BRITISH ISLES

Brian Nixon

University Tutorial Press

Published by
UNIVERSITY TUTORIAL PRESS LIMITED
842 Yeovil Road,
Slough SL1 4JQ.

ISBN 0 7231 0760 2
Published 1978
Reprinted 1978 (twice), 1979, 1980 (twice), 1981, 1982

Set by The Pitman Press, Bath.
Printed and bound in Great Britain by
Butler & Tanner Ltd, Frome and London

Contents

Preface

The main aim of this book is to provide a comprehensive coverage of the geography of the British Isles, suitable for pupils taking examinations at sixteen plus. Such an aim has naturally imposed certain constraints, the most important of which is the need to group material in well defined, self contained topics, relevant to the requirements of the various examination boards. These topics have been built into a general framework based upon the theme of change, and, in keeping with modern examination needs, information has been expressed in visual form wherever possible and exercises have been included to help develop the skills of interpreting such information. Finally, because it is unashamedly an examination text, no attempt has been made to extend the boundaries of secondary school geography and the techniques of the 'new' geography are included only in certain limited and well-tried circumstances.

I would like to thank John Light for the time and trouble which he has taken with the illustrations, and the editorial and production staff of UTP for the speed and efficiency with which they have produced this book. I am also grateful to P Travers, T Purdy and I G Knights for supplying me with information on agriculture and to the many organisations which have supplied photographs, the acknowledgements for which appear on page 312.

Introduction

The British Isles are situated on the continental shelf off the west coast of Europe. Politically, they are divided into two countries—the United Kingdom and the Republic of Ireland, and the former is itself divided into the major units of England, Scotland, Wales and Northern Ireland. The combined area is more than three hundred thousand square kilometres and the population in 1971 was 58,465,000. This makes the islands one of the most densely populated parts of the earth's surface and the United Kingdom, at least, one of the most densely populated countries.

UNIT	AREA (SQ KM)	POPULATION (000's)
England	131,700	46,454
Wales	19,300	2,765
Scotland	78,700	5,206
Great Britain	229,700	54,425
Northern Ireland	14,100	1,537
Islands	1,000	175
United Kingdom	244,800	56,137
Republic of Ireland	68,900	2,978
British Isles	313,700	59,115

The problems of supporting such a large population on such a small land area are obvious. In fact, this became possible only with the emergence of Britain as the world's first modern industrial nation during the eighteenth and early nineteenth centuries. Then the growth of coal based industries allowed manufactured goods to be produced which could be exported to pay for the food and raw materials needed to support the population and the expanding industries. It was also during this period that Britain acquired a vast overseas empire which immeasurably strengthened her trading position and enabled her to become the wealthiest nation on earth.

During the twentieth century Britain has lost this position and her economy has faced increasing problems which have produced a serious decline in relationship to other countries in the world. As a result, it is difficult to view the present situation realistically and it is easy to forget that Britain is still one of the richest nations in the world and still one of the leading industrial and trading countries. The following figures may help to put things in perspective.

COUNTRY	CALORIES PER DAY	INFANT MORTALITY (PER 000)	% OF POP. IN TOWNS	% OF LAB. FORCE IN FARMING	ENERGY USED (PER HEAD)	GROSS NATIONAL PRODUCT ($)
The World	100	87	17	47	100	—
United Kingdom	130	20	71	5	275	3,375
United States	130	25	50	8	560	6,597
India	84	139	9	68	10	120
Liberia	97	40	1	74	21	334
Sweden	129	13	25	18	282	6,876
Japan	100	18	41	21	186	4,152
Peru	86	91	15	55	31	473
USSR	125	27	—	34	255	—

Note: The statistics for diet and consumption of energy are measured from the world average which is said to represent 100.

It is clear from these figures that there is a great divide between the developed nations of the West and East and the undeveloped nations of the Third World. It is also clear that on any yardstick Britain belongs to the former group and is indeed one of the wealthiest members of it.

Using information given in Table 2, complete the following exercise:

a) Assuming that the Gross National Product is a measure of the wealth of a nation, work out the correlation between the GNP and each of the other factors given in the table.

(For a simple method of calculating the correlation coefficient see page 308.)

b) i) Which of the factors mentioned show the strongest correlation?

ii) Using the factor which has the strongest correlation, divide the countries into two groups—those which are developed and those which are underdeveloped. (Assume that Britain is a developed nation.)

At the same time it is important to remember that Britain does have serious economic problems and that these problems underlie many of the changes which have taken place in recent years. It is not surprising, therefore, that they form an important part of this book and are examined in some detail in the individual sections.

Trade and World Links

From the late eighteenth century onwards Britain was faced with the problem of feeding a rapidly growing population and, in order to do this, she was forced to import ever increasing amounts of food. These foodstuffs were paid for by exporting manufactured goods and, by the end of the nineteenth century, dependence on trade had become so great that, with little more than two percent of the world's population, Britain was conducting more than one third of world trade in manufactured goods. Since that time Britain's share of world trade has declined enormously, largely on account of competition from other countries, but, in spite of this, she remains one of the leading trading nations of the world.

The Changing Pattern of Trade

THE CHANGING PATTERN OF BRITAIN's TRADE 1935–1975.

TYPE	1935		1975	
	IMPORTS	EXPORTS	IMPORTS	EXPORTS
	(EXPRESSED AS % OF TOTAL)			
Food	45·0	7·4	16	7·6
Raw materials	28·0	7·9	12	3·8
Fuel	4·7	9·0	20	4·8
Manufactures	22·3	75·7	51	83·8
BRITAIN'S TRADE (Million U.S. Dollars)	2,042	1,239	45,642	52,009
WORLD TRADE (Million U.S. Dollars)	12,093	11,457	903,000	871,000

Study the table above and page 6 and complete the following exercise:

a) Complete the following statements by using one or more of the alternatives given:

i) In terms of value, Britain's trade has increased/decreased/ stayed the same since 1935.

ii) During this time Britain's share of world trade has increased /decreased/stayed the same.

iii) This development is the result of high unemployment in Britain/ increased competition from other industrial countries/the sinking of British ships during two world wars/failure to modernise British industry.

iv) In 1935 foodstuffs and raw materials for industry (excluding fuels) made up 25/50/75 over 90% of Britain's imports.
v) Today this proportion has risen/declined to 25/50/75 over 90%.
vi) This change is a result of a decline in the demand for food/an increase in the import of manufactured goods/an increase in food production in Britain/an increase in the import of oil/ an increase in the production of raw materials at home/a decline in the demand for raw materials for industry.

b) Write a brief description of the pattern of Britain's trade and how it has changed during the twentieth century.

It is clear from the table that Britain's trade has become much more complicated and that the traditional pattern of importing foodstuffs and raw materials and exporting manufactured goods, although still present, is less easy to detect. This is largely the result of a steady increase in the import of manufactured goods from the EEC, Japan and the United States. One of the consequences of this development has been that Britain finds it increasingly difficult to export enough goods to pay for the imports. This produces a *deficit* which has become a serious problem for Britain, particularly since the oil price rises of the early nineteen seventies.

Changes have also taken place in the distribution of Britain's trade with the rest of the world. The development of Britain as the world's leading trading nation in the eighteenth and nineteenth centuries went hand in hand with the acquisition of a vast overseas empire. Countries in the empire were expected to supply raw materials to Britain's industries and to provide markets for the manufactured goods which they produced. It is not surprising, therefore, that for most of this period, empire trade overshadowed trade with other countries in the world, particularly since this trade was conducted on special terms which made it very attractive. By 1913, this pattern had already begun to break down and today trade with the Commonwealth (which replaced the Empire) is small in comparison with trade with other areas.

Study the information given on page 6 and complete the following exercise:

a) Trade in 1913.
i) Name three areas with which Britain had a trade surplus.
ii) Name two areas with which Britain had a trade deficit.
iii) In which group are most Commonwealth countries likely to be found?

b) Trade in 1974.
i) List the four main markets for Britain's exports.
ii) List the four main sources of imports.

iii) Name two areas with which Britain has a trade surplus.
iv) Name two areas with which Britain has a trade deficit.

c) Write a brief account of the changes which have taken place in Britain's trade since 1913.

d) Refer to the map below.
   i) Name the six countries with the largest merchant fleets, arranging them in order of importance.
   ii) Four of these countries are major trading partners of Britain. Two are not. Name these two and try to find out why they have such large merchant fleets.

e) The following are trade routes which are of great importance to Britain: the North Atlantic, the Cape of Good Hope, the Suez Canal, Cape Horn, The Panama Canal. For each:
   i) Using an atlas, identify the routes on the map.
   ii) Name one trading partner to which Britain is linked by each route.
   iii) Name one major port on each route.
   iv) Arrange the routes in order of importance according to the volume of trade carried on them.

WORLD TRADE ROUTES (showing relative volume of trade)

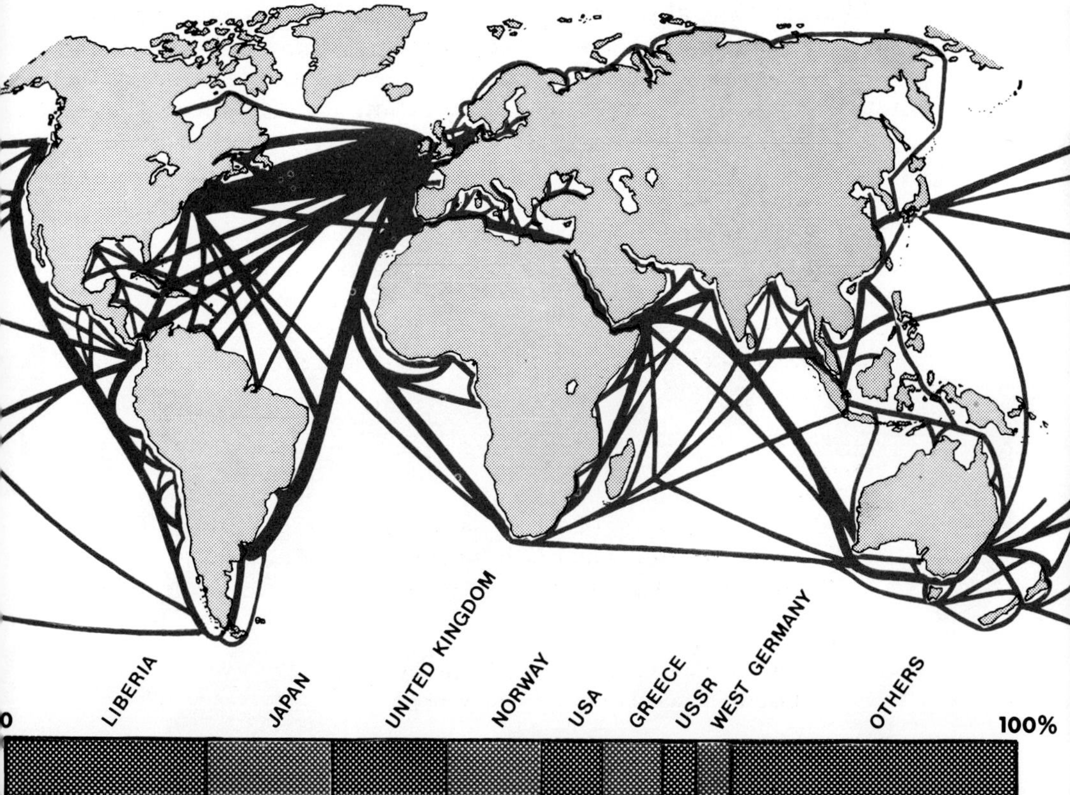

WORLD MERCHANT FLEETS

TRADE BY AREA 1974

EXPORTS

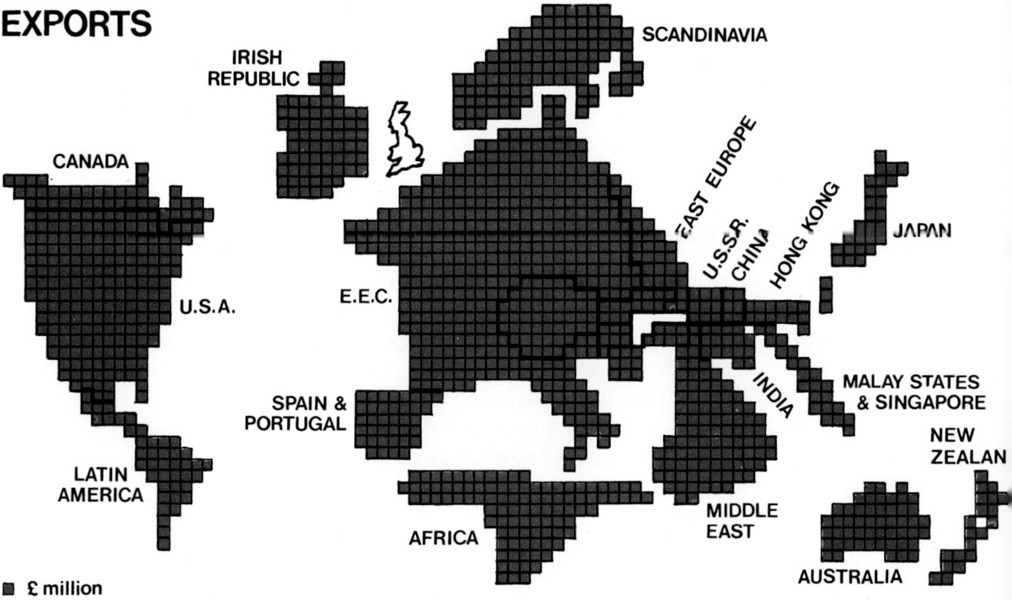

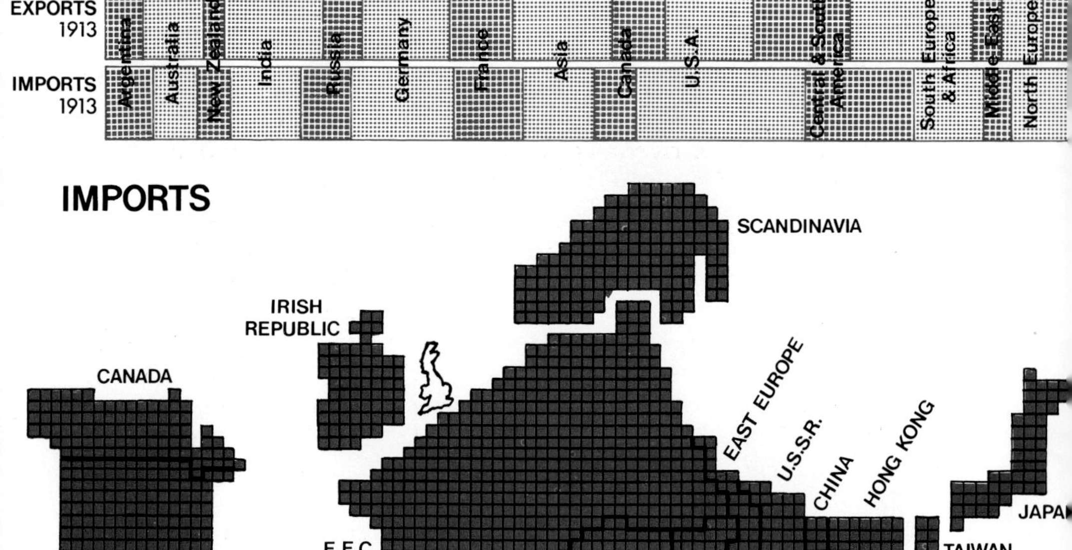

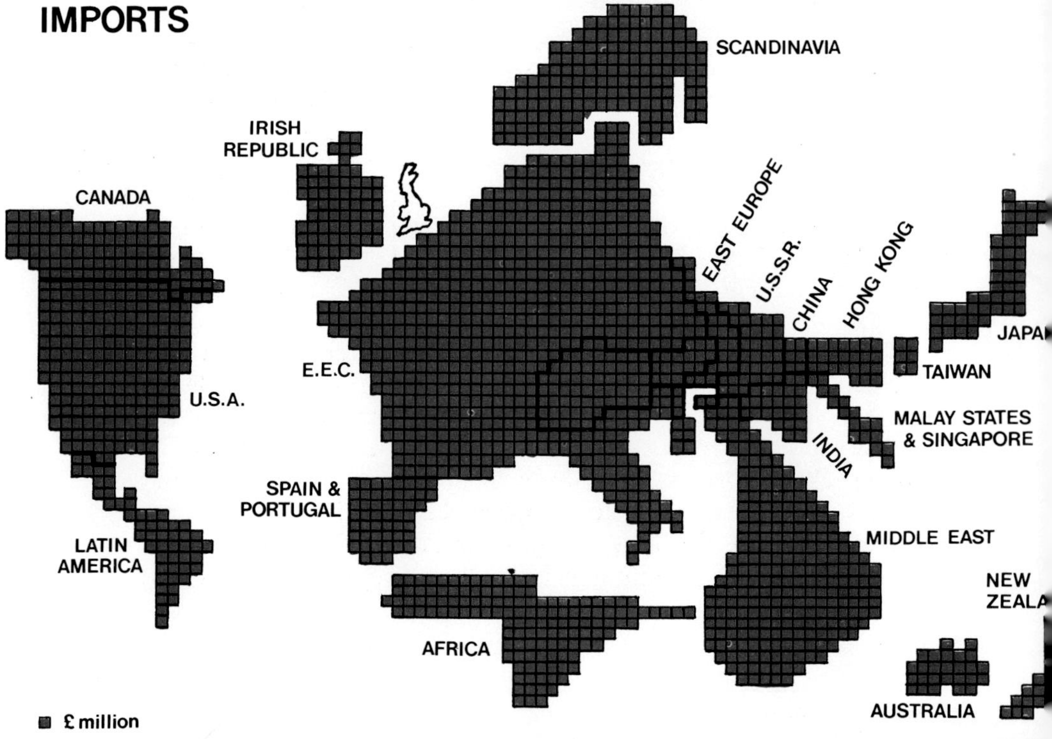

Since trade involves the movement of goods from one place to another, often over great distances and in great quantities, it is not surprising that the major trading countries of the world also own the largest merchant fleets. This is particularly true of Britain which, as an island, is totally dependent upon seaborne trade and, for almost two hundred years until the Second World War, Britain's merchant fleet was by far the largest in the world. Since then Britain has lost ground to her rivals and both Liberia and Japan now have larger fleets. (Although it is important to remember that Liberia is used as a flag of convenience and is not a major trading nation.)

Ports

Almost all of Britain's trade is handled at a comparatively small number of ports. Most of these ports are old established and have been involved in trade for several hundred years. Reasons for their development are numerous and complex.

1. The majority of ports have grown up at the mouths of rivers which give sheltered water, deep enough to take the comparatively small ships which were common before super-tankers came into use.

2. Such sites were usually tidal and, from the eighteenth century onwards it became usual to construct dock basins which could be isolated from the sea or river by closing their gates. This meant that, as the tide ebbed and the water level in the estuary began to fall, the gates could be closed and the water level in the dock could be maintained at a high level, so that loading or unloading could continue regardless of the state of the tide. Prior to this ships had been forced to anchor out in the main channel and small boats (lighters) were needed to bring the cargoes ashore.

3. Many of the dock systems built during the nineteenth and early twentieth centuries became too small to handle the larger vessels afloat today and this resulted in the abandonment of old port areas and the building of new docks nearer the open sea, or even the construction of entirely new ports (called *outports*).

4. It is at these new sites that most of the modern facilities have been installed. The time spent by a ship in a port may be regarded as time wasted and the efficiency of a port can be measured by the speed with which cargoes are loaded and unloaded. Many of Britain's older ports are relatively inefficient in this respect and attempts have been made to improve the situation by installing specialist handling facilities, such as ore and grain docks, oil terminals with massive pumping facilities, container terminals and roll

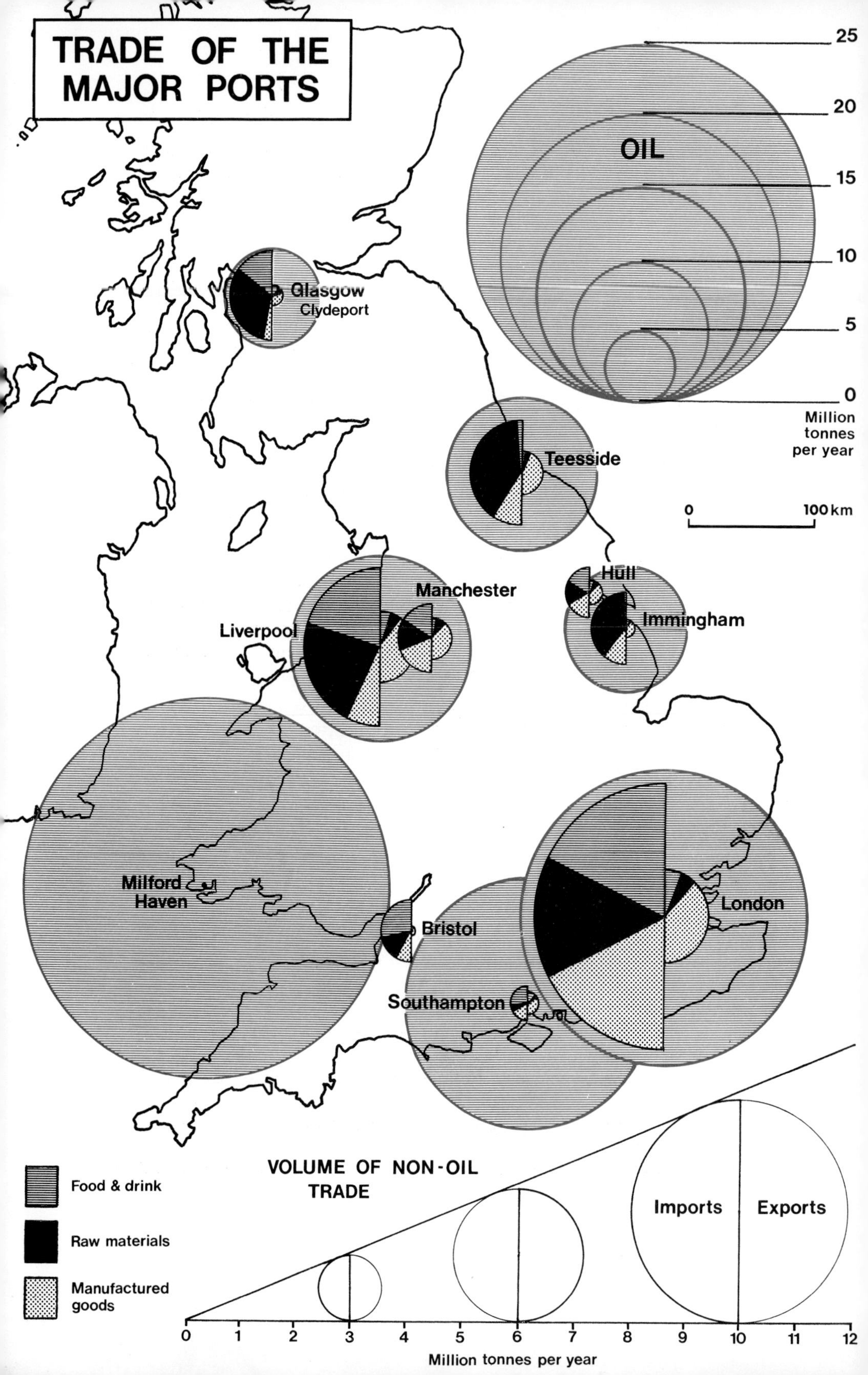
TRADE OF THE MAJOR PORTS
OIL
25
20
15
10
5
0
Million tonnes per year
0
100 km
Glasgow
Clydeport
Teesside
Hull
Immingham
Manchester
Liverpool
Milford Haven
Bristol
Southampton
London
Food & drink
Raw materials
Manufactured goods
VOLUME OF NON-OIL TRADE
Imports
Exports
0 1 2 3 4 5 6 7 8 9 10 11 12
Million tonnes per year

on/roll off berths (see page 10). Such facilities are expensive and they have, therefore, been installed down river from the old docks, at sites where the water is deep enough to take the largest of modern vessels.

5. Apart from site, the most important factor in the growth of a port is its accessibility to a large and prosperous area of the country. Such an area—the area served by a port—is called the *hinterland* and it can vary in size from a few hundred square kilometres in the case of a small local port to virtually the whole of Britain in the case of London.

The effects of these factors can be seen in the present distribution of ports in Britain and in the size and nature of their trade.

Using information given on page 8, complete the following exercise:

a) i) Name two ports which have a well balanced pattern of trade, ie a large volume of imports and exports and a large oil trade.
   ii) Name two ports which have virtually no oil trade.
   iii) Name one port which has virtually no general trade.

b) One method of examining the effect of one factor on another is to work out the *correlation coefficient.* This can be done quite simply, using the method described on page 308. In this case it is useful to examine the relationship between the depth of water in the ports and the nature of their trade. Refer to the table below.

APPROXIMATE DEPTH OF THE LEADING PORTS

PORT	DEPTH OF GENERAL BERTHS (METRES)	DEPTH OF OIL TERMINAL (METRES)
Glasgow	12·6	11·8*
Teesport	10·0	13·8
Hull	12·7	12·7
Immingham	10·0	17·0
Liverpool	13·9	13·9
Manchester	8·0	8·0
Milford Haven	9·0	20·0
Bristol	10·9	10·9
London	13·8	14·6
Southampton	12·8	12·8

* The Finnart terminal is 24 metres deep.

i) List the ports in rank order according to the depth of water. (Number them from 1–10).

ii) List the ports according to the volume of their imports. Again number them from 1–10.

iii) Calculate the correlation between depth of water and volume of imports.

iv) Repeat the procedure and calculate the correlation between depth of water and exports.
v) Repeat the procedure for oil trade and calculate the correlation between depth of water and the volume of oil imports.

c) i) Which type of trade is most strongly influenced by the depth of the channel?
ii) Why should this be so?
iii) Why is the volume of exports so small when compared with the volume of imports? An examination of the make-up of the trade of the ports will help in this.

d) Describe the patterns of trade at Teesside, Manchester and Milford Haven and explain how and why they differ.

AREAS OF TRADE OF THREE PORTS.

Areas of trade (as a % of the total tonnage handled)

PORT	EIRE	EEC	SCANDIN-AVIA	MEDITER-RANEAN	AFRICA
Manchester	7	26	13	6	6
Teesport	0·5	14	11	35	7
Milford Haven	12	6	7	5	8

PORT	FAR EAST	MIDDLE EAST	NORTH AMERICA	SOUTH AMERICA	AUSTRAL-ASIA
Manchester	0·1	1	18	13	11
Teesport	0·5	11	7·5	5	0·2
Milford Haven	0·7	80	0·8	0·3	0

It is clear from this exercise that the oil trade in particular is dependent upon the depth of water, and that modern terminals have been built where deep water is available. This is to be expected in view of the size of modern tankers which are now approaching half a million tonnes dead weight.

Other trends are less obvious but equally important. One of the most important developments in port activity in recent years has been the introduction of container traffic.

Using information given on pages 11 and 12, complete the following exercise:

a) i) Explain what is meant by the term container.
ii) Describe how container traffic is organised.
iii) Explain how this method differs from the traditional methods of handling goods in ports and explain why it is more efficient.

b) i) List the main container ports, ranking them according to the volume of container traffic handled.
ii) Rank the ports according to the total volume of non-oil trade handled.

Felixstowe docks showing, in the foreground, the car terminal; and in the background, the storage area for containers and quayside container handling facilities.

iii) Name three ports which are both major general ports and major container ports.
iv) Name three ports which are major container ports but which are not among the ten major ports.
v) Calculate the correlation between container traffic and general trade in the ports mentioned on page 8.
vi) Calculate the correlation between container traffic and depth of water in these ports. What does this tell you about container traffic?

The introduction of container traffic has led to an enormous reduction in the amount of time spent loading and unloading ships. This increased efficiency has been achieved, however, only by building new and expensive handling facilities at many ports. Several factors have influenced the choice of site for these facilities. For example:

1. Although container ships are not particularly large by modern standards, most terminals have been built where deep water is available.

2. The old port areas have also been avoided because they are often situated near to the centre of cities where congestion is a serious problem and where there is little room for development.

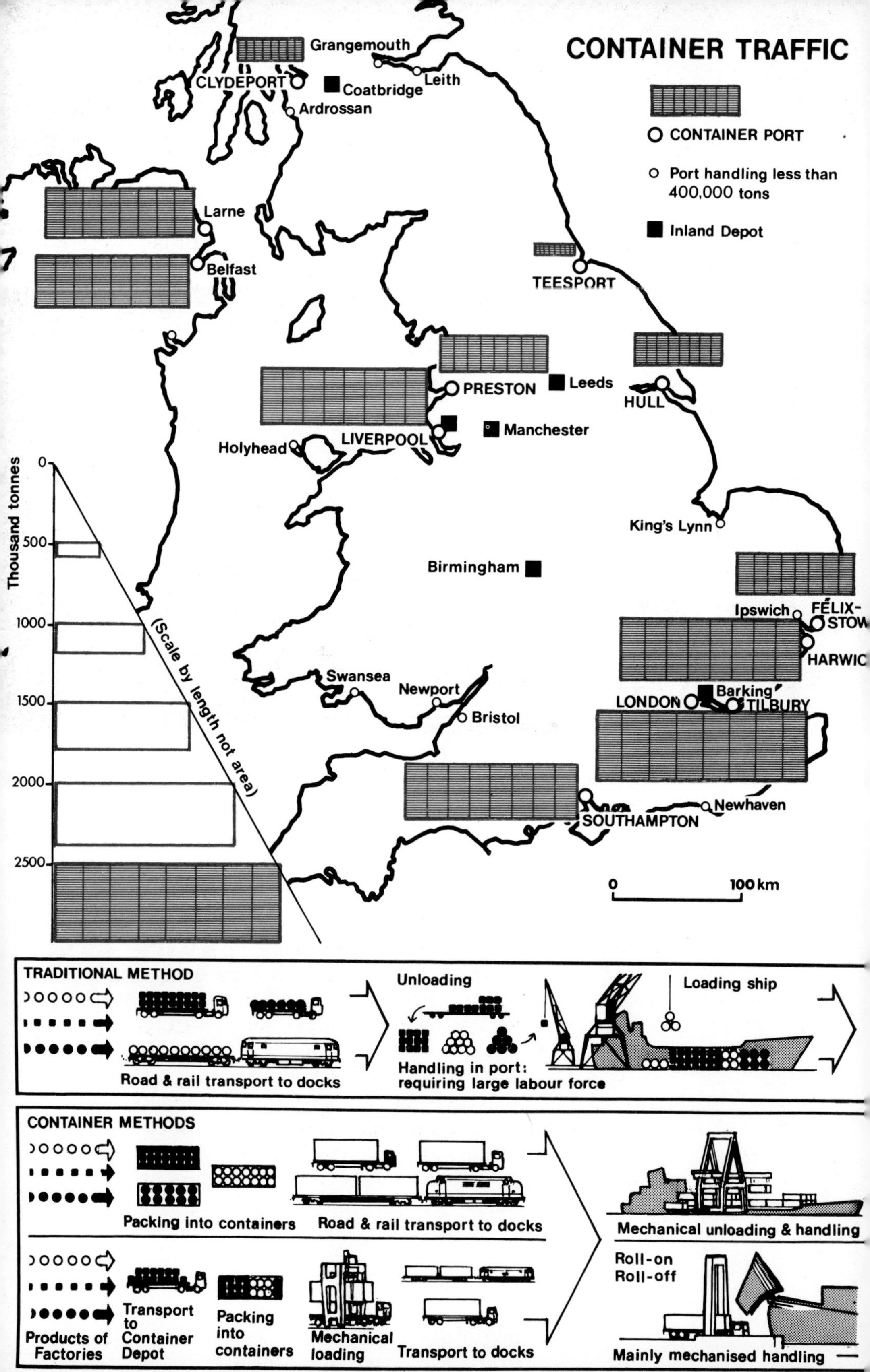
CONTAINER TRAFFIC
CONTAINER PORT
Port handling less than 400,000 tons
Inland Depot
Grangemouth
Leith
CLYDEPORT
Coatbridge
Ardrossan
Larne
Belfast
TEESPORT
PRESTON
Leeds
HULL
LIVERPOOL
Manchester
Holyhead
King's Lynn
Birmingham
Ipswich
FELIX-STOW
HARWIC
Swansea
Newport
Bristol
Barking
LONDON
TILBURY
SOUTHAMPTON
Newhaven
0
100 km
Thousand tonnes
0
500
1000
1500
2000
2500
(Scale by length not area)
TRADITIONAL METHOD
Road & rail transport to docks
Unloading
Handling in port: requiring large labour force
Loading ship
CONTAINER METHODS
Packing into containers
Road & rail transport to docks
Mechanical unloading & handling
Products of Factories
Transport to Container Depot
Packing into containers
Mechanical loading
Transport to docks
Roll-on Roll-off
Mainly mechanised handling

Felixstowe Ferry Terminal with a roll on/roll off car ferry alongside. Note the area of land needed to handle the vehicles outside the docks.

3. Labour relations in the old ports were often poor and this has encouraged the development of facilities in smaller ports where there is less danger of strikes affecting traffic. Felixstowe and Harwich, for example, have grown rapidly, capturing trade from the Port of London, and benefiting from the enormous expansion of trade with Europe which has taken place in recent years.

Another type of port which has grown rapidly in recent years is the *ferry port*. Originally designed to handle passengers and mail, the leading ferry ports provide regular sailings, often at very short intervals. For this reason, the ports have to be very efficient and turn around time is reduced to a minimum. It is not surprising, therefore, that many of the techniques now used at the container terminals were first developed at the ferry ports, eg roll on/roll off berths.

Most ferry ports are old established. They grew up to provide services to Europe and Ireland and the sites chosen were usually those controlling the shortest crossing points, eg Dover and Folkestone on the Channel coast; Harwich on the east coast; Southampton on the south coast; Fishguard, Holyhead and Stranraer on the west coast. Equally important, since the ports concentrated on passengers rather than freight, communications with the rest of Britain had to be good. Land transport is generally faster than sea travel and, for this reason, most passengers prefer to keep voyage times to a minimum and to transfer to the road or rail network as soon as possible. For example, passengers destined for London, itself a major port, have traditionally disembarked at Southampton, Tilbury or the Channel ports and continued their journeys by land.

Britain's ferry ports have grown rapidly in recent years to meet the increased demand for passenger berths to Europe and Ireland. Even more important, however, they have had to provide facilities for motorists; hence the introduction of roll on/roll off ferries. Much of this traffic consists of holiday makers and is, therefore, seasonal. The facilities are used throughout the year, however, to move freight and Britain's ferry ports are emerging as major ports, handling large amounts of container traffic.

The extent of these changes in the distribution and organisation of Britain's ports can perhaps best be seen by studying a single example.

Port Study: Bristol

Four hundred years ago Bristol ranked second only to London among Britain's ports. Situated on the Avon, above the tidal waters of the Severn Estuary, the port had grown rapidly, depending upon the continued prosperity of the woollen industry in the Cotswolds and the expansion of world, as opposed to European, trade. The latter was particularly important and Bristol, along with other west coast ports, became involved in trade with newly established colonies in America and, as a result of this, in the infamous slave trade. The slave trade developed in the seventeenth century and reached its peak towards the end of the eighteenth century. Merchants from Bristol carried manufactured goods from the Midlands and South Wales to West Africa where they were exchanged for slaves. This new cargo was then shipped direct to the colonies in North America and the West Indies where it was sold and where new cargoes of sugar, cotton and tobacco were bought

Bristol showing the position of Avonmouth (a), the Clifton Gorge (b), the lock gates of the city dock system (c) and the floating harbour (d).

for the return journey to England. The risks of this triangular trade were great but profits were enormous and Bristol prospered.

By the beginning of the nineteenth century this period was coming to an end and Bristol's problems were becoming more apparent. Some of these problems can be attributed to the abolition of the slave trade but even more stemmed from the fact that the port was too far from the sea and ships were already experiencing difficulties in navigating the Avon. Improvements carried out during the century staved off total collapse. The old dock system, situated where the Frome meets the Avon, was improved and new docks were built near the city centre. Even more important, the channel linking the port to the sea was improved and made navigable for larger vessels. Little could be done about the river through the Avon Gorge, however, and, by the end of the century, Bristol's docks were totally inadequate and more drastic solutions were sought. This involved building a completely new outport at Avonmouth on the Severn estuary. Here the channel was deep enough to take large vessels and there was a large area of low flat land which could be used for dock installations and for future industrial development. A second outport was built at Portishead and today these two dock complexes handle virtually the whole of Bristol's trade; the City Docks falling into virtual disuse.

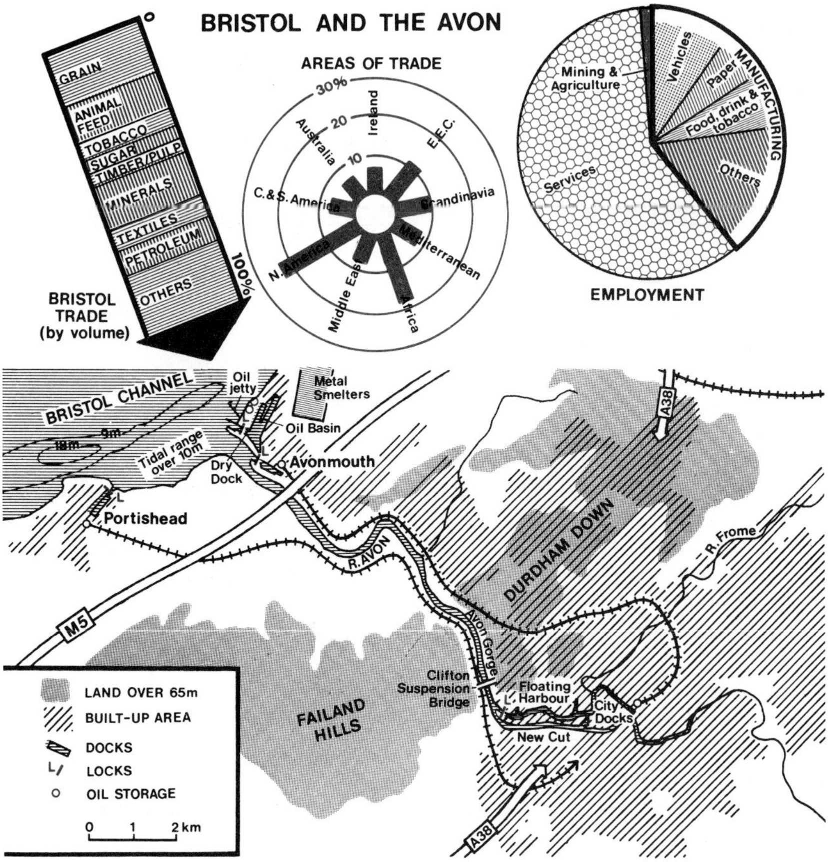

BRISTOL AND THE AVON

BRISTOL TRADE (by volume)

0

GRAIN
ANIMAL FEED
TOBACCO
SUGAR
TIMBER/PULP
MINERALS
TEXTILES
PETROLEUM
OTHERS

100%

AREAS OF TRADE

10
20
30%
Ireland
E.E.C.
Scandinavia
Mediterranean
Africa
Middle East
N. America
C.& S. America
Australia

EMPLOYMENT

Mining & Agriculture
Vehicles
Paper
Food, drink & tobacco
Others
MANUFACTURING
Services

BRISTOL CHANNEL
18m
9m
Tidal range over 10m
Oil jetty
Metal Smelters
Oil Basin
Dry Dock
Avonmouth
Portishead
M5
R. AVON
A38
DURDHAM DOWN
R. Frome
Avon Gorge
Clifton Suspension Bridge
Floating Harbour
City Docks
New Cut
FAILAND HILLS

LAND OVER 65m
BUILT-UP AREA
DOCKS
LOCKS
OIL STORAGE
0 1 2 km

Portishead docks.

Although Bristol has changed out of all recognition during the twentieth century, the influence of the past can still be traced both in its pattern of trade and in its industrial structure.

Using information contained in the map and photographs, complete the following exercise:

a) i) Why have the City Docks declined?
   ii) Why is Avonmouth a good location for an outport?
   iii) Why are lock gates necessary at the entrance to the main docks there?

b) i) List the three main areas with which Bristol trades, arranging them in order of importance.
   ii) Compare this with the pattern for Hull (see page 168).
   iii) How do the two patterns reflect the history of the two ports?

c) i) List the main imports of Bristol, arranging them in order of importance.
   ii) Name two commodities which were once very important to Bristol and which are still imported.

d) Using the statistics for British industries given on page 80, calculate the *location quotients* for the following industries in the Bristol area:
   Mining and Agriculture
   All manufacturing industries
   Paper
   All service industries
   Food, drink and tobaccco
   Vehicles
   (For the method of calculation see page 311)

e) The pattern which emerges is typical of British ports. Describe it and try to explain:
   i) the importance of service industries
   ii) the sources of raw materials for the manufacturing industries.

Avonmouth docks.

The actual distribution of industries in the Bristol area also reflects these developments. Many of the industries make use of imported raw materials or are connected with shipping. Such industries are called '*port industries*' and they tend to be located near to the major docks. In Bristol some of the traditional port industries such as tobacco processing and the bottling of wine and spirits still remain in or near to the city centre. Other large scale port industries such as metal refining and the manufacture of chemicals have been attracted to the outports where the cargoes are unloaded and where there is room for large factories. A third group of industries has developed on the outskirts of the city and these are usually less dependent upon the port.

The success of Bristol, both as a port and as an industrial centre, is reflected in the continued prosperity of the city. Two factors have contributed to this in recent years:

1. Many of the industries of the area are growth industries. (See the list on page 81.)

2. The extension of the motorway network has made the port more accessible and has enormously extended its hinterland. Industrial areas as far afield as Lancashire, Yorkshire and London are now within an easy day's drive of the port. Furthermore, the building of the Severn Bridge, improving links with South Wales, has enabled Bristol to become the commercial and business centre for the whole of the Severn estuary. The significance of this improvement can be seen from the following exercise:

a) Trace the map on page 143 which shows the motorway network of Britain. On it mark Bristol and the following cities: London, Birmingham, Liverpool, Leeds, Glasgow.
b) Refer to page 144. Assuming average travelling speeds of 45 kph by main road and 80 kph by motorway (for goods vehicles):
   i) calculate the travel time to each of the cities by road
   ii) calculate the travel time by motorway
   iii) plot the area within four hours travel time of Bristol by main road.
   iv) plot the area within four hours travel time by motorway. This can be done by plotting the appropriate distance along each of the roads and motorways serving the city and by joining these points together.
c) Describe the effect this improvement in communications has had on the hinterland of the port.
d) Using the method described on page 309, calculate the index of directness of the Bristol–Cardiff route
   i) before the building of the Severn Bridge (distance 145 km)
   ii) after the building of the bridge.

e) Using the above speeds for freight traffic, what would be the travel time between the two cities, in each case?

Air Traffic

The most striking development in the field of transport in recent years has undoubtedly been the growth of air traffic. The extent of this growth can be seen from the following figures.

GROWTH OF AIR TRAFFIC

	1954	1974
Number of flight's	49,000	403,000
Passengers (millions)	0·7	43
Freight (000 tonnes)	48	700

Impressive as these figures may seem, they do in fact represent a very slight change in the overall pattern of transport in Britain, and in 1975 only 0·03% of freight traffic and 0·5% of passenger traffic was by air (see page 133).

This failure to challenge sea and land transport on all but the long distance passenger routes stems from the nature of air transport itself. For example, the largest of modern aircraft can move only small amounts of freight and, as a result, running costs are much higher (per tonne/km) than those for other forms of transport. Against this, however, air transport is much faster than any of its rivals and this can offset higher costs, particularly in the field of passenger transport. Even here there are problems and the advantages of speed in the air can be lost by delays on the ground. This can be seen most clearly in the movement of freight (see below).

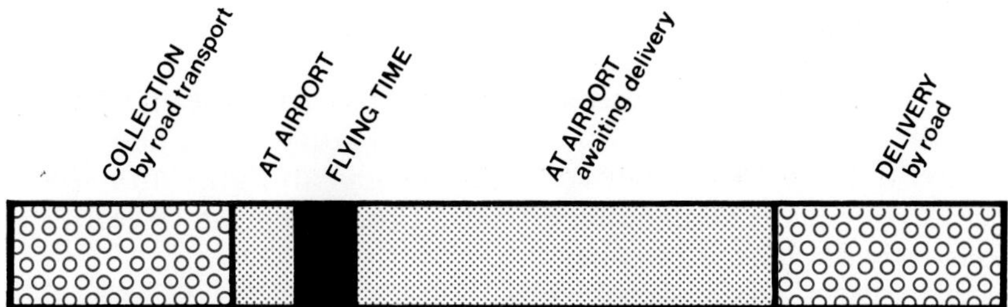

THE MOVEMENT OF FREIGHT BY AIR showing the time taken for each stage of a typical journey from London to Glasgow

a) Work out the proportion of the total journey time spent:
   i) in the air
   ii) waiting at the airport
   iii) in collection and delivery.

b) Assuming that for passenger the time spent travelling to and from the airport and waiting at the airport is less, arrange the following journeys in rank order according to the advisability of using air transport:
   i) as a passenger to Sydney (Australia)
   ii) moving one thousand tonnes of coal from Immingham to Copenhagen
   iii) as a passenger from London to Rome
   iv) moving one thousand watches from Zurich to London
   v) as a passenger from London to Birmingham.

   Explain why you have ranked them in the order chosen.

c) i) Concorde has reduced travel time between London and the eastern seaboard of the USA from seven hours to three hours. What effect does this have on total journey time (from door to door)?
   ii) Give other possible methods of reducing total travel time for the journey.

It is clear from the exercise that the efficiency of air transport can be increased either by improving aircraft or by improving links between airports and the main centres of population. The former has taken place at a rapid rate and aircraft are now both larger—although not large enough to compete in the bulk freight market—and faster than they have ever been. The second alternative has proved much more difficult. Travel time between cities and airports can be reduced either by improving transport facilities between the two or by locating airports nearer to the city centres. Of these, improvement in transport has been widely tried and rail, motorway, mono-rail and even helicopter links have been established. The second alternative has again proved impossible, largely on account of the many factors which limit the choice of site for an airport. These include:

1. The space required, which is enormous in the case of a modern airport. Runways have to be over three kilometres in length, terminal buildings for passengers and freight are large and service areas are equally demanding of space. Consequently a large area of relatively flat land is essential.

2. Noise is a serious problem with modern jet aircraft, particularly on take-off and landing. Anyone living near a major airport will suffer from this and it is obviously better to site airports away from settlements. This, however, serves only to increase the problems of access to city centres.

3. Another reason for avoiding built-up areas is the risk of accident which are also greatest during take-off and landing.

Given such constraints, it has not only proved impossible to locate airports near to city centres where land values are high and building densities at their greatest, it has also proved increasingly difficult to find sites for any new airports at all in a country as overcrowded as the United Kingdom. Opinion about the need for new airports is divided but it is generally agreed that the need is greatest in the London area where existing airports are already under strain.

Airport Study: London Heathrow

London is served by two major airports—Heathrow and Gatwick. Of these, Heathrow is by far the most important and its dominant position among Britain's airports can be seen below. This one airport handles seventy-three per cent of all passenger traffic and sixty-five per cent of all freight passing through British airports. It is already one of the largest airports in the world and its problems are typical of most major airports today.

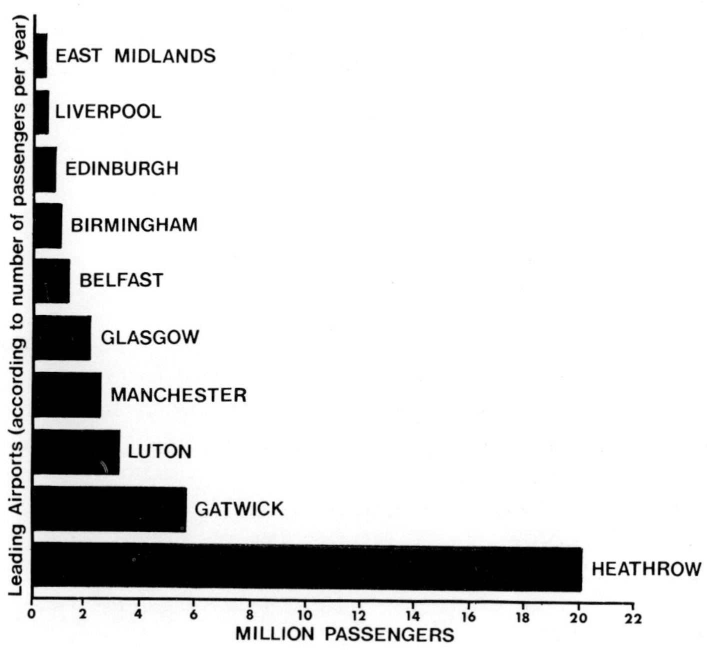

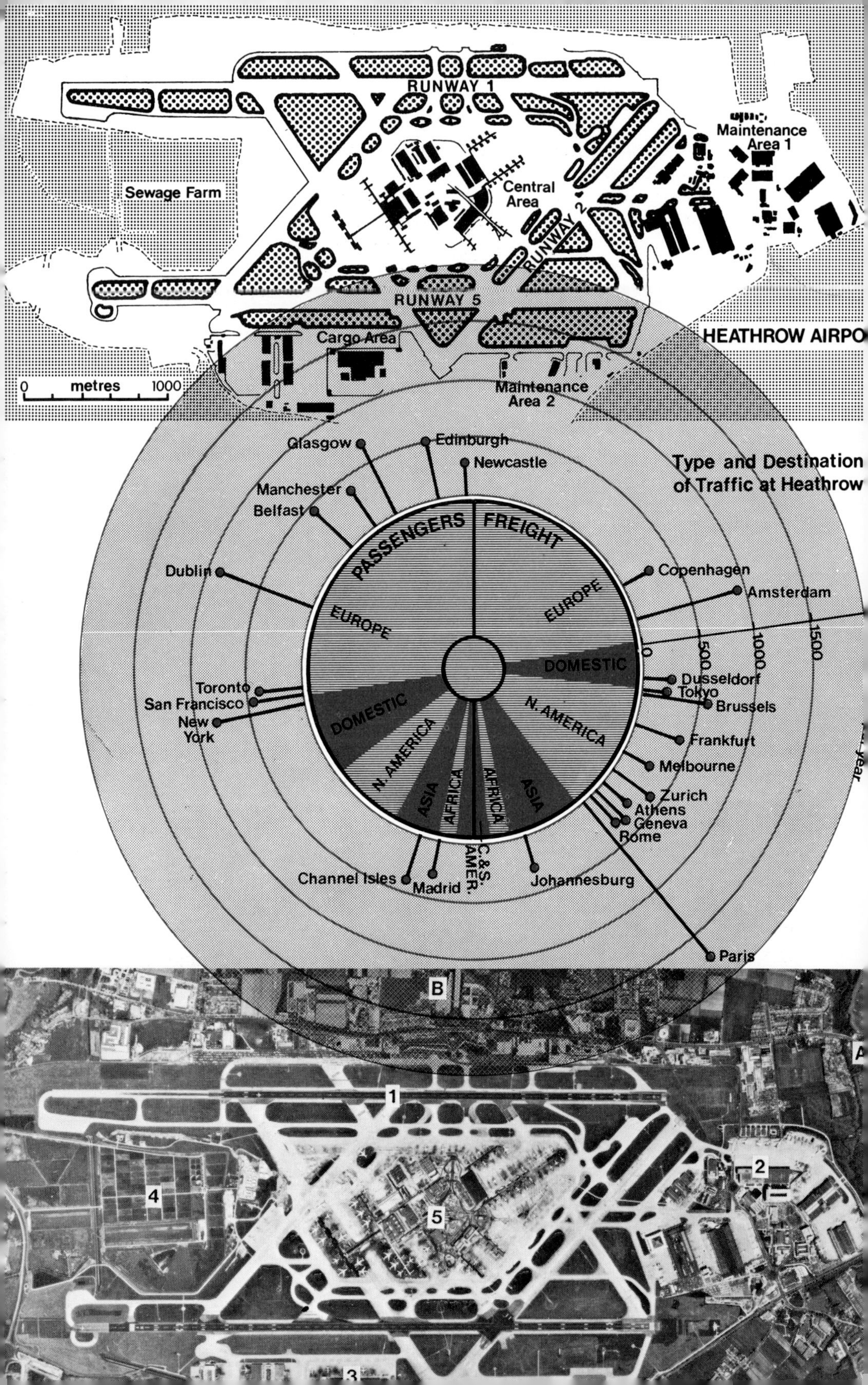
HEATHROW AIRPO
RUNWAY 1
RUNWAY 2
RUNWAY 5
Maintenance Area 1
Central Area
Sewage Farm
Cargo Area
Maintenance Area 2
0 metres 1000
Type and Destination of Traffic at Heathrow
PASSENGERS
FREIGHT
EUROPE
EUROPE
DOMESTIC
DOMESTIC
N. AMERICA
N. AMERICA
ASIA
AFRICA
AFRICA
ASIA
C.&S. AMER.
0
500
1000
1500
Glasgow
Edinburgh
Newcastle
Manchester
Belfast
Dublin
Toronto
San Francisco
New York
Channel Isles
Madrid
Johannesburg
Copenhagen
Amsterdam
Dusseldorf
Tokyo
Brussels
Frankfurt
Melbourne
Zurich
Athens
Geneva
Rome
Paris
year
B
A
1
2
3
4
5

Using information given on page 22 complete the following exercise:

a) Refer to the map of London on page 242 and locate the airport.
   i) How far from the city centre is the airport?
   ii) What links exist between the airport and the city centre?
   iii) Why is Heathrow in a good position to capture much of the traffic of the Midlands? Why is Gatwick less well situated?

b) Describe the advantages of Heathrow as a site for a major airport.

c) i) Name the features marked 1–5 on the photograph.
   ii) Why is such a complicated pattern of runways necessary?
   iii) Which of the settlements A or B will be most seriously affected by noise?
   iv) What is the approximate area of the airport?
   v) Why would it be difficult to extend the airport?

d) i) List the six airports which have the strongest links with Heathrow, ranking them according to the number of flights per year.
   ii) List the main areas of the world which have links with Heathrow, ranking them according to the number of passengers and the amount of freight.
   iii) Describe the pattern of traffic at Heathrow airport and explain how it related to the advantages and disadvantages of air transport described earlier.

Heathrow covers an area of more than twenty square kilometres, handles more than twenty million passengers and nearly half a million tonnes of freight per year, and has aircraft landing or taking off at two or three minute intervals at peak periods. It is already the busiest international airport in the world and, if traffic continues to grow, it will not be able to cope. As we have seen, extensions to Heathrow would be difficult and the only solution would appear to be the diversion of aircraft to other airports. Gatwick, as London's second airport, is an obvious choice but it is already busy and would be equally difficult to extend. The diversion of traffic to airports in other parts of Britain is also possible but, since the majority of passengers and the bulk of freight are destined for the London area, the increased travel times involved make this development highly unlikely. It is not surprising, therefore, that for the last decade, sites have been examined for development as a third airport for London and the South East. In every case opposition has been enormous, usually from people living in the areas concerned, and the project has now been abandoned in view of the possibility that air traffic may not grow as quickly as was anticipated and that in the future aircraft will be even larger and fewer will be required.

Agriculture

In 1975 Britain imported more than one third of her total food supply. Of these imports less than one quarter were tropical products which cannot be economically grown in the British Isles. The remainder were products from temperate areas which could be grown in Britain but which, for a variety of reasons, are not (see below).

BRITAIN'S FOOD SUPPLY (SHOWING THE PROPORTION OF MAJOR FOODS PRODUCED AT HOME)

PRODUCT	1938	1975
Wheat and flour	23	52
Barley	46	95
Oats	94	99
Potatoes	?	93
Milk	100	100
Butter	9	22
Cheese	24	54
Eggs	61	97
Beef	49	85
Lamb	36	43
Pork	78	93
Bacon	29	44
Poultry	80	99
Sugar	33	31

Among the most important of these reasons are:

1. The small area of land available in Britain.
2. Physical limitations such as relief, drainage and soil type which further restrict the area of land available for farming.
3. The climate and weather conditions.
4. The structure of farming in Britain.

A closer examination of the table does, however, show that the pattern of food imports has changed and that Britain is now much less dependent upon imports than it was forty years ago. This means that the factors which limit food production in Britain are themselves not rigid, and that they have changed during this period. Naturally, changes have been greatest in the structure of farming and the physical limitations still remain.

The Land

Few areas are as complex or as varied as the British Isles, and, over very short distances, great differences can occur in both structure and scenery. To summarise such variety is difficult and exceptions can be found to every generalisation.

There does, however, appear to be a marked divide between *lowland* and *highland* areas. In the north and west of Britain there are large areas of land over three hundred and fifty metres in height and in North Wales, the Lake District and Northern Scotland altitudes of nine hundred metres are reached. These highland areas are related to geological conditions, for it is in the north and west that the most ancient rocks occur. Such rocks have been extensively folded and faulted and are often resistant to erosion.

To the south east the land is much lower, rarely reaching three hundred metres in height. Once again the influence of the underlying rock formations is important, and, over most of the area, rocks are younger, less disturbed by folding and faulting and less resistant to erosion. This does not mean that the land surface is uniform. For example, outcrops of limestone and chalk form ridges or *escarpments* which extend across England from the south coast to the north east. These escarpments are separated by large areas of lowland which are usually based on clays of various kinds.

The underlying structure, although important, is only one of the factors which have contributed to the development of the natural landscape of Britain. A second, and often more striking, contribution has been made by the pattern of erosion. Although erosion by running water is most important today, it is the effects of glaciation which have left the greatest imprint on the land. In the upland areas this is seen in a series of striking landforms such as U shaped valleys and corries. On the lowlands the effects are more marked, for the ice sheets deposited vast amounts of debris which completely mask the underlying rocks. In East Anglia, for example, these deposits are more than sixty metres thick and, without them, large areas of land would now be below sea level.

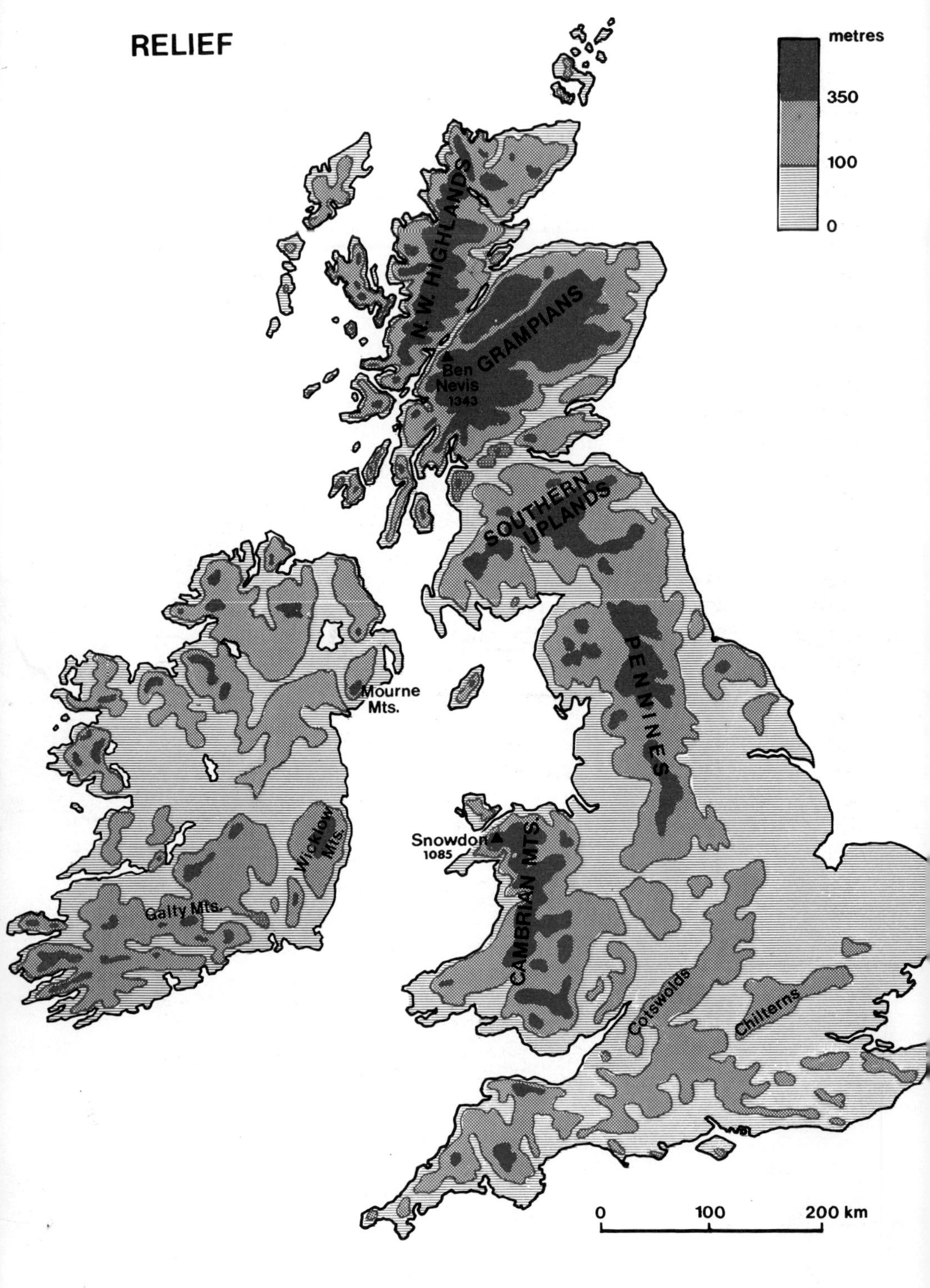
RELIEF
metres
350
100
0
N.W. HIGHLANDS
GRAMPIANS
Ben
Nevis
1343
SOUTHERN
UPLANDS
PENNINES
Mourne
Mts.
Wicklow
Mts.
Galty Mts.
Snowdon
1085
CAMBRIAN MTS.
Cotswolds
Chilterns
0
100
200 km

This combination of the underlying rock structure and a discontinuous cover of deposited material has important effects on the pattern of agriculture in Britain. For example:

1. Altitude has an important effect on farming. At 250 metres land becomes 'marginal' in terms of general farming and, although even cereals can be grown at heights of 500 metres, arable farming tends to become uneconomic. This is the result partly of the deterioration in climatic conditions, which takes place with increased height, and partly of the more difficult surface conditions which exist at altitude.

2. Relief is important, particularly in highland areas where slopes can be so steep as to prevent ploughing or at least make it uneconomic. In lowland areas the effects are usually less dramatic but slopes can cause local changes in farming.

3. Soils are formed when the rocks on the surface of the land are broken down by weathering. Because rates of weathering differ and because the land surface is varied, soils themselves vary enormously, often over very short distances. Generally speaking, however, in the highland areas soils are poorly developed and the heavy rainfall has washed out (*leached*) much of the mineral content. As a result, they tend to be thin and infertile. The pattern over the lowlands is even more complicated, particularly on the glacial deposits which are very varied. Overall, however, they do tend to be deeper and more fertile.

4. Drainage depends upon two main factors—the rate at which water percolates into the rock and the rate at which it runs off the surface. Where both take place at a slow rate drainage presents serious problems. This occurs over large areas of lowland Britain, particularly where clays outcrop or where the glacial deposits are very fine. In some places, the lowlands around the Wash for example, drainage was so poor that marshland occurred and peat was deposited. Such areas have been drained and now provide good farmland.

Although physical factors such as these influence farming, it is important to remember that during the last three hundred years many changes have been made which have modified their effects. Among the most important of these changes are the drainage of the clay lands and the improvement of light soils.

Climate and Weather

The British Isles lie in the cool temperate latitudes, off the west coast of Europe and on the eastern margins at the Atlantic Ocean. This position has strongly influenced the climate which is marked by the following general characteristics:

1. Cool summers with average temperatures reaching 15°C.
2. Mild winters with averages exceeding 5°C.
3. Rainfall which ranges from 60 cm in the south east to 400 cm in the western mountains.
4. No marked dry season.

The factors which have helped to produce this general pattern are relatively clear.

1. *Latitude*

Temperatures on the earth's surface tend to decrease from the Equator to the poles. This is known as the *temperature gradient* and it can be most clearly traced in the summer temperatures in Britain. As can be seen from the map on page 29, in July northern areas are generally cooler than those in the south. During the winter this pattern is obscured by the influence of the surrounding seas.

2. *The Sea*

Sea and land heat up and cool down at different rates. This has a striking effect on temperature conditions in Britain. During the summer the sea heats up more slowly than the land and has a cooling effect on many coastal areas. This is less important than the effect of latitude and, as we have seen the temperature gradient is from north to south. During the winter the sea retains its heat longer and has a warming effect on adjoining land areas. As a result the north–south gradient is replaced by a gradient from west to east. This reflects the fact that the North Atlantic has a much greater warming effect than the smaller North Sea. The contrast between east and west is made more pronounced by the influence of the North Atlantic Drift, a warm current which raises temperatures on the west coast by as much as 3°C.

3. *Prevailing Winds*

For more than one hundred and fifty days each year, on average, the British Isles come under the influence of westerly winds. These winds blow from the Atlantic Ocean and they spread the influence

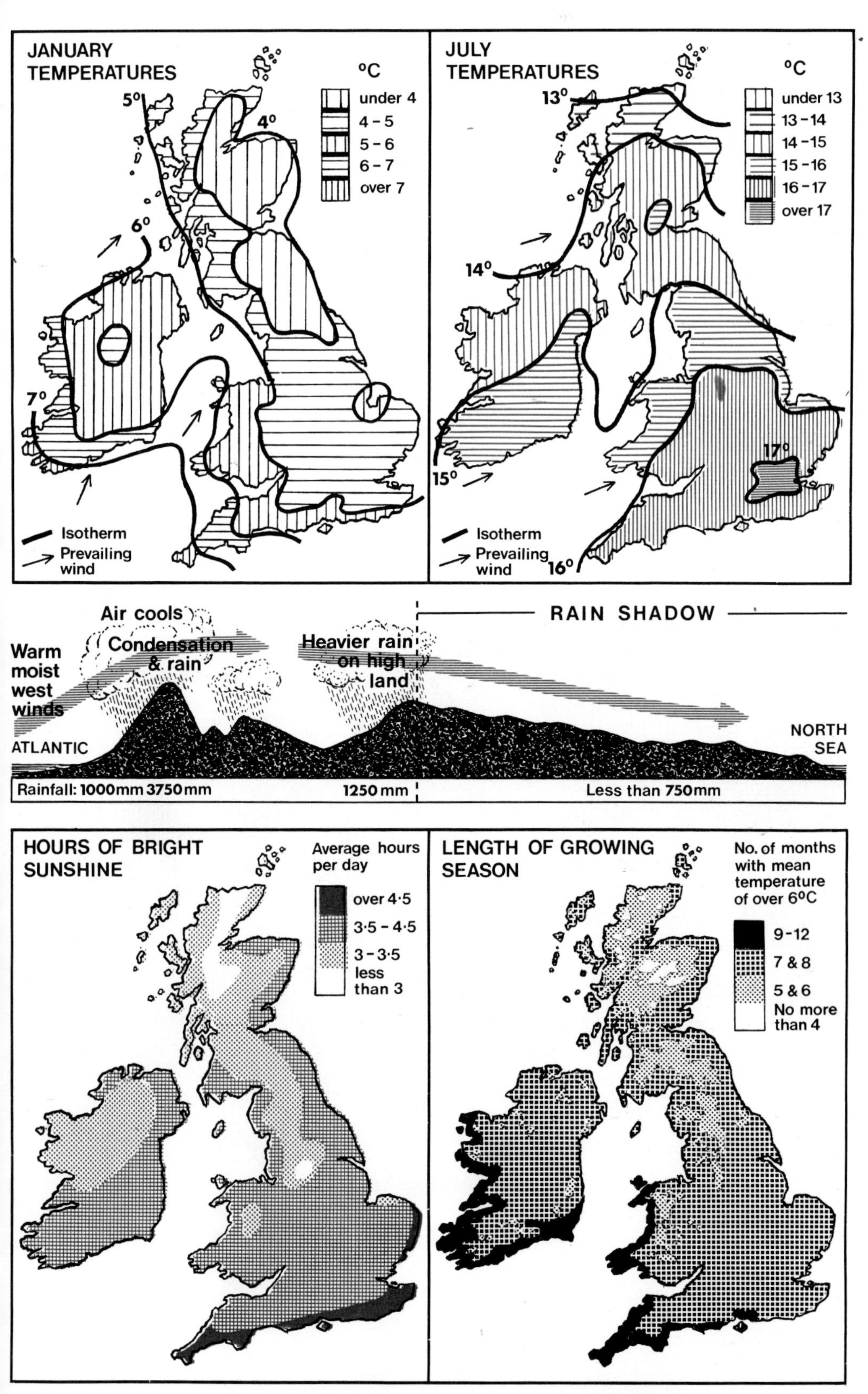

JANUARY TEMPERATURES
°C
under 4
4 - 5
5 - 6
6 - 7
over 7
5°
4°
6°
7°
Isotherm
Prevailing wind
JULY TEMPERATURES
°C
under 13
13 - 14
14 - 15
15 - 16
16 - 17
over 17
13°
14°
15°
16°
17°
Isotherm
Prevailing wind
RAIN SHADOW
Air cools
Condensation & rain
Heavier rain on high land
Warm moist west winds
ATLANTIC
NORTH SEA
Rainfall: 1000mm 3750mm
1250 mm
Less than 750mm
HOURS OF BRIGHT SUNSHINE
Average hours per day
over 4·5
3·5 - 4·5
3 - 3·5
less than 3
LENGTH OF GROWING SEASON
No. of months with mean temperature of over 6°C
9-12
7 & 8
5 & 6
No more than 4

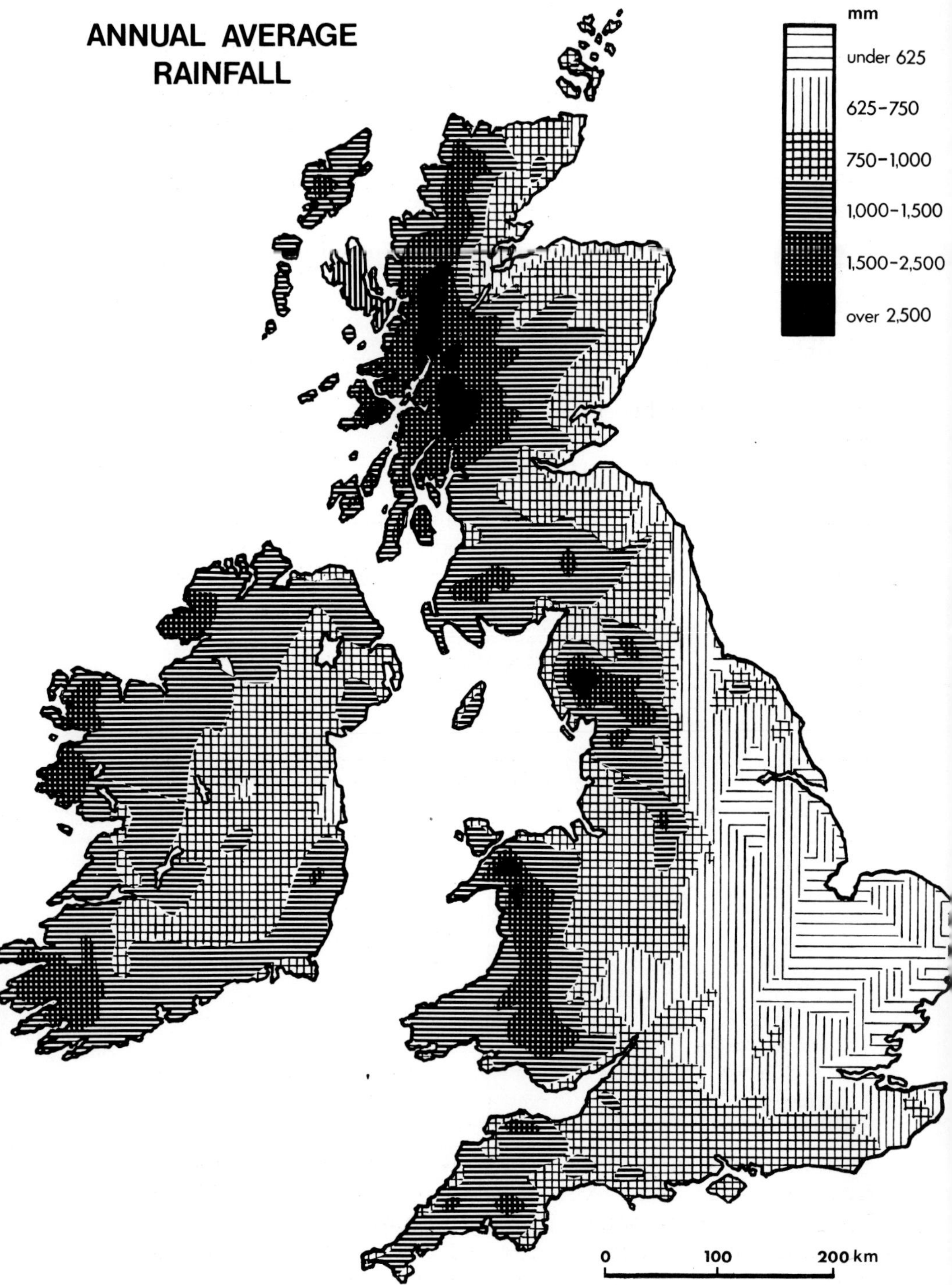
ANNUAL AVERAGE RAINFALL
mm
under 625
625–750
750–1,000
1,000–1,500
1,500–2,500
over 2,500
0
100
200 km

of the sea far inland. As a result, conditions in all parts of Britain tend to be equable and continental influences are limited even in those areas nearest to Europe.

The combined effects of the influence of the sea and prevailing winds are equally apparent in the general pattern of rainfall over the country. The westerly winds are onshore and, having crossed a large expanse of ocean, they contain moisture, which is released as rain when the air is forced to rise on reaching land. Furthermore, rainfall in the west is increased by the fact that the coast is bounded by high land which creates strong upward movements of air and increases its rate of cooling. The areas to the east of the highlands tend to be much drier because they lie in a *rainshadow*. These eastern areas also tend to show the effects of continental influences in that summer rainfall is often higher than might be expected.

CLIMATIC STATISTICS

PLACE	JULY TEMP (°C)	JANUARY TEMP (°C)	TOTAL (mm) RAINFALL	SEASON OF MAXIMUM RAINFALL
A	13	5	1800	Winter
B	17	4	610	Slight summer and autumn maximum
C	16	7	925	Winter
D	14	3	625	Slight summer maximum

Using information given on pages 29 and 30, complete the following exercise:

a) Trace an outline map of the British Isles. On it mark:
   i) the 5°C January isotherm
   ii) the 15°C July isotherm.

b) This will divide the British Isles into four divisions. Using the following words—warm/cool/mild/cold/wet/dry—describe:
   i) summer temperatures
   ii) winter temperatures
   iii) rainfall

   in each of the four divisions.

c) The above table contains climatic statistics for four places. In which of the four subdivisions does each place occur?

d) Account for the differences in climatic conditions in the four places.

Concentrating on general factors makes it difficult to appreciate that the British climate is among the most complex in the world and that its weather conditions are among the most varied. This variety and complexity can be attributed to two main causes—the incursion of *airmasses* and the passage of *depressions* over the country.

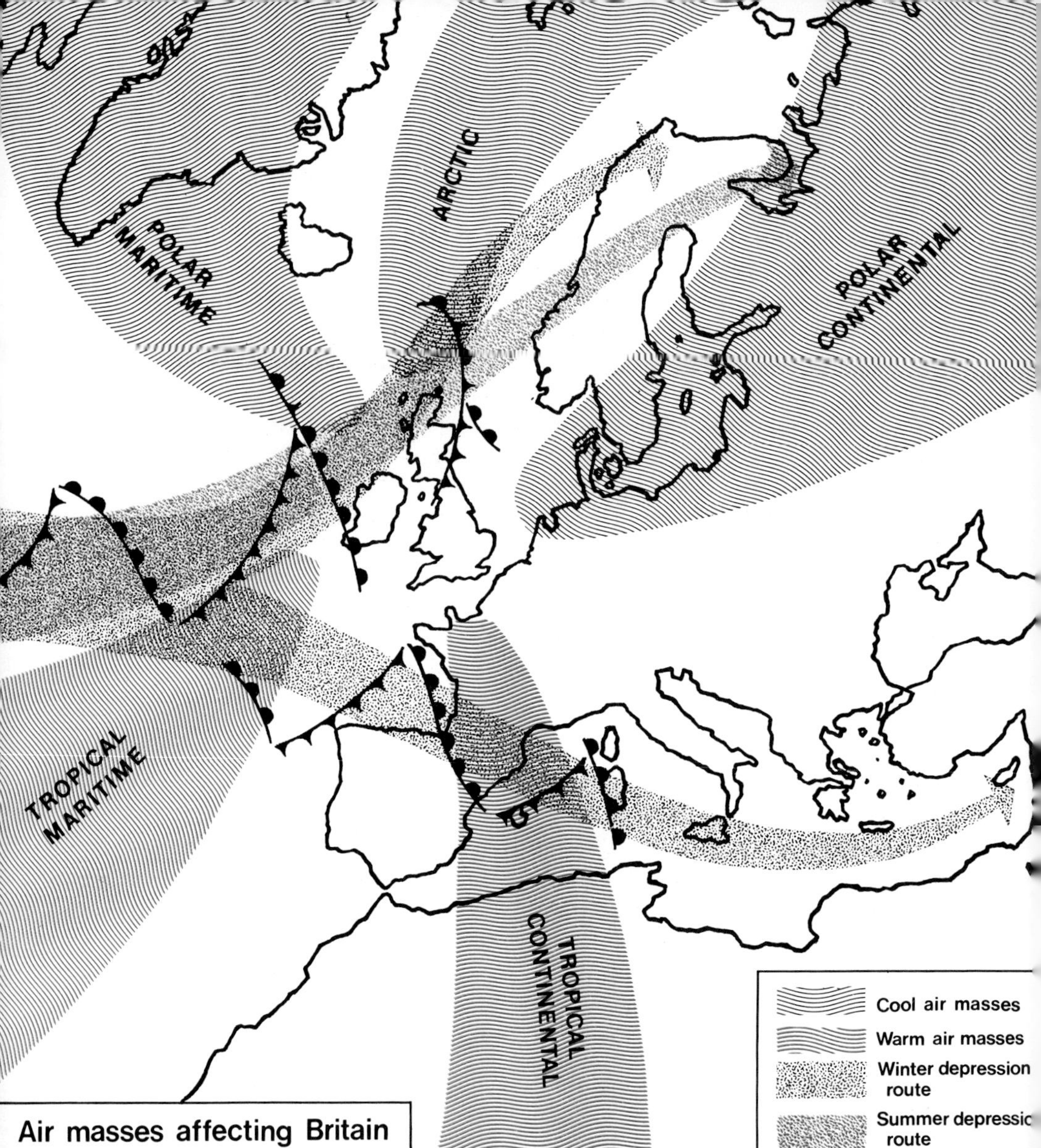

Air masses affecting Britain

1. Air Masses

An air mass is a vast body of air which is more or less similar in temperature and humidity throughout its area. Since a single air mass can cover an area of more than one million square kilometres and since its movement across the earth's surface is slow, its influence on the weather of regions which it crosses can be very striking. Britain is affected by five major air masses, each with a different character. They are:

1. Polar Maritime air which brings cool, rainy weather to Britain during the summer, and colder, wet conditions during the winter. Since this air mass crosses the warm Atlantic Ocean it is less cold than might be expected during the winter.

2. The Arctic air mass influences Britain less frequently but when it does it produces similar conditions with lower temperatures.

3. If Polar Continental air reaches Britain during the winter very severe weather is experienced, with low temperatures and snow. During the summer continental air gives dry sunny conditions.

4. The Tropical Continental air mass brings heat wave conditions during the summer. It rarely reaches Britain during the winter.

5. Tropical Maritime air penetrates more frequently, bringing hot, humid weather during the summer and mild, rainy weather during the winter.

It is the interplay of these airmasses which produces the changing pattern of weather in the British Isles. This in itself would be difficult to decipher but the picture is further complicated by the fact that Britain lies in the zone of contact between tropical and polar air—a zone which produces depressions.

2. Depressions

Depressions are large areas of low pressure which generally move towards Britain from the North Atlantic Ocean. They are formed in the zone of contact between warm tropical air and cool polar air, and, as a result, most temperate latitude depressions contain a wedge of warm air virtually surrounded by cooler air. It is the presence of this *warm sector* which gives rise to the complex weather pattern associated with depressions, including rapid changes in temperature and rainfall.

Using information given on pages 34 and 35, complete the following exercise:

a) What is the name given to:
   - i) the segment of warm air
   - ii) the rear edge of the warm air
   - iii) the leading edge of the warm air?

b) i) Why does rain fall along the zone of contact between warm and cold air?
   - ii) Describe how the two fronts differ and explain the effect this has on the pattern of rainfall in the depression.

c) Write a description of the sequence of weather which will be produced as the depression moves across Britain from west to east. Include changes of pressure, wind direction and cloud cover, as well as temperature and rainfall.

A depression over Britain -showing the weather associated with it

SATELLITE PHOTOGRAPH

SECTION THROUGH THE DEPRESSION

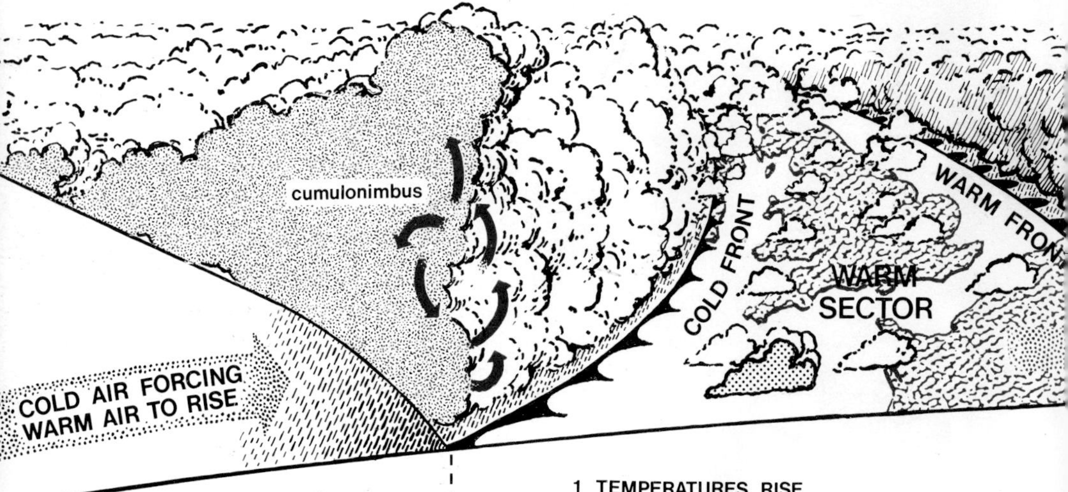

1 TEMPERATURES FALL
2 HEAVY CLOUD COVER
3 HEAVY RAINFALL
4 WINDS VEER
5 PRESSURE RISES
6 SKIES CLEAR

COLD FRONT PASSES

1 TEMPERATURES RISE
2 RAIN STOPS
3 CLOUD COVER BREAKS
4 SKIES CLEAR GIVING SUNSHINE WITH SCATTERED SHOWERS
5 WINDS VEER

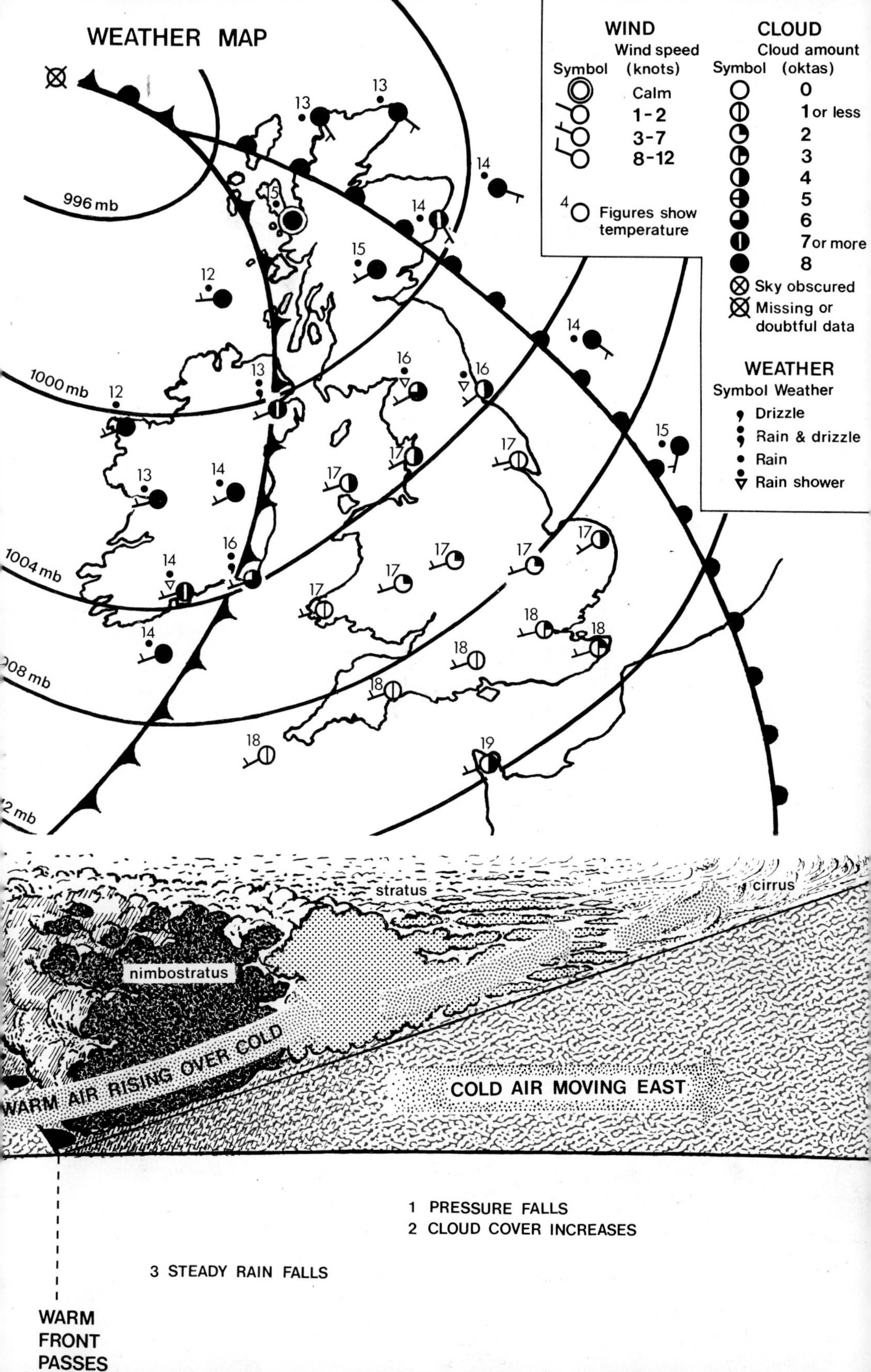

WEATHER MAP
996 mb
1000 mb
1004 mb
008 mb
2 mb
WIND
Wind speed
Symbol (knots)
Calm
1-2
3-7
8-12
Figures show temperature
CLOUD
Cloud amount
Symbol (oktas)
0
1 or less
2
3
4
5
6
7 or more
8
Sky obscured
Missing or doubtful data
WEATHER
Symbol Weather
Drizzle
Rain & drizzle
Rain
Rain shower
stratus
cirrus
nimbostratus
WARM AIR RISING OVER COLD
COLD AIR MOVING EAST
1 PRESSURE FALLS
2 CLOUD COVER INCREASES
3 STEADY RAIN FALLS
WARM FRONT PASSES

Depressions dominate the British weather and, on average, two frontal systems cross the country each week. They vary in intensity and, as a result, in the weather which they produce. Some are very deep and contain a well developed warm sector, with its attendant fronts. Others are filling in and the warm sector has been lifted off the ground to form an *occluded front*, which is usually much weaker than the fronts described above.

3. **Anticyclones**

Depressions are separated by ridges of higher pressure which give more settled drier weather. Such features are usually short lived and are succeeded by new frontal systems. Occasionally, however, large areas of high pressure, known as *anticyclones* develop. These are usually associated with incursions of tropical or continental air and, if they become well established, they can produce long spells of settled weather—hot and dry in the summer; cold and dry in the winter, with frost and fog.

It is the variety and changeability of the British weather which presents most problems to farmers. Average conditions are remarkably good and temperature, rainfall, hours of sunshine and length of growing season combine to make possible a wide range of farming in all parts of Britain. But the weather in any particular year can be unfavourable and this can cause serious losses. Furthermore, the variations in climate across the British Isles, whilst not great enough to exclude farming activities, do tend to encourage changes in emphasis from one place to another.

The Pattern of Farming

The Factors influencing Farming

Physical Factors (see previous sections)

1. Rainfall
2. Temperature
3. Length of growing season
4. Amount of sunshine
5. Altitude
6. Relief and slope
7. Drainage
8. Soils
9. Aspect

Aspect is also important because it causes local changes in climate. For example, south facing slopes tend to be warmer than north facing slopes, and valley floors, although sheltered, often experience more severe frosts than the sloping valley sides.

Economic Factors

Most farmers aim to make a profit and farming is therefore, strongly influenced by economic factors. Among the most important of these factors are:

1. Availability of markets for the produce.
2. Access to these markets.
3. Prices obtained for the produce.
4. Production costs which are in turn determined by factors such as the size of farm and the efficiency of its organisation.

Changes in either costs or prices can produce changes in the pattern of farming.

Government Policy

The present pattern of farming in Britain owes a great deal to decisions taken by the national government. During the nineteenth century Britain became increasingly dependent upon imported food. The danger of this situation became apparent during the two world wars of the twentieth century when the country was almost starved into defeat by the German blockade. As a result, it was decided to support agriculture by paying subsidies to farmers which would enable them to compete with overseas producers. This meant that food could be sold in shops at prices which did not cover production costs and that the British farmer depended for his profit upon subsidies paid out of the revenue from taxes. By changing the subsidies the government could change the pattern of farming. Entry into the EEC ended this system but agriculture is still protected, by an artificial price structure and by duties imposed on imported food.

Personal Choice

Every farmer decides how he will use his own land and it is this, more than anything else which produces the variety of farming found within very small areas of Britain.

The combination of so many factors has produced a very complex pattern of farming in Britain. It is possible, however, to identify certain general trends.

Using information given on pages 26, 29, 30, 38, and 40, complete the following exercise:

a) There is a marked difference in farming and farming landscape between western and eastern Britain (see photographs on page 39). Details of the differences can be obtained by choosing the correct

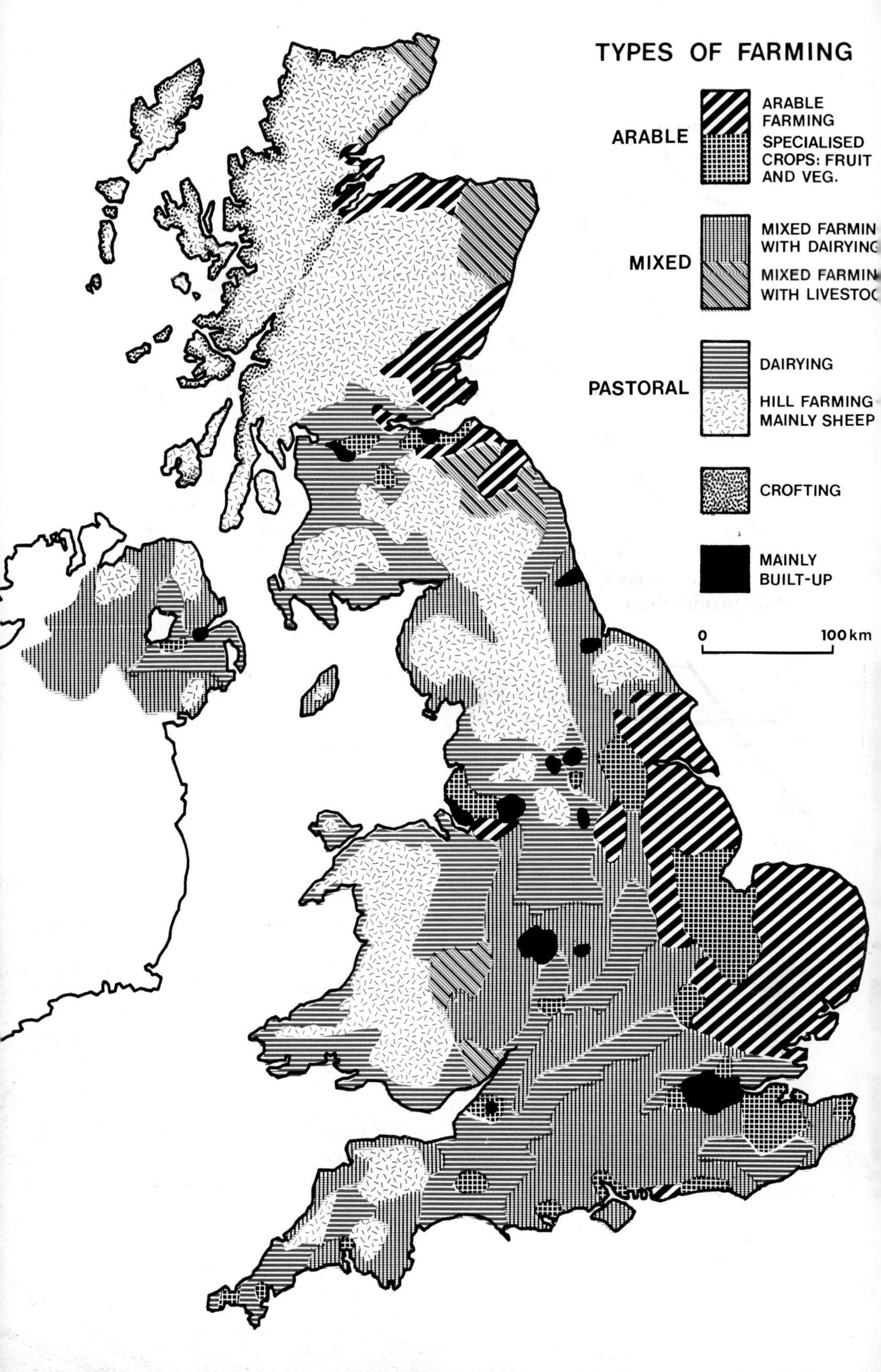
TYPES OF FARMING
ARABLE
ARABLE FARMING
SPECIALISED CROPS: FRUIT AND VEG.
MIXED
MIXED FARMIN
WITH DAIRYING
MIXED FARMIN
WITH LIVESTOC
PASTORAL
DAIRYING
HILL FARMING MAINLY SHEEP
CROFTING
MAINLY BUILT-UP
0
100 km

statements from the following i) for the west, ii) for the east.

1. Large areas of land over/under 250 metres.
2. Relief rugged/gentle, making the use of large machines easy/difficult.
3. Rainfall more than/less than 750 mm.
4. January temperatures over/below 5°C.
5. July temperatures over/below 15°C.
6. Average sunshine more than/less than 3·5 hours per day.
7. Growing season above average/below average/varied.
8. Conditions favour pasture/arable farming.

b) Describe the main changes which have taken place in the pattern of British agriculture during the twentieth century. (see page 40).

c) Referring to the factors listed on pages 36 and 37, explain why these changes have taken place.

The picture of British farming which emerges from the maps and statistics is fairly clear. The industry is highly efficient, a remarkably large proportion of the total land area is used for farming, a wide range of food is produced, productivity per hectare is high, mechanisation has spread rapidly and the labour force has declined while productivity has increased.

Farmland in Cornwall.

Farmland in East Anglia.

Changes in British Agriculture

LAND USE (AS A % OF LAND AREA)

1938 ENGLAND & WALES	1938 SCOTLAND	1938 N. IRELAND	TYPE OF LAND	1975 ENGLAND & WALES	1975 SCOTLAND	1975 N. IRELAND
30	19	36	Arable	48	20	24
51	10	44	Pasture	33	5	49
17	71	20	Rough grazing	16	71	21

MAJOR CROPS (AS A PERCENTAGE OF TOTAL CROPLAND)

CROP	1938	1975
Wheat	14·8	14
Barley	7	33
Oats	18	2·3
Sugar beet	2·7	2·9
Potatoes	5·3	3·7
Root crops	8·1	1·7
Vegetables	1·9	2·7
Rotational grass	31	28

LIVESTOCK (MILLIONS)

TYPE	1938	1975
Cattle	8·6	14·4
Sheep	25	27·9
Pigs	4·5	8·9
Poultry	76	144
Horses	1·1	NA.

YIELDS (IN TONNES PER HECTARE)

	1938	1975
Wheat	2·2	4·4
Barley	2·0	3·8
Oats	1·9	3·9
Potatoes	17·5	28·5
Sugar beet	16·5	34·1

WORKFORCE

	1938	1975
Workforce (000's)	825	409

MACHINES USED ON FARMS (000's)

	1938	1975
Tractors	101	517
Combined harvesters	15	61

SIZE OF FARMS (AS A PERCENTAGE OF TOTAL NUMBER OF FARMS)

	UNDER 2 Hect.	2–20	20–40	40–120	OVER 120 Hect.
1938	16	50	16	15	3
1975	18	44	16	17	5

But such a picture would not be a true reflection of the complexity of British farming. Even the distinction between eastern and western Britain is far from clear and it is important to remember that arable and pastoral farming are practised in all parts of the country and that all that the map shows is a change of emphasis. Equally important, by concentrating on general trends, it is easy to underestimate the immense variety of farming which is found in Britain, even within very small areas. And finally, of course, general trends can give no indication of the complexity of individual farms or of farming in individual regions. This can be seen only from more detailed case studies.

Farm study: Hill Farming in the Lake District

Hill farming is the large scale rearing of stock, particularly sheep, using upland pastures. It is found throughout highland Britain and, of all the types of farming in the country, it is the most obviously influenced by environmental factors such as altitude, slope, rainfall and quality of soils. In many places these factors combine to make large areas of land '*marginal*', ie barely capable of producing a worthwhile return for the farmer, and it is not surprising that the last hundred and fifty years has seen a gradual drift of farmers away

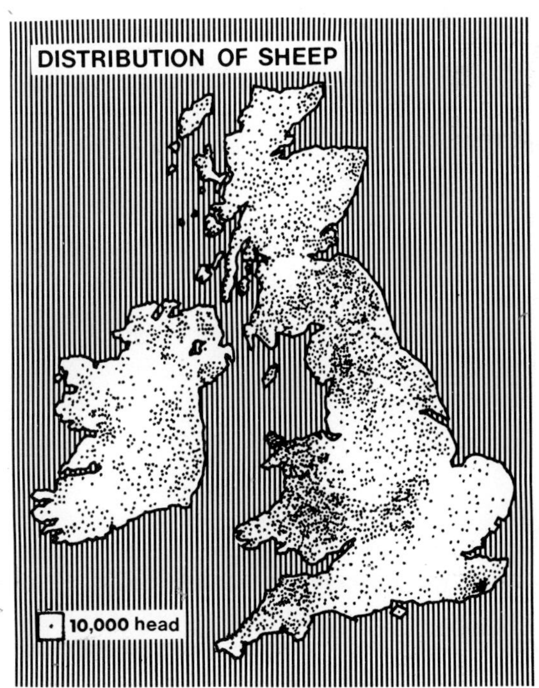

HILL FARM

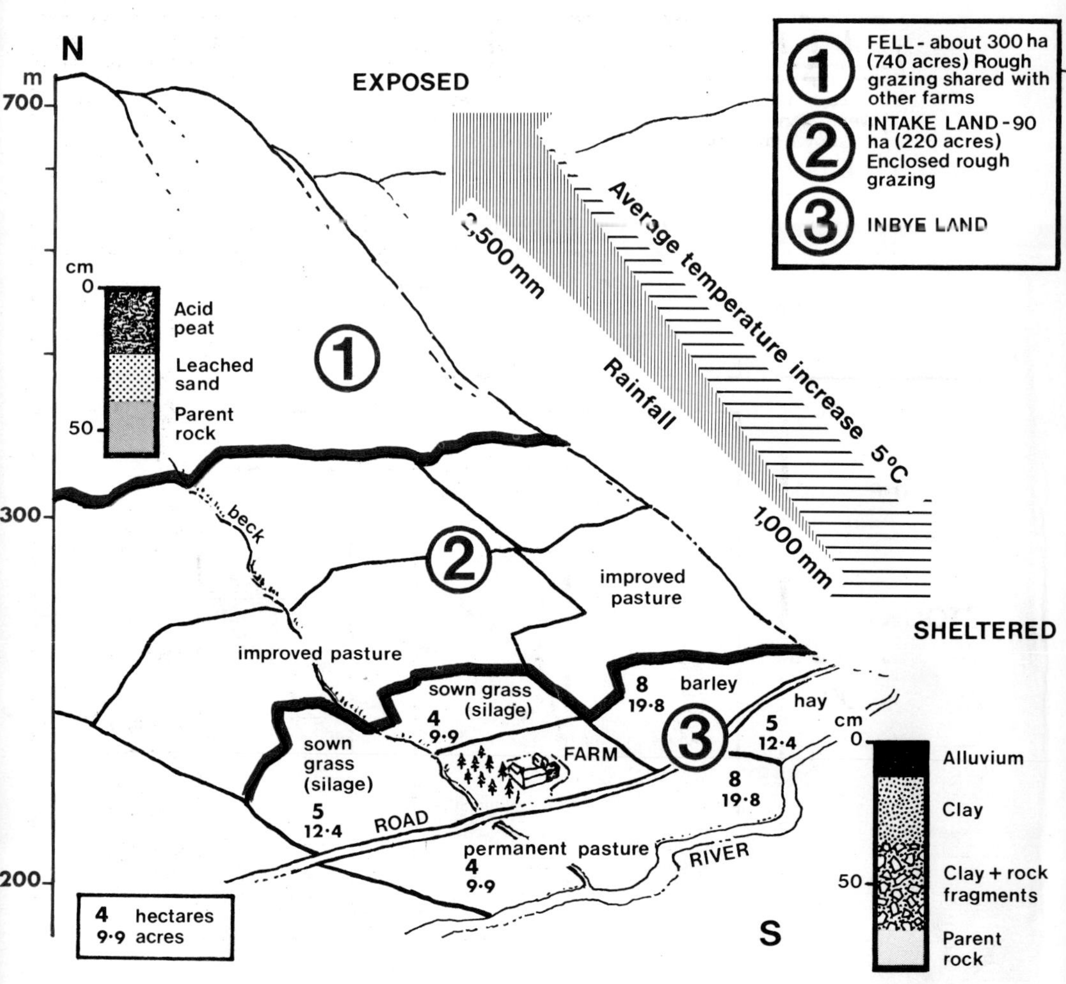

from such areas and the amalgamation of their small holdings to produce the large hill farms that we see today.

The nature and problems of hill farming are best seen by studying one farm.

Grange Farm

Grange Farm is a holding of one hundred and twenty five hectares. This would appear to be very small for a hill farm but, as is usual in the Lake District, the farmer has grazing rights on three hundred hectares of open *fell* a large part of which is held in common by all of the farmers in the parish. A smaller area of fell is actually owned

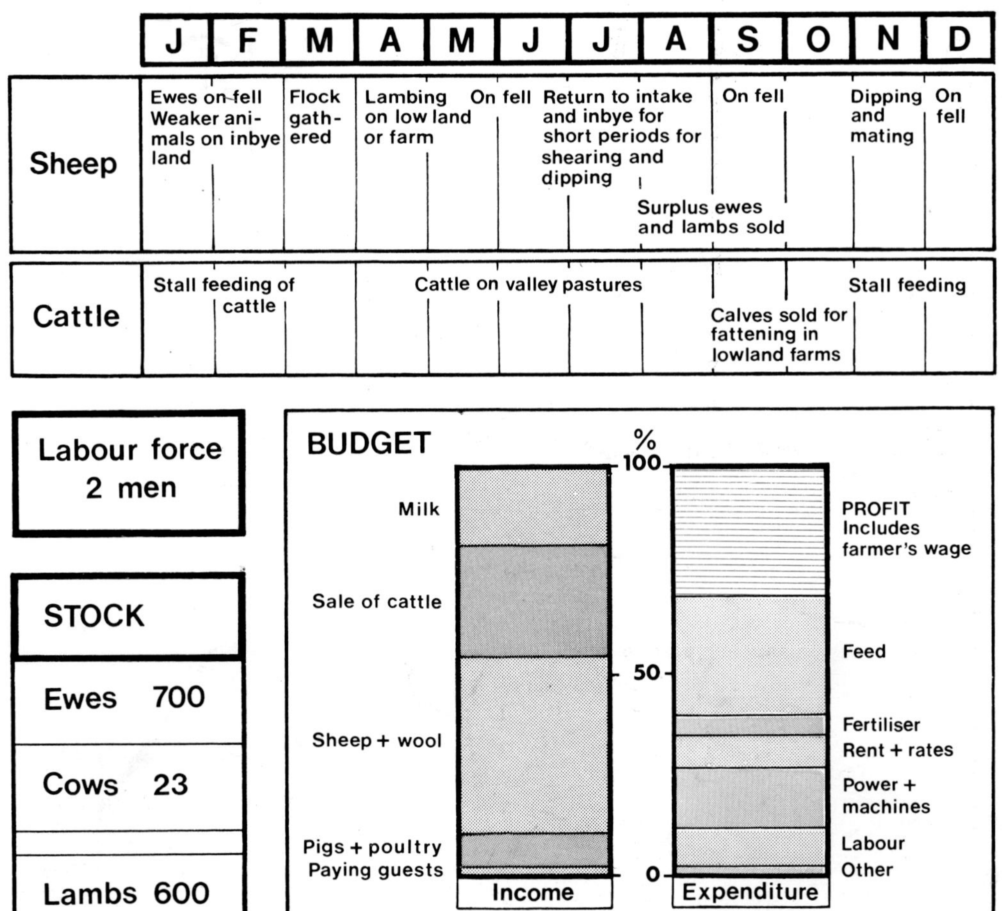

by the farmer. This area, known as the *intake* has been enclosed and parts of it have been improved to provide better pasture. Both the fell and the intake suffer from exposure, steep slopes and thin acid soils which have been leached by the heavy rains. The best land is on the valley floor and this, known as the *inbye*, is where the main crops are produced. Here soils are better and the climate is more sheltered. The inbye does, however, make up a very small proportion of the farm and arable farming is of minor importance, being practised largely to provide winter feed for the stock. In fact the entire organisation of the farm centres on maintaining the sheep and cattle which together contribute more than ninety per cent of the total farm income.

Refer to the farm plan and complete the following exercise:

a) Calculate, as a percentage of the total land available (including fell):
   i) the area of inbye
   ii) the area of intake
   iii) the area of fell.

b) i) List the crops grown on the farm.
   ii) On what type of land are they grown?
   iii) What are the advantages of this land?
   iv) Where is the permanent pasture located? Why is it located there?

c) i) For how many months are animals kept on the open fell?
   ii) Why are they brought down from the fell?
   iii) Why are cattle not reared on the fell?

d) i) How does the intake differ from the open fell?
   ii) What is it used for?
   iii) Part of the intake has been improved. What are the factors which caused the farmer to improve this area while leaving the remainder as rough pasture?

e) i) What proportion of the farm income comes from sheep?
   ii) What proportion comes from cattle?
   iii) List two other contributors to the farm income.
   iv) Would you say that farming is intensive or extensive?

f) Write a brief account of the hill farm, showing how its organisation is influenced by environmental factors.

Productivity per hectare is low and stock rearing on this type of land remains marginal. Many steps have been taken to improve the situation.

1. Land has been enclosed and improved to give better pasture. The importance of this can be seen from the fact that on the intake three ewes can be reared on each hectare of land, while on the open fell this figure drops to less than one. Unfortunately there are limits to the extent of these improvements. The higher land is colder and less hospitable and slopes are much steeper. As a result, soils are poorly developed and incapable of supporting improved pasture. Even more important, most of this land is held in common and this makes improvement very difficult.

2. The introduction of cattle has brought great benefits to the hill farm. Once again, however, further development is limited because pasture is in short supply, particularly the lowland pasture needed for cattle.

3. New breeds of sheep have been introduced to replace the hardy mountain sheep. This has led to better quality meat and wool and generally improved yields.

4. The government has given grants and subsidies to encourage

Farmland in the Lake District showing the contrasts between the fell, the intake and the valley floor.

improvement schemes and to help overcome the problems of a difficult environment.

In spite of this, hill farming remains marginal and the farmer, like many others in the Lake District and other parts of highland Britain, has turned to the tourist trade. Paying guests are taken, a cottage is let during the summer and plans are in hand to develop camping and caravan facilities near the river.

The Regional Setting

Similar trends can be seen in the Lake District as a whole. Agriculture is important in the regional economy and, as might be expected in an upland area, its patterns and problems are strongly influenced by environmental factors.

Large areas of land are over six hundred metres above sea level and, as a result, altitude severely limits farming activities. Furthermore, the valleys of the region have been deepened by glaciation and this makes slope and aspect very important. Add to this the fact that the Lake District is one of the wettest areas of Britain and it is

General view of the Lake District at Helvellyn showing the effects of glaciation eg Corrie (a), Arête (b), Tarn (c), U-shaped Valley and Finger Lake (d).

not surprising that arable farming is of negligible importance or that the bulk of the land is either permanent pasture or rough grazing. Stock rearing is by far the most important farming activity and, as was the case at Grange Farm, it is only possible when valley pastures are available to supplement the poor, rough grazing of the fells. In fact, in some parts of the Lake District links have been established between hill farms and farms on the surrounding lowlands, and animals are moved from the highland pastures during the winter or for fattening. This movement, known as *transhumance*, was once very important and, although it has declined, it still persists in some areas.

Because the land is marginal there has been a steady movement of farmers away from hills. A few managed to find employment in other local industries such as slate quarrying, but the majority were forced to move out of the area. Of these, many travelled the short distance to the coastal lowlands of west Cumbria. Here they were attracted to two areas in particular:

1. The Whitehaven–Workington area where the occurrence of coal and iron ore had given rise to an important iron and steel industry.

2. The Furness district around Barrow which contained some of the richest iron ores in Britain. These were developed during the late nineteenth century and Barrow became a boom town with major iron making, engineering and shipbuilding industries.

During the twentieth century both areas have suffered a long period of decline and many of the early basic industries have virtually disappeared. The Cumberland coalfield is almost worked out and mining has ceased over most of its area. Iron ore production has been revived but is on a very small scale, and the last major steelworks closed during the 1960's. Replacement industries have been slow to arrive, largely on account of the remoteness of the region, and the coastlands no longer attract population from the hill country. In fact, in recent years there has been some revival of the economy of the upland area, based largely on the growth of the tourist industry.

The tourist industry became established in the Lake District during the nineteenth century but the numbers involved were very small. This situation changed completely after the Second World War when the number of visitors increased enormously. Two factors have contributed to this:

1. The designation of the area as a national park which gave people increased access to an area of great scenic beauty.

2. The increase in car ownership and the building of the

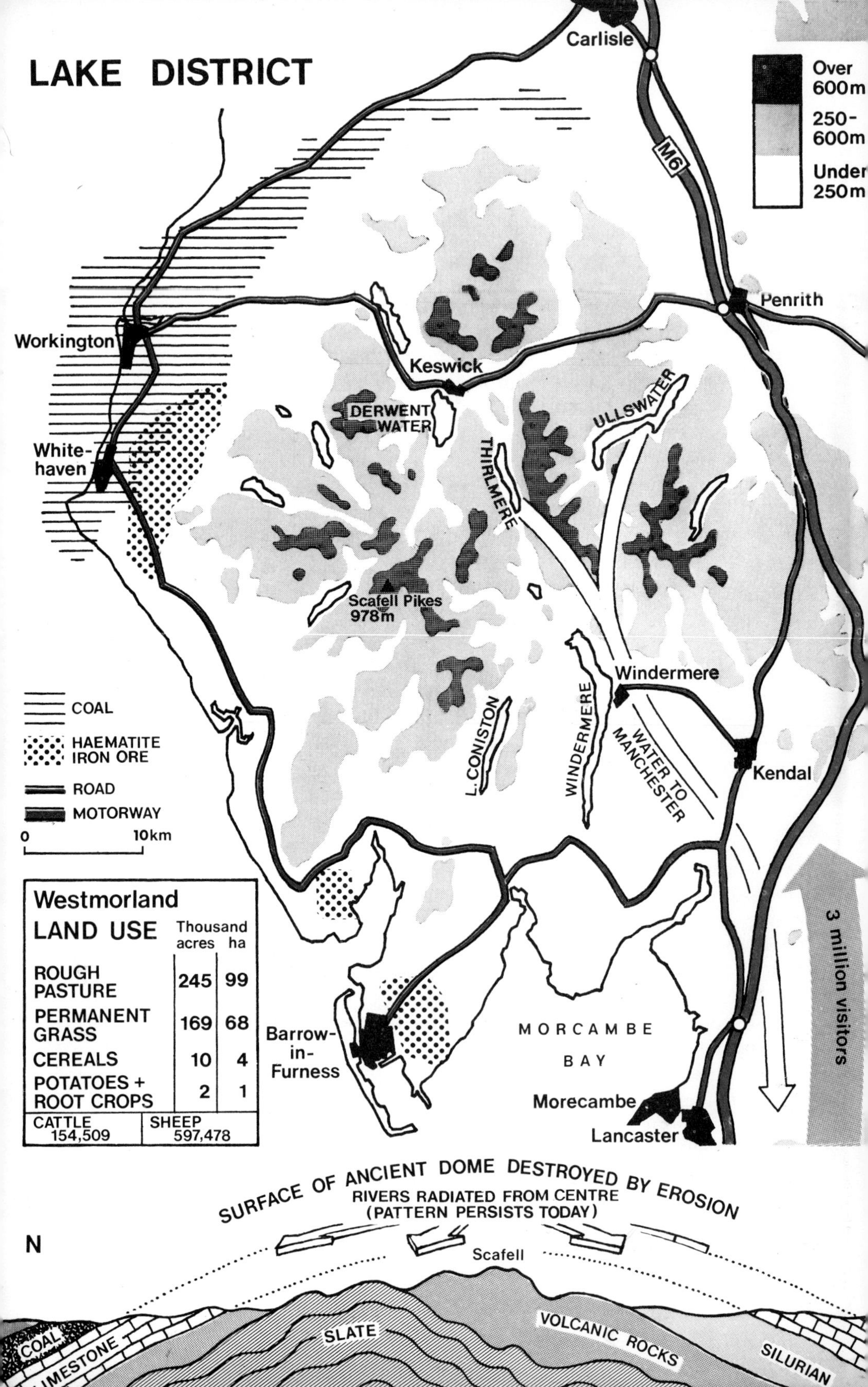

Westmorland LAND USE	Thousand acres	ha
ROUGH PASTURE	245	99
PERMANENT GRASS	169	68
CEREALS	10	4
POTATOES + ROOT CROPS	2	1
CATTLE 154,509	SHEEP 597,478	

motorway network which brought a large part of northern England within easy reach of the areas.

The effects of tourism are seen in the amount of money brought into the region and the number of jobs created.

Farming Study: Dairy Farming

The bulk of Britain's agriculture is carried out in lowland areas and here the pattern is much more complex and the relationship between farming and the environment much less clear. Nowhere is this more true than in the case of dairy farming where the interplay of environmental and economic factors has produced a very complicated pattern of distribution.

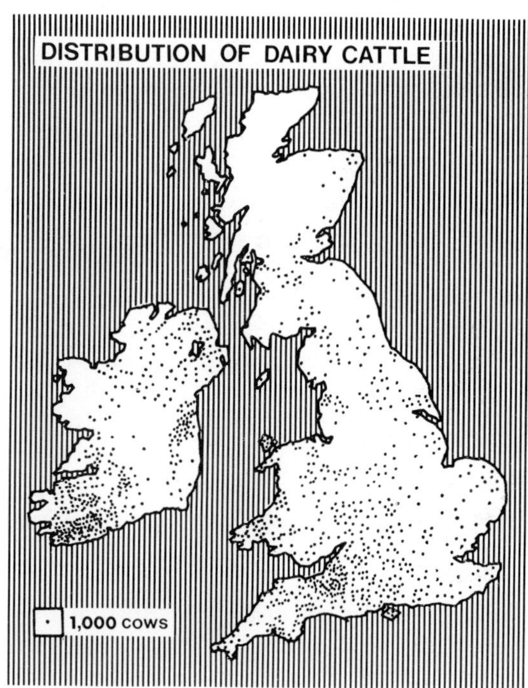

1. Environmental Factors

Since dairy cattle require good quality pasture if they are to produce high yields of milk, dairy farming has tended to develop in areas which favour grass growth.

Using information given on pages 26, 29, 30 and 39, complete the following exercise:

a) List the factors which have favoured the development of pasture in western Britain.
b) The following are important dairy farming regions:
the Plain of Somerset
South Devon
South West Wales
North Lancashire
Ayrshire
Cheshire
the Belfast Region
Kent and Sussex
   i) Trace a blank outline map of Britain.
   ii) On it mark and name the dairying regions mentioned above.
c) Using the climatic factors mentioned on page 39:
   i) Name the regions which conform to all of the factors.
   ii) Name the regions which conform to 3 or 4 of the factors.
   iii) Name the regions which conform to 2 or less of the factors.
d) Mark on the map the main conurbations in Britain (include Belfast, p. 284).

It is clear that there is a strong relationship between the distribution of dairy farming and environmental factors, and that the most important regions are in the west, where the heavy rainfall, mild winters and cool summers all favour the development of lush pasture land. At the same time, it is important to remember that dairy farms are numerous outside this region. Often their development can be explained in terms of the environment. For example, in Kent and Sussex the most important dairying districts are on heavy clay soils which retain moisture and which are difficult to cultivate. Similarly relief can be important locally, and pasture land is often found on land which is too steep to be ploughed. But such factors are of minor importance when compared with the influence of economic factors, particularly the availability of markets.

2. Economic Factors

Most of the milk produced in Britain is consumed in liquid form and it is, therefore, of vital importance that it reaches the consumer while it is fresh. As a result dairy farming tended to develop within one day's journey of the main towns and cities, almost regardless of environmental considerations. The building of the railways during the nineteenth century, however, meant that most parts of lowland Britain lay within these zones, and it was at this time that dairy farming became concentrated in the areas best suited to it. In spite of this, traces of the old pattern can be seen, and the availability of markets has played an important part in the development of dairying in south east England and East Anglia which supply London, and in Humberside and the Vale of York, which supply the Yorkshire cities.

As the distance between dairy farms and their markets increased, so the need for an efficient system of collecting milk grew. Today this is done by tanker and the milk is transported to a nearby dairy or collecting centre where it is processed and then shipped in bulk to the markets. Here, after further processing, the milk is put into containers for delivery to the consumers. Few dairying regions lie outside this network and milk is now transported by road and rail to distant markets. For example, London obtains milk from farms in the West Country which are more than five hundred kilometres away. In the more remote areas, however, a larger proportion of the milk yield is taken to dairies for conversion into cheese and butter.

The effects of these factors can be seen in the organisation of the individual dairy farm.

New Farm

Refer to the information given on pages 26, 29, 30 and 39, and complete the following exercise:

a) The farm is located on the Cheshire Plain. Locate this area on page 38 and, using the list of factors given on page 39, explain why dairying is important there.

b) Which markets are likely to be served by the area?

c) i) What is the area of the farm and how many cattle are kept?
  ii) What is the area of permanent pasture? Express this as a percentage of the total farm area.
  iii) What is the area under other forms of grass? Express this as a percentage of the total area.
  iv) What is this grass used for?
  v) What is the area under crops?
  vi) Draw a pie graph to show land-use on the farm.
  vii) One area is not used for pasture or cultivation. What is it used for and why is it left uncultivated?

d) During the winter grass growth ceases and the food value of the pasture is low. Alternative sources of feed are required.
  i) Name two types of winter feed produced on the farm.
  ii) Imported feed is bought in. What proportion of the farm expenditure is taken up by this?
  iii) How is the farm organised to provide food for the animals during the winter?
  iv) How does this differ from the method used during the summer?
  v) The farm has recently been modernised to improve winter feeding. What evidence is there to support this statement?

e) i) Why is there always a surplus of cattle for sale on dairy farms?

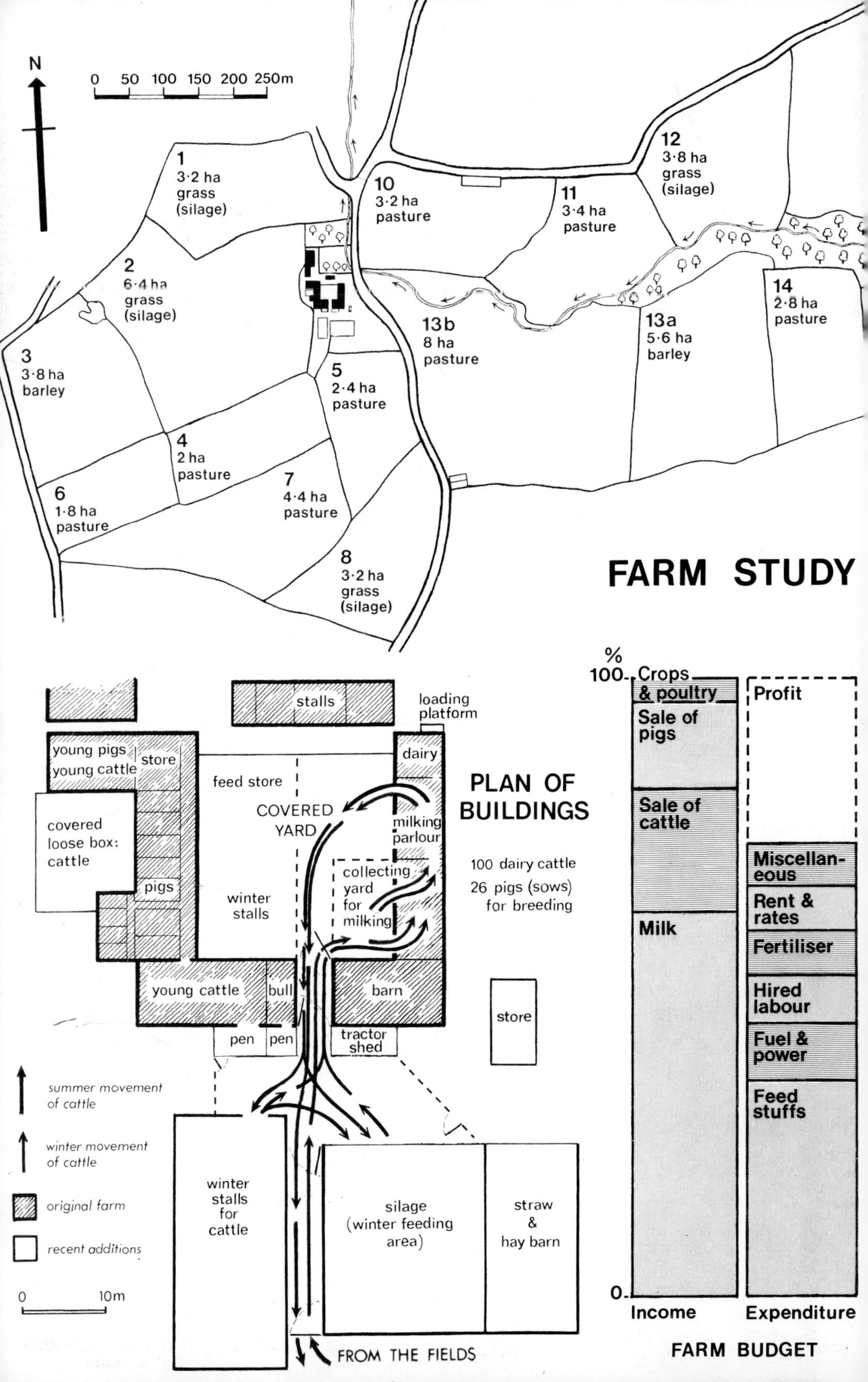

N
0 50 100 150 200 250m
1
3·2 ha
grass
(silage)
2
6·4 ha
grass
(silage)
3
3·8 ha
barley
4
2 ha
pasture
5
2·4 ha
pasture
6
1·8 ha
pasture
7
4·4 ha
pasture
8
3·2 ha
grass
(silage)
10
3·2 ha
pasture
11
3·4 ha
pasture
12
3·8 ha
grass
(silage)
13b
8 ha
pasture
13a
5·6 ha
barley
14
2·8 ha
pasture
FARM STUDY
stalls
loading platform
young pigs
young cattle
store
feed store
dairy
COVERED YARD
milking parlour
covered loose box: cattle
pigs
winter stalls
collecting yard for milking
young cattle
bull
barn
pen
pen
tractor shed
store
PLAN OF BUILDINGS
100 dairy cattle
26 pigs (sows) for breeding
summer movement of cattle
winter movement of cattle
original farm
recent additions
0 10m
winter stalls for cattle
silage (winter feeding area)
straw & hay barn
FROM THE FIELDS
%
100
0
Crops & poultry
Sale of pigs
Sale of cattle
Milk
Income
Profit
Miscellaneous
Rent & rates
Fertiliser
Hired labour
Fuel & power
Feed stuffs
Expenditure
FARM BUDGET

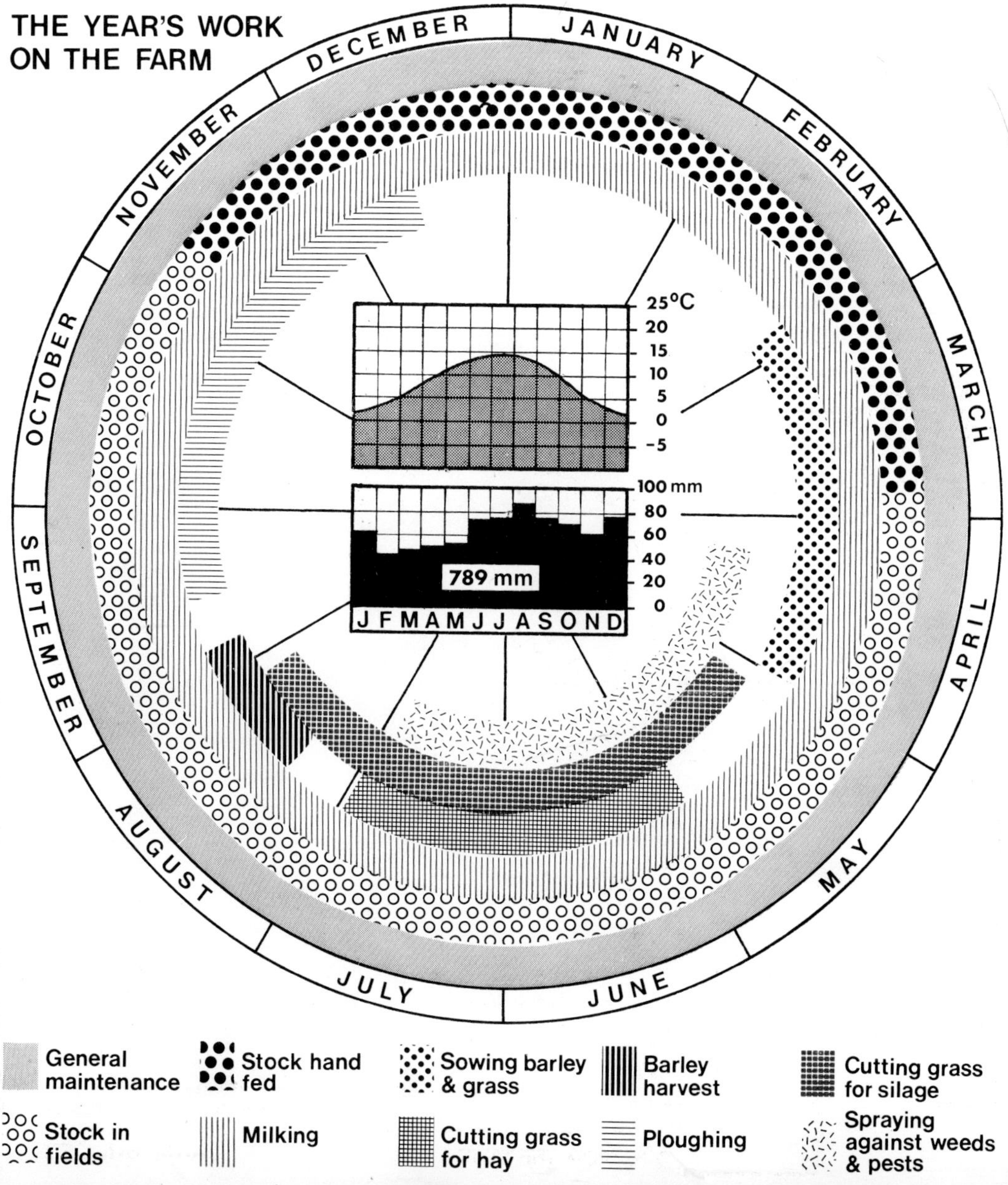

THE YEAR'S WORK ON THE FARM

JANUARY
FEBRUARY
MARCH
APRIL
MAY
JUNE
JULY
AUGUST
SEPTEMBER
OCTOBER
NOVEMBER
DECEMBER

25°C
20
15
10
5
0
−5

100 mm
80
60
40
20
0

789 mm

J F M A M J J A S O N D

- General maintenance
- Stock hand fed
- 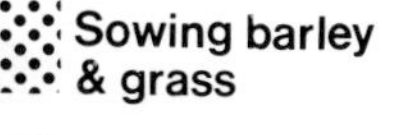 Sowing barley & grass
-   Barley harvest
- 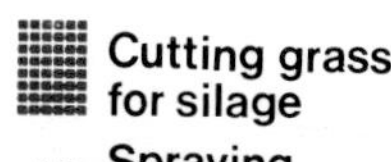 Cutting grass for silage
- Stock in fields
-  Milking
- Cutting grass for hay
- Ploughing
- Spraying against weeds & pests

ii) What proportion of the farm income does this supply?

f) i) What evidence is there that the dairy farm is more intensively farmed and requires a larger labour force than the hill farm?

ii) Write a brief description of the dairy farm and explain how it differs from the hill farm which also specialises in the rearing of animals.

The organisation of the dairy farm centres on the need to milk the cows twice each day. This, more than anything else, imposes limits upon the size of herd which can be handled and, in turn, restricts the size of farm. It also determines the way in which the farm is organised. Whenever possible animals are kept near to the farm. This means that the fields near the farm are usually put down to grass while outlying fields are used for crops, particularly feed crops such as barley. During the winter the problems are even greater since the animals spend a large proportion of the time in small yards or paddocks, entering the feeding area whenever they like. Because it is situated in one of the main dairying regions, the collection and processing of the milk is very efficient. In fact, the milk is taken to a dairy less than ten kilometres away from the farm and it is processed for sale in the nearby industrial area of North Staffordshire.

In recent years the farmer has faced increasing problems, most of which stem from the rapid rise in the price of imported feedstuffs and fuel.

A milking Parlour: taken with a 'fish eye' lense.

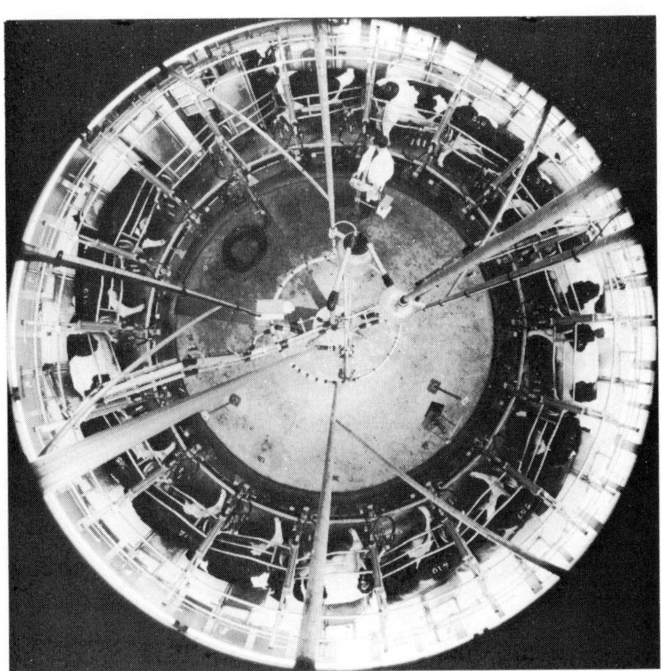

Farming Study: Cereal Farming in East Anglia

The second major type of farming in lowland Britain is arable farming and, as was seen on page 39 this tends to be most important in eastern areas.

In both organisation and lay-out the arable farm differs considerably from the dairy farm described in the previous section.

Green Farm

Green Farm is a holding of almost two hundred hectares, situated some twenty five kilometres north of Norwich in north east Norfolk. It is almost entirely given over to arable farming.

Refer to pages 56 and 57 and complete the following exercise:

a) i) Describe the relief of the area in which the farm is located.
   ii) Give the maximum and minimum temperatures, total rainfall and season of greatest rainfall in the area.
   iii) Using the list of factors given on page 39, explain why conditions favour the development of arable farming.

b) i) What is the total area of the farm? How does this compare with the dairy farm?
   ii) What is the average size of the five largest fields? How does this compare with the five largest fields of the dairy farm?

Combine Harvester at work on a modern arable farm.

CEREAL FARM: in East Anglia

	O	N	D	J	F	M	A	M	J	J	A	S
Temperature °C	11	7	4	4	4	6	8	12	15	17	17	14
Rainfall mm TOTAL: 650	64	69	61	61	46	38	48	43	43	66	53	58
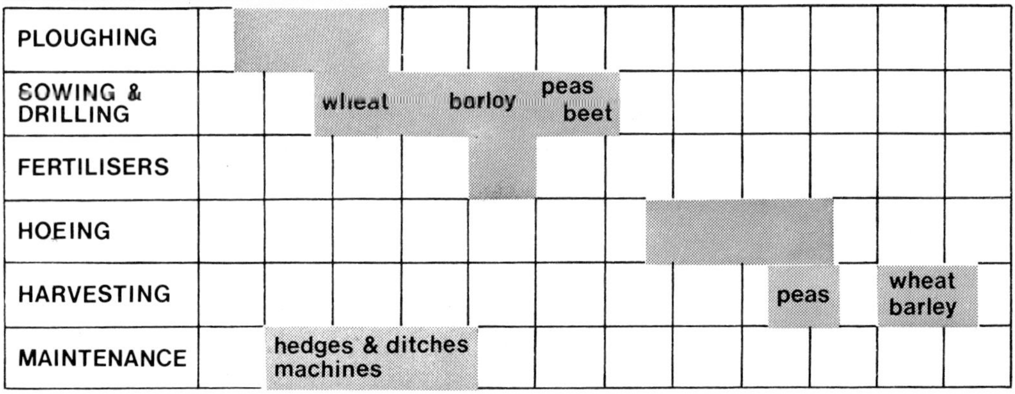 PLOUGHING												
SOWING & DRILLING			wheat		barley	peas beet						
FERTILISERS												
HOEING												
HARVESTING									peas		wheat barley	
MAINTENANCE		hedges & ditches machines										

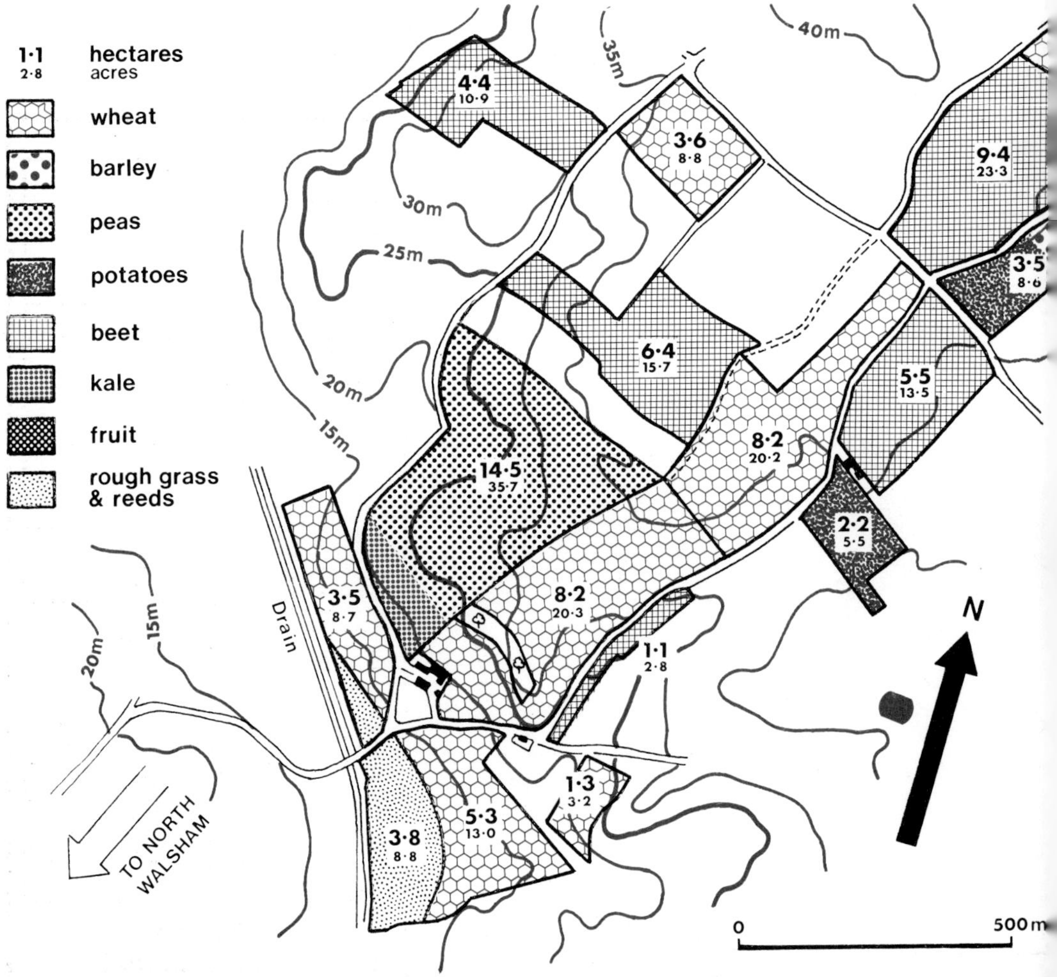

1·1 hectares
2·8 acres

- wheat
- barley
- peas
- potatoes
- beet
- kale
- fruit
- rough grass & reeds

40m
35m
30m
25m
20m
15m
4·4 10·9
3·6 8·8
9·4 23·3
3·5 8·6
6·4 15·7
5·5 13·5
8·2 20·2
14·5 35·7
2·2 5·5
3·5 8·7
8·2 20·3
1·1 2·8
Drain
N
1·3 3·2
5·3 13·0
3·8 8·8
TO NORTH WALSHAM
0 500m

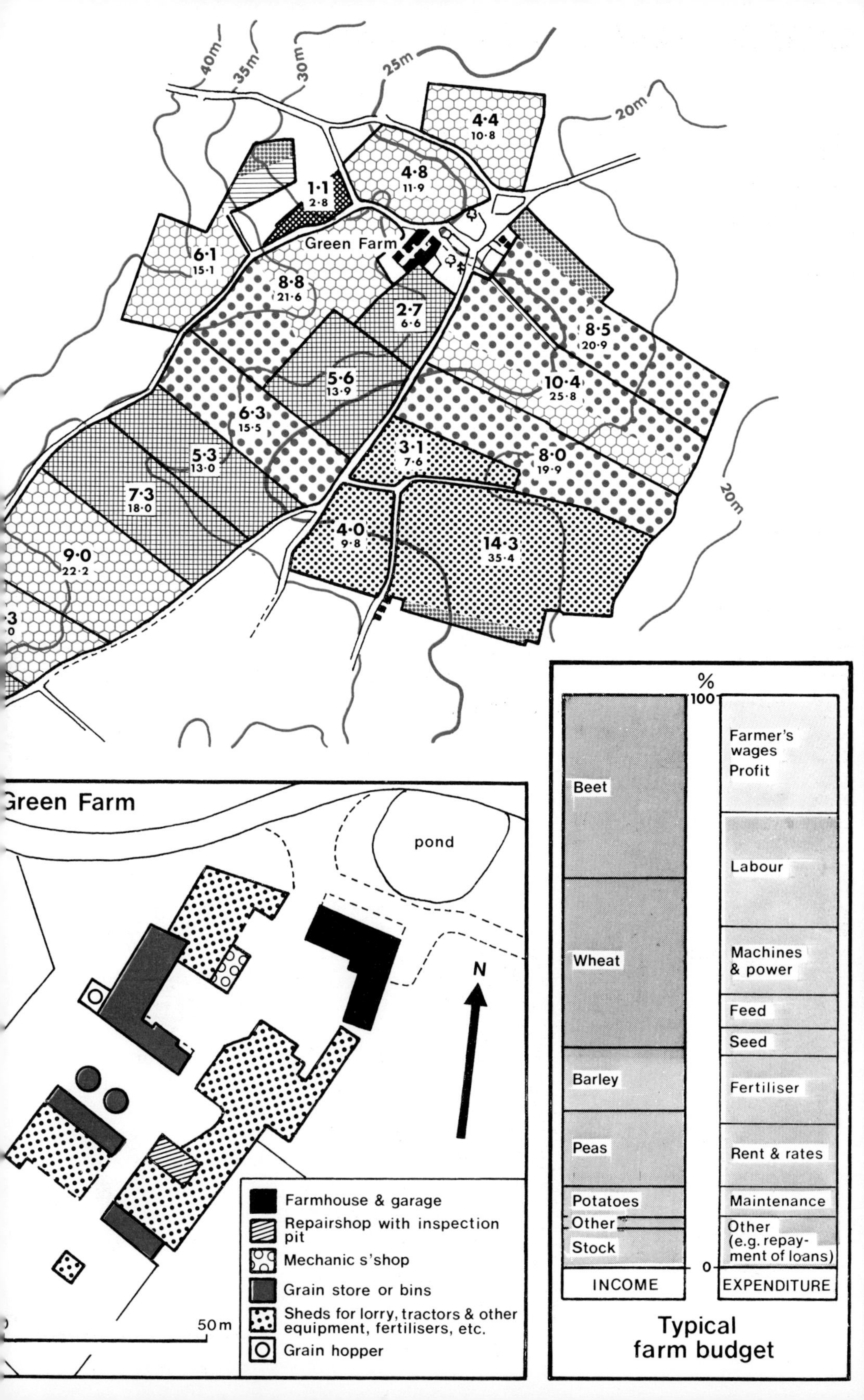
40m
35m
30m
25m
20m
20m
4·4
10·8
4·8
11·9
1·1
2·8
6·1
15·1
Green Farm
8·8
21·6
2·7
6·6
8·5
20·9
5·6
13·9
10·4
25·8
6·3
15·5
5·3
13·0
3·1
7·6
8·0
19·9
7·3
18·0
4·0
9·8
14·3
35·4
9·0
22·2
Green Farm
pond
N
50m
Farmhouse & garage
Repairshop with inspection pit
Mechanic s'shop
Grain store or bins
Sheds for lorry, tractors & other equipment, fertilisers, etc.
Grain hopper
%
100
0
Beet
Wheat
Barley
Peas
Potatoes
Other
Stock
INCOME
Farmer's wages
Profit
Labour
Machines & power
Feed
Seed
Fertiliser
Rent & rates
Maintenance
Other (e.g. repayment of loans)
EXPENDITURE
Typical farm budget

CROP ROTATION ON THE ARABLE FARM

YEAR	ACTUAL TYPE OF CROP	IDEAL TYPE OF CROP
1	Peas	Peas
2	Sugar beet	Wheat
3	Cereal	Beet
4	Cereal	Wheat
5	Peas	Peas
	etc	

Changed to give:

1. Longer period between crops of peas.
2. Longer period between crops of sugar beet (both exhaust the soil).
3. Longer period between cereal crops (limits spread of disease).

c) i) List the main crops grown on the farm, arranging them in order according to area.
   ii) In each case write down the contribution made by the crop to the farm income.
   iii) What do these two sets of figures tell you about the value of the crops?

d) What is crop rotation? Explain why it is necessary and describe how it is carried out on this farm.

e) What proportion of the farm is under grass? Where is this land located? Why is it left under grass?

Many of the features of Green Farm are characteristic of arable farms in eastern England.

1. The farm is large, covering an area of 200 hectares.

2. Fields are large and regular in shape. This allows large machines to be used during ploughing and harvesting.

3. Hedges have been replaced with wire fences which can be moved to give access to machines.

4. Most of the main operations on the farm have been mechanised and this enables a large farm to be run by a small labour force.

5. Crop rotation, while still practised, has become less important and crops such as barley are grown for several years in succession. This has been made possible by the increased use of chemical fertilisers.

6. Stock rearing—in this case beef cattle—does not depend on permanent pasture. Instead the animals are kept in relatively small yards or paddocks where they are hand fed on imported feedstuffs

and on crops produced on the farm eg barley, sugar beet tops etc. In other respects Green Farm is not typical. For example, the fields are smaller and more dispersed than is usual in a highly mechanised arable farm. And stock rearing is less important here than on most farms.

The Regional Setting

Although the map below only shows the distribution of barley production in Britain, the pattern which emerges is a fairly accurate reflection of the distribution of most of the major crops grown in the country. The concentration in eastern England is obvious and this can be explained largely in terms of environmental factors. This emerges quite clearly from the following exercise, as does the concentration of crop production in the area as a whole and in East Anglia in particular.

a) Trace an outline map of the British Isles. On it:
   i) Shade using horizontal lines, the land under under 90 metres (see map on page 26).
   ii) Shade, using vertical lines, the area with less than 750 mm of rain (see map on page 30).
   iii) Shade, using diagonal lines, the area with an average of more than 3·5 hours of sunshine each day (see page 29).

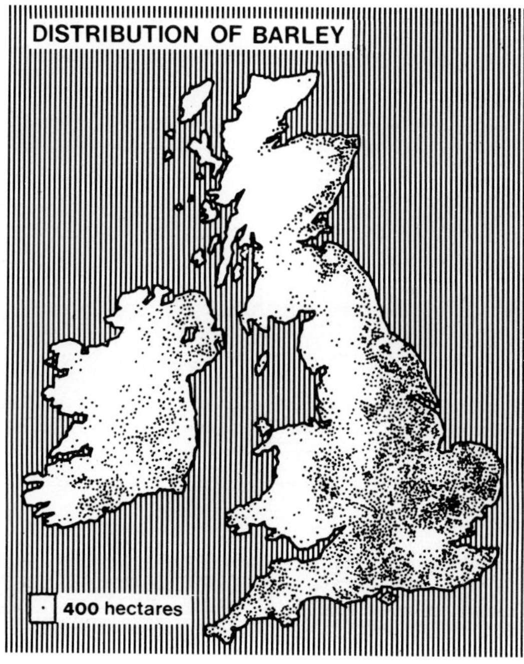

iv) Shade, using opposite diagonals, the area with July temperatures over 15°C (see page 29).

b) The area most densely shaded will be the area most suited to arable farming.

i) Compare this area with the distribution of barley growing in Britain. Point out major differences between the two patterns.

ii) Describe in your own words the environment in that part of Britain best suited to barley cultivation.

c) i) Refer to the statistics given below and calculate the location quotient for each of the crops in Norfolk and for arable farming as a whole. (see page 40 for the figures for Britain and page 311 for the method.)

ii) List the crops in order of concentration in the area.

iii) Which crops are not well represented in Norfolk? Why are they unimportant?

FARMING IN NORFOLK

TOTAL AREA OF AGRICULTURAL LAND	420,000 ha
AREA OF ARABLE LAND	345,000 ha
Proportion of arable area under following crops:	
Wheat	23%
Barley	40%
Oats	2%
Sugar beet	15%
Potatoes	5%
Vegetables	8%

(To obtain the area under each crop as a % of the total land area multiply the above percentages by 0·75.)

Although there is a strong emphasis on arable farming, and on cereal production, in particular, in East Anglia, this does not mean that agriculture is by any means uniform throughout the region. In fact, there is considerable variety both in terms of individual farms and in terms of general trends. Two factors have contributed to this variety.

1. Geology and Soils

The rocks underlying East Anglia are comparatively simple. An area of chalk which has produced low hills with very gentle relief, is bounded in the east and west by broad belts of clay. Here the land is very low lying—in places it is below sea level—and the relief is even less marked. This comparatively simple pattern has been complicated by the fact that the entire region has been covered by material deposited by successive ice sheets. This material ranges from heavy clays to light sands and gravels, and great changes can occur within very short distances. Furthermore, the soils which have developed on these deposits vary enormously and this has affected the land use of the area.

For example, a large part of East Anglia is made up of chalk, covered by a thick layer of fairly heavy boulder clay. Here arable farming is highly developed and barley, wheat and sugar beet are grown on a large scale. In the middle of this boulder clay plateau, however, lies an area of very poor sand. This is known as the Breckland and, on it, the land-use is completely different, with sheep farms and forestry plantations replacing the rich arable farms of the surrounding area. Similar contrasts can be seen in north Norfolk where the pattern of glacial deposits is very complicated and very varied. Once again the soils tend to be sandy but here they have been improved by centuries of careful cultivation and farming is highly developed, with sugar beet cereals and vegetables particularly important. This is a characteristic shared by large parts of East Anglia and in many places, poor soils have been improved to give the prosperous farmland that we see today.

2. The Influence of Markets

In general the produce of arable farming does not deteriorate rapidly. As might be expected, therefore, the individual arable farm is less dependent upon access to markets than almost any other kind of farm. But, in spite of this, in East Anglia as a whole, the influence of markets, particularly the London market, is very strong and has had important effects on the general pattern of farming. For example, the development of dairying and market gardening in Essex, from the nineteenth century onwards depended almost entirely upon the fact that the area was very near to London. Furthermore improvements in transport have seen the influence of London spreading throughout the region and dairy farming is important as far north as Norfolk, while vegetables have become one of the major crops throughout East Anglia. In the latter case, however, it is important to remember that many of the vegetables are sold to canneries or freezing plants and are not supplied direct to the market. In fact, there has been a growing tendency in East Anglia for farms to be bought by these food companies and this has increased the trend towards larger farms.

EAST ANGLIA

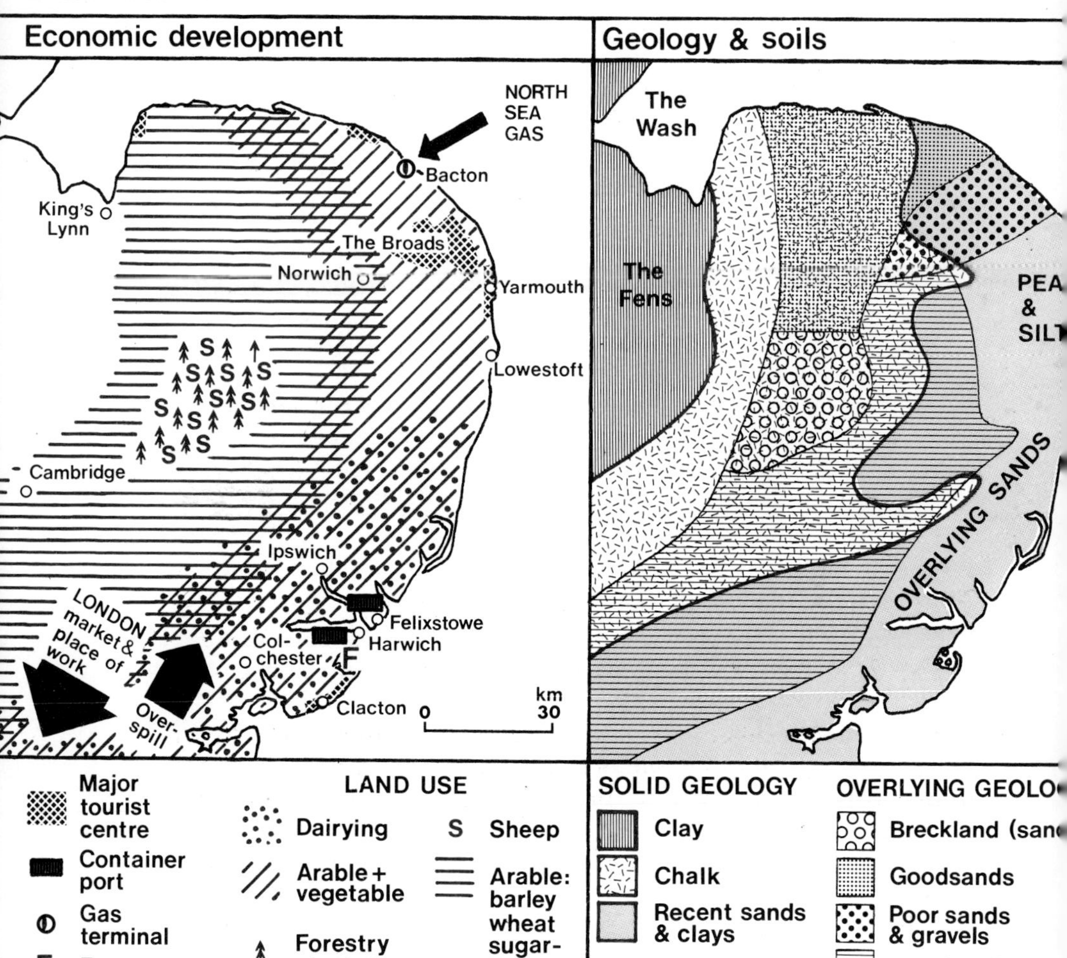

Agriculture also plays an important part in the entire economic development of East Anglia. For example, recent employment statistics show that the proportion of the population employed in agriculture is five times greater in East Anglia than in Britain as a whole and that the pattern of industrial employment is much more dependent upon agriculture than in any other major region of Britain. The nature of these links can be seen from the following table and agriculture-based industries are found in most of the market towns of the region. They are particularly important in the larger regional centres such as Norwich and Cambridge, as are the service industries upon which the farmers depend.

AGRICULTURE-BASED INDUSTRIES IN EAST ANGLIA

INDUSTRIES WHICH SUPPLY GOODS TO FARMS	FARM PRODUCE	INDUSTRIES BASED ON FARM PRODUCE
Animal feed	Sheep	Textiles
	Beef cattle	Leather and shoes
		Meat processing
Fertilisers	Cereals	Flour milling
		Brewing
Manufacture of farm machinery	Sugar beet	Sugar refining
	Vegetables	Canning and freezing
	Soft fruits	
	Mustard	Vinegar

Widespread as this traditional pattern of industry became in East Anglia, exceptions did occur. In the coastal towns, for example, fishing, tourism and even port industries developed, albeit on a much smaller scale than in other parts of Britain.

During the twentieth century, however, the pattern of industry began to change and the rate of change has become rapid since the Second World War. Much of this can be attributed to the influence of London, eg

1. An increasing proportion of the population of East Anglia works in London.

2. Both industries and people have been encouraged to move out of London to reduce growth there and to ease congestion (see page 246). Some of this 'overspill' has been accommodated in East Anglian towns and this has led to an increase in growth industries such as light engineering, in the region.

3. East Anglian ports have also benefited from the problems of the port of London and, as we have seen (page 10), Felixstowe, Harwich and Ipswich have become important ferry and container ports.

4. Tourism has become important along large stretches of the coastline and resorts such as Yarmouth and Clacton are among the largest in Britain. In addition, the Norfolk Broads—a series of flooded medieval peat diggings—have become a major tourist attraction.

Such developments have helped to offset the decline which has taken place in the traditional industries such as leather, textiles and fishing.

Farming Study: Market Gardening

Market gardening is the large scale production of fruit and vegetables. Both types of crop are grown on other farms but only as a subsidiary activity; market gardens specialise in their production. As the name implies, market gardening is strongly influenced by access to markets. This is largely because both fruit and vegetables deteriorate rapidly after picking and must, therefore, be moved to market as quickly as possible. As a result, market gardens have tended to grow up near to the large industrial cities, often in areas which are not ideally suited to the production of such crops. In recent years improvements in transport and changes in the methods of food processing have resulted in vegetable growing spreading to new areas. This has complicated the pattern of market gardening in Britain.

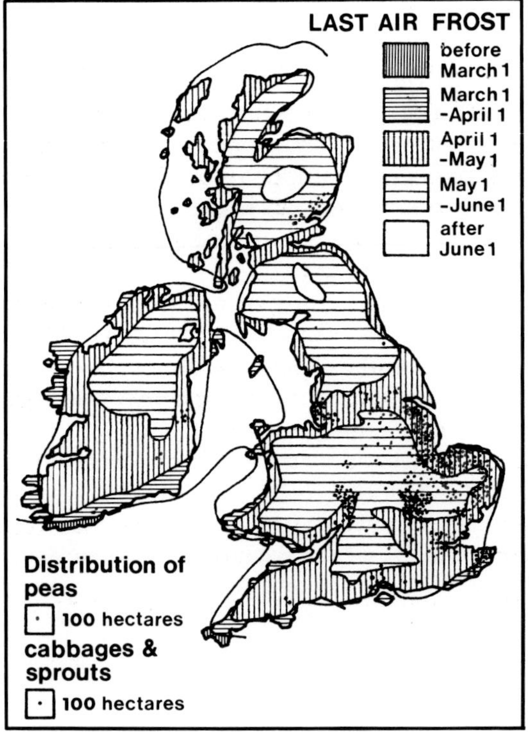

Complete the following exercise:

a) Draw an outline map of the British Isles using the above map. On it mark:

i) The following major market gardening regions:
South East Lancashire
Bedfordshire
Vale of Evesham
The Fens
Lea Valley
Strathmore

ii) The major conurbations (see page 284).

b) Around each conurbation draw a circle, the radius of which represents 50 km.

c) i) Name the market gardening areas which fall within the circles.
ii) Alongside each write the name of the nearest conurbation.
iii) Name the market gardening areas outside the circles.

Using only a single factor—distance from markets—therefore, it is possible to identify the two major types of market gardening, namely the traditional type which has grown up near to the cities and the new type which is located in less accessible areas.

The Vale of Evesham: a traditional market gardening area

The Vale of Evesham is part of the lower Avon valley. Situated less than forty kilometres to the south of Birmingham, the valley developed as a market gardening region during the nineteenth century. Nearness to the city was an important factor in this development but the area did possess certain environmental advantages. The most important of these were:

1. The area is sheltered, lying as it does between the hills of the Welsh borders and the escarpment of the Cotswolds.

2. The Severn Estuary allows warm air from the Atlantic to penetrate during the winter and this lengthens the growing season and reduces the risk of frosts (see Map opposite).

3. The sloping valley sides allow cold air to 'drain' away and this further reduces the risk of frost.

The farms which have developed in the area are among the most highly specialised in Britain and among the most distinctive.

Using information given on pages 66 and 67, complete the following exercise:

a) What are the advantages of the site as the location for a market garden?

b) i) What is the total area of the farm?
ii) What is the area of the largest field?
iii) How do these figures compare with those for the dairy farm and the arable farm?

c) i) List the main crops, arranging them in order, according to area occupied.

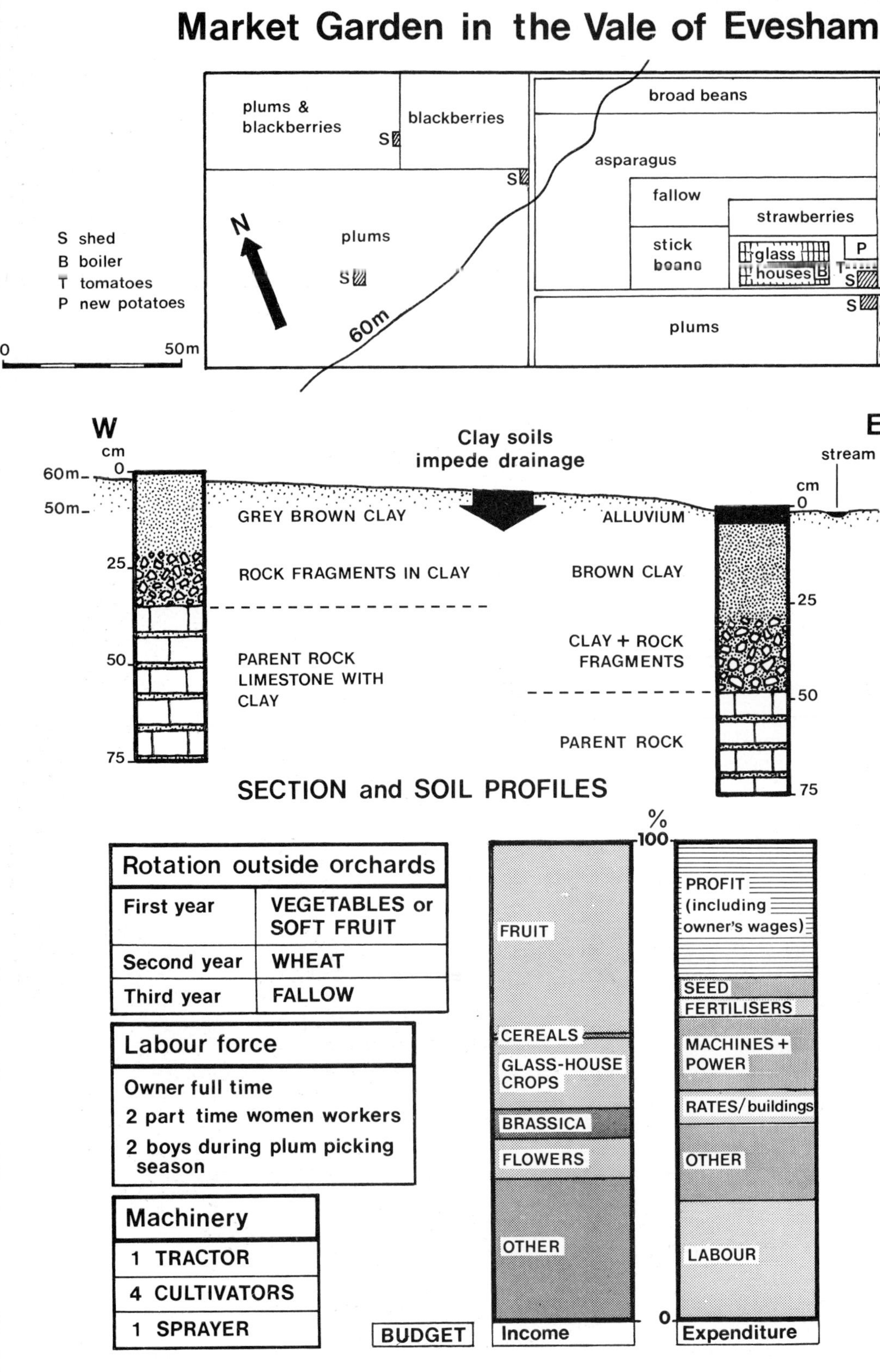
Market Garden in the Vale of Evesham
plums & blackberries
blackberries
S
S
plums
S
60m
broad beans
asparagus
fallow
strawberries
stick beans
glass houses
B
T
P
S
S
plums
Damson trees
N
S shed
B boiler
T tomatoes
P new potatoes
0
50m
W
E
cm
0
60m
50m
Clay soils impede drainage
stream
GREY BROWN CLAY
ALLUVIUM
25
ROCK FRAGMENTS IN CLAY
BROWN CLAY
25
50
PARENT ROCK
LIMESTONE WITH
CLAY
CLAY + ROCK
FRAGMENTS
50
PARENT ROCK
75
75
SECTION and SOIL PROFILES
Rotation outside orchards
First year | VEGETABLES or SOFT FRUIT
Second year | WHEAT
Third year | FALLOW
Labour force
Owner full time
2 part time women workers
2 boys during plum picking season
Machinery
1 TRACTOR
4 CULTIVATORS
1 SPRAYER
%
100
0
FRUIT
CEREALS
GLASS-HOUSE CROPS
BRASSICA
FLOWERS
OTHER
PROFIT (including owner's wages)
SEED
FERTILISERS
MACHINES + POWER
RATES/buildings
OTHER
LABOUR
BUDGET
Income
Expenditure

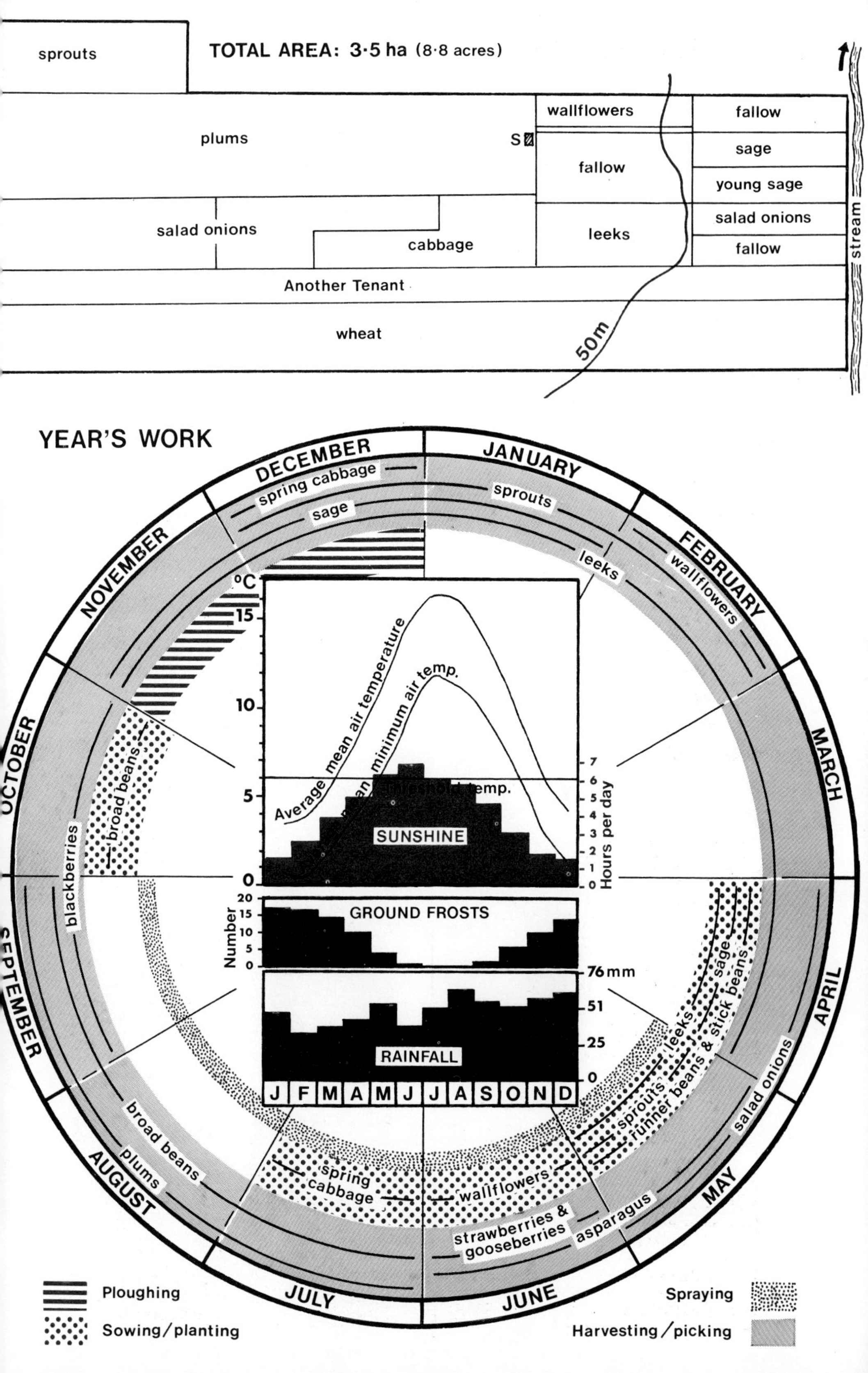

sprouts
TOTAL AREA: 3·5 ha (8·8 acres)
plums
S
wallflowers
fallow
fallow
sage
young sage
salad onions
fallow
salad onions
cabbage
leeks
Another Tenant
wheat
50m
stream
YEAR'S WORK
DECEMBER
JANUARY
FEBRUARY
MARCH
APRIL
MAY
JUNE
JULY
AUGUST
SEPTEMBER
OCTOBER
NOVEMBER
spring cabbage
sprouts
sage
leeks
wallflowers
broad beans
blackberries
sage
leeks
runner beans & stick beans
sprouts
salad onions
spring cabbage
wallflowers
strawberries & gooseberries
asparagus
broad beans
plums
°C
15
10
5
0
Average mean air temperature
mean minimum air temp.
Threshold temp.
SUNSHINE
Hours per day
7
6
5
4
3
2
1
0
GROUND FROSTS
Number
20
15
10
5
0
76 mm
51
25
0
RAINFALL
J F M A M J J A S O N D
Ploughing
Sowing/planting
Spraying
Harvesting/picking

ii) Compare this list with the contribution of each crop to the farm income. What does this tell you about the value of the main crops on the farm?
iii) How has the need to produce a variety of crops influenced the way in which the farm is laid out?
iv) Describe how the crops are rotated and explain why this is necessary.

d) i) What are the main items of expenditure in the farm budget (excluding the farmers income)?
ii) Compare this with expenditure on the dairy farm and arable farm. What does this tell you about the nature of market gardening?

The farm is typical of farms in the area. It is small, and more than half the farms in the Vale of Evesham are less than twenty hectares in area. The labour force is large for such a small farm and part time labour is employed during planting and harvesting. And, in addition, the farm is highly mechanised. This involves the investment of a large amount of money so that, in order to make the farm pay, it is

Market Gardens in the Vale of Evesham.

necessary to use the land almost continuously and to obtain high yields per hectare. Because the area involved is so small, many techniques can be used to make this possible. For example, soils can be improved by the application of fertilisers and dressings, irrigation can be used to overcome the problem of drought, and glass-houses can be built to extend the growing season and to allow the production of crops which require temperatures higher than those which occur in Britain. On the other hand, operating on a small scale has disadvantages and farmers in the Vale of Evesham have joined together to form co-operatives for buying materials and marketing produce.

At one time most of Britain's vegetables were produced on farms of this kind but, in recent years, their importance has tended to decline. In the Vale of Evesham this has been reflected in a twenty per cent decline in the area under vegetables and a thirty-five per cent decline in the area under fruit. Much of this decline is the result of the growing popularity of frozen foods and the development of a new kind of horticulture to supply them.

The Fens: a 'new' market gardening area

The Fens is an area of drained marshland bordering the Wash. Draining the area was both difficult and expensive but it has provided some of the most fertile farmland in Britain. Because it is situated in the heart of Eastern England, the pattern of farming on the Fens is similar to that seen in East Anglia. But the fertility of the soil and the expense of maintaining drained land has encouraged farmers to specialise in crops, such as vegetables, which bring high returns and this has produced important differences.

LAND-USE ON THE LINCOLNSHIRE FENS

CROP	% OF AREA	CROP	% OF AREA
Wheat	29	Potatoes	10
Vegetables	22	Sugar beet	10
Barley	11	Soft fruit	7
		Bulbs	4
		Grass	6

FARM SIZE ON THE LINCOLNSHIRE FENS

UNDER 20	20—40	40–120	120–200	200–400	OVER 400 (HA)
72%	12%	10%	3%	2%	1%

a BEFORE 1650

1 River meanders across the low flat clay vale to the sea.
2 Silt is deposited on the river bed, causing the water level to rise. This causes floods.
3 Marshland or Fen produced.

---

b 1700

1 Main drains cut to straighten river
2 This increases the rate of flow an silt is washed into the sea.
3 Land drains built into the main drai and the river.
4 Produces fertile agricultural land.

---

c 1750

1 As the peat Fen dries out the pea tends to shrink and is blown away
2 The rivers flow on the silt which they have deposited. This does n shrink.
3 The rivers now flow above the lev of the peat Fen and flooding star

---

d TODAY

1 Water in the drains is pumped into the river which is now flowing 20 above the level of the Fen.
2 The level of the peat is still bein lowered and in places the peat ha disappeared exposing the under lying clay

---

STAGES IN THE RECLAMATION OF THE FENS

The Fens near Wisbech.

Using information given above, complete the following exercise:

a) i) Write a brief account of the draining of the Fens, pointing out the special problems which were faced.
ii) Describe the present landscape of the Fens and explain how drainage has affected it.
iii) What are the advantages of the area as an arable farming region? (Refer to the factors given on page 39).

b) Compare the statistics on page 69 with those for Norfolk on page 60.
i) Name three crops which are important in both areas.
ii) Name three crops which are more important in the Fenland area.
iii) In what ways does the pattern of farming resemble that of Eastern England?
iv) In what ways does it differ?

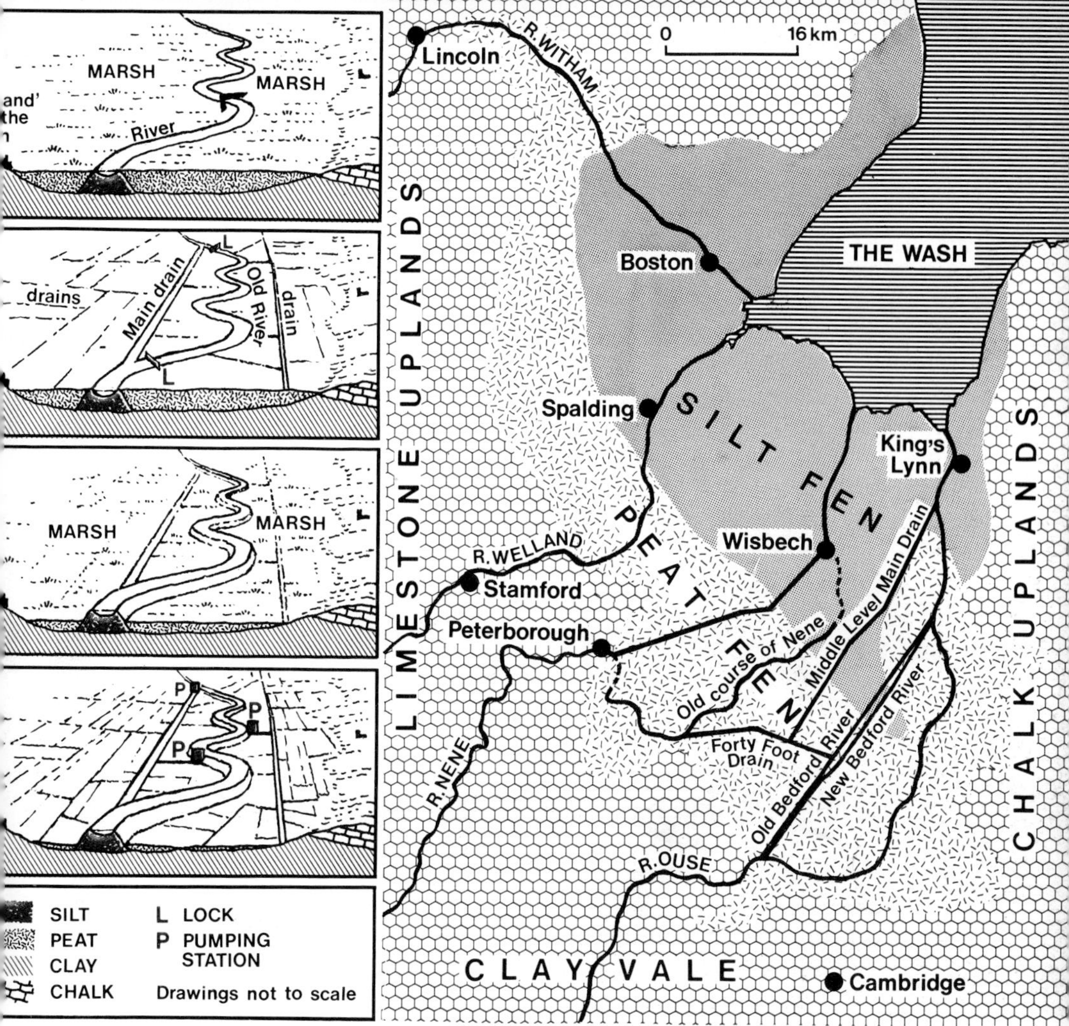

Another factor which has encouraged this concentration on the large scale production of fruit and vegetables has been the development of new food processing techniques and the building of factories in the areas where food is grown. As a result, instead of supplying the cities direct, the Fenland farms supply canneries and freezing plants which process the food before it reaches the consumer. This means that access to markets is no longer of vital importance. It also means that the demands of the food manufacturers have to be met and this has had important effects on the organisation of the farms, eg.

1. The main crops are peas, green beans, broad beans and cauliflowers, all of which are widely used in freezing plants and canneries. In the Vale of Evesham, on the other hand, there is a concentration on vegetables such as cabbage which are supplied direct to the market.

2. The crops are grown on a much larger scale than in the Vale of Evesham.

3. New varieties have been developed to suit this type of cultivation, eg peas with very small leaves which will not clog picking machines.

Most of the vegetable producing districts of Britain fall into one or other of the categories described above. There are, however, one or two exceptions. Perhaps the most important of these are in the West Country and the Channel Islands. Here the small, intensively cultivated farms, producing a wide range of vegetables and cut flowers resemble those in the Vale of Evesham. But, instead of serving a nearby city, they serve the entire British market. This involves high transport costs which would cripple most market gardens. The industry survives, however, by concentrating on the production of early crops which fetch high prices in the markets and so offset the higher transport costs. In this case, therefore, it is a combination of environmental factors—namely the mild winters and early springs—which have encouraged the development of a type of farming not usually dependent upon environment.

The Fishing Industry

Britain's second major source of food is the surrounding sea. In 1975 more than one million tonnes of fish were landed by British ships and further quantities were imported. Impressive as this may seem it marks a decline from landings earlier this century. Reasons for this decline will be examined later in this section.

Although fish are widespread in the oceans of the world, it is only in certain limited areas that they occur in sufficient numbers to make large scale fishing an economic proposition. Such areas are called *fishing grounds* and they are usually found where the following conditions occur:

1. Shallow waters where light and oxygen are readily available to the fish.

2. Cold waters which contain minerals in solution.

3. Large quantities of *plankton* which feed on the minerals and which, in turn, provide food for the fish.

For centuries the British fishing industry depended upon fishing grounds near Britain, particularly those in the North Sea. By the fifteenth century however, fishermen were already searching for new grounds and it is known, for example, that they had reached the Grand Banks of Newfoundland within a few years of Columbus' first voyage to America. By the nineteenth century fishing in distant waters was highly developed and several distinct methods of fishing had been developed to cope with different kinds of fish and with different fishing grounds. Two main types of fish are caught.

1. *Pelagic Fish*

Pelagic fish such as herring, mackerel and pilchards are free swimming fish which live near to the surface and move in shoals. They have traditionally been caught by a method known as *drifting*. This involves releasing a long net which is suspended in a vertical position from a line of buoys. The net is then allowed to drift with the tide and the wind across the path of a shoal and the fish become trapped in the mesh. Drifting is still practised on the *inshore* fishing grounds off the coast of South West England and Wales, ie the fishing grounds near to the coast. Other variations have been developed eg seine fishing but all have declined in recent years, partly because of the introduction of new methods and partly because the North Sea herring fisheries, which were Britain's main source of pelagic fish, have declined. For centuries the herring moved southwards from northern Scotland in June to East Anglia in October and November. Overfishing and pollution have reduced the

Modern Stern Trawler.

shoals and many east coast ports have seen their fishing fleets decline and, in some cases, disappear.

2. *Demersal Fish*

Demersal fish such as cod, haddock, hake and plaice live near to the sea bed. Such fish are caught by *trawling*. The trawl is a bag shaped net which is drawn along the sea bed by a ship known as a trawler. It is held open by two boards known as 'otter boards', and the fish swim into the open end and are trapped either by the net or by the currents which are created by the trawl. When the net is hauled in, the otter boards close and escape is impossible. Since trawling was introduced at the beginning of the nineteenth century it has become by far the most important method of fishing and has replaced other methods on most fishing grounds. It is particularly important, however, in the deep water fisheries which provide a large proportion of the total catch.

The relative importance of these types of fishing can be seen from the map on page 76 and the following table.

FISH LANDINGS IN BRITAIN 1974

TYPE	CATCH (AS A % OF TOTAL)
Cod	27
Herring	15
Haddock	14
Skaithe	6
Plaice	4
Whiting	3
Mackerel	2
All demersal	66
All pelagic	28
All shellfish	6

Study the map and the statistics and complete the following exercise:

a) i) Name the main fish caught, arranging them in order of importance.
   ii) In each case state whether the fish are pelagic or demersal.

b) i) Name the three most important fishing areas near to Britain, arranging them in order of importance.
   ii) Name four distant water fisheries, arranging them in order of importance.
   iii) Assuming a total catch of one million tonnes, what proportion will be obtained from home waters and what proportion from distant waters?

Although home waters remain more important, distant water fishing has developed rapidly in recent years. Many factors have contributed to this growth.

1. Overfishing and pollution have reduced catches in the seas near Europe.

2. Technical improvements have made long voyages more economic, eg.

a) Modern stern trawlers are larger and can remain at sea for long periods.
b) Refrigeration units allow the catch to be stored on board ship for longer periods.
c) Some countries use factory ships which process the catch of an entire fleet of trawlers thus allowing them to concentrate on fishing and to stay at sea for very long periods.
d) Modern devices such as echo sounders are used to locate shoals of fish.

3. Supplies of top quality 'white' fish, particularly cod, are limited in home waters.

The very success of these methods has produced its own problems and overfishing is now threatening even the most remote fishing grounds. Even more important, in some areas breeding fish are now being caught and the future of certain species is in doubt. Many attempts have been made to reverse this trend. They include:

1. Limiting the catch of each country using the fishing grounds. Such 'quotas' are difficult to enforce.

2. Excluding foreign trawlers from an ever larger area of sea around the coast. These *fishing limits* have been extended from five kilometres to their present three hundred and twenty kilometres, and each change has produced serious conflicts between the fishing nations and the countries imposing the limits. As a result of the most recent extension, Britain has lost access to most of the major Icelandic and Norwegian fishing grounds.

3. Using larger gauge nets to as to allow young fish to escape.

For Britain, the loss of distant water fishing grounds means that more attention will have to be paid to home waters, and that resources there will have to be more wisely used. This could involve two major developments:

1. The opening up of new fishing grounds beyond the limit of the continental shelf. Here the waters are deeper and the fish less numerous. This means that fishing is more difficult and more expensive. Furthermore, many of the fish are unusual and, therefore, difficult to market.

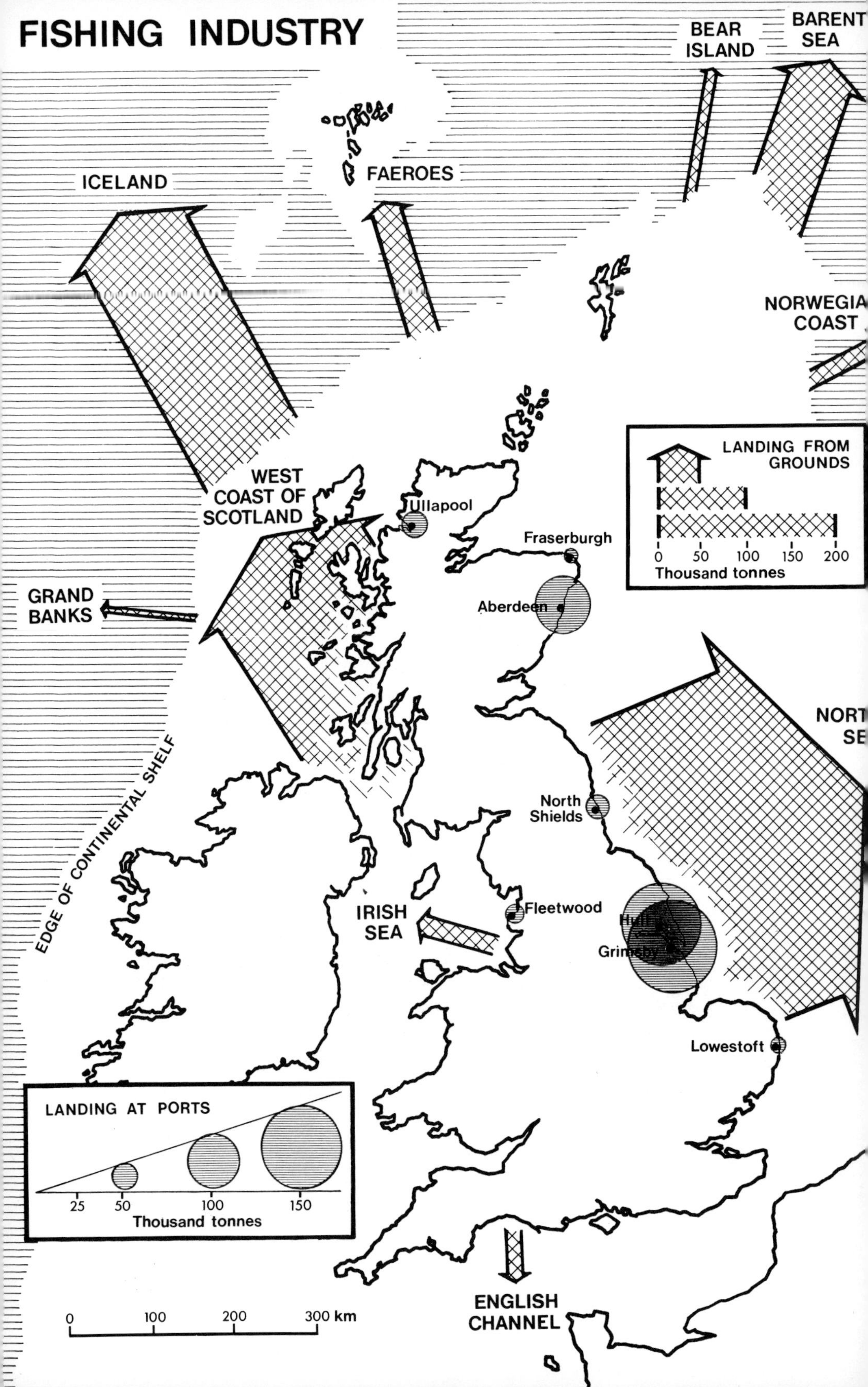
FISHING INDUSTRY
ICELAND
FAEROES
BEAR ISLAND
BARENT SEA
NORWEGIA COAST
WEST COAST OF SCOTLAND
GRAND BANKS
Ullapool
Fraserburgh
Aberdeen
LANDING FROM GROUNDS
0 50 100 150 200
Thousand tonnes
NORT SE
EDGE OF CONTINENTAL SHELF
North Shields
Fleetwood
IRISH SEA
Hull
Grimsby
Lowestoft
LANDING AT PORTS
25 50 100 150
Thousand tonnes
0 100 200 300 km
ENGLISH CHANNEL

2. Fish farming which involves rearing fish in controlled areas. Again costs are high and there are problems of disease but small schemes are now working and further development seems likely in view of the growing problems of the deep sea fishing industry; problems which include the loss of fishing grounds, competition from other fleets and the rising costs of maintaining trawlers at sea.

Fishing Ports

Developments in the fishing industry have had important effects on the ports which depend on fishing.

Complete the following exercise:

a) From the map opposite:
   i) Name the major fishing ports, arranging them in order of importance.
   ii) What proportion of the total catch is landed, a) at Hull and Grimsby, b) at all North Sea ports?
   iii) Why are so many of the major fishing ports located on the east coast?

b) Refer to the photograph and to the map on page 168:
   Which of the following factors may have contributed to the growth of Grimsby as a fishing port:
   i) A natural harbour on a rocky coast. ii) Convenient for the North Sea fishing grounds. iii) Situation where the deep water channel of the Humber approaches the south shore. iv) Near to London. v) Central situation in relationship to the main industrial cities. vi) Early rail links with markets. vii) No tides to hinder handling of catch. viii) Freezing and canning plants can also handle the produce of the nearby Lincolnshire vegetable farms.

For a long time the fishing industry was scattered through a large number of ports, many of them no more than villages. In recent years, however, it has tended to become concentrated in a few large ports. This is partly due to the increase which has taken place in the size and cost of the trawlers needed for deep sea fishing, and partly to changes in the markets for fish. For example, the sales of fresh fish have declined and an increasing proportion of the catch is taken for freezing, canning and processing into pet food or fertiliser. This can be carried out more cheaply in large factories which have been built in the main ports.

Industry

Britain was the birthplace of modern industry and, by the late nineteenth century, the Industrial Revolution had produced a large scale movement of people from the countryside into the towns. Here they formed the industrial work force most of which was initially employed in *manufacturing industry*, actually producing goods for sale. Within a short time, however, growth in this area of industry slowed down and the twentieth century has seen an ever increasing proportion of the population employed in *service industries*, such as local government, office work and education.

	AGRICULTURE AND FISHING	MANUFACTURING INDUSTRY	SERVICE INDUSTRY
1851	22	33	42
1971	1·9	34	62

Industrial activity is not evenly distributed throughout Britain, and an examination of page 79 clearly shows that there are marked differences from one region to another. For example, the industrial population tends to be concentrated in the South East, the West Midlands, Lancashire and Yorkshire. And manufacturing industry tends to be relatively more important in the Midlands and the North, while service industries predominate in the South. Attempting to explain these patterns forms an important part of any geography of the British Isles and it involves a careful study of the factors which might influence the *location* of an industry in a given area. Most important among these are:

1. Access to raw materials or components.
2. Availability of fuel and power.
3. An efficient transport system to allow the cheapest possible assembly of materials and to give access to markets.
4. Labour supply.
5. The existence of related industries in the area.

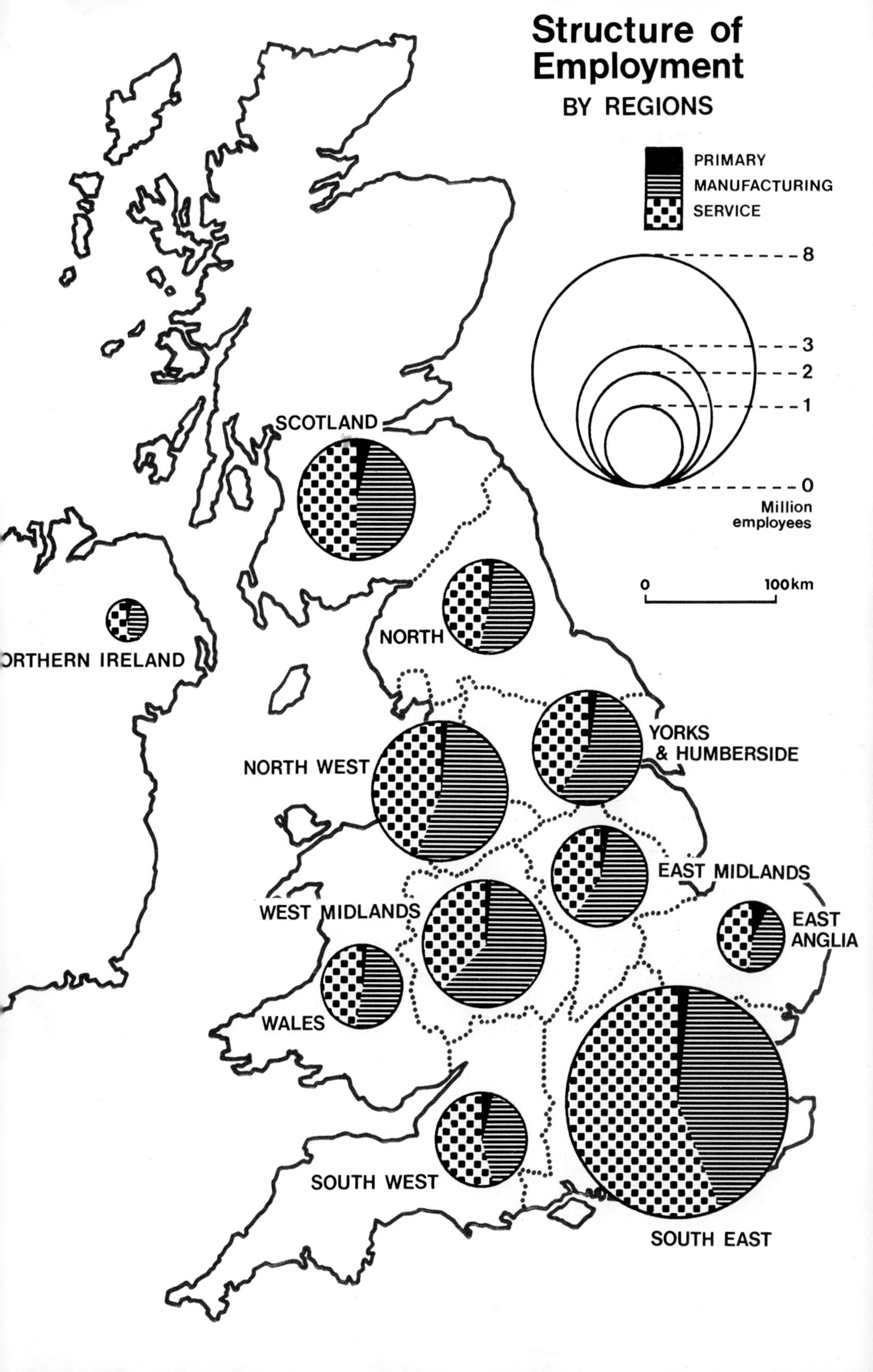

Structure of Employment
BY REGIONS
PRIMARY
MANUFACTURING
SERVICE
8
3
2
1
0
Million employees
0
100km
SCOTLAND
NORTH
ORTHERN IRELAND
YORKS & HUMBERSIDE
NORTH WEST
EAST MIDLANDS
WEST MIDLANDS
EAST ANGLIA
WALES
SOUTH WEST
SOUTH EAST

6. The personal decision of a business man.

7. Government policy which might encourage or discourage development in an area.

In considering the location of a single factory, it may be that one of these factors has been of overwhelming importance. But, when an industry becomes concentrated in an area, it is probable that several factors have contributed to its location there. Identifying these factors is not easy, particularly since the present pattern of industry in Britain owes much to earlier periods, when the factors influencing location may have been very different from those which exist today. Indeed, in some cases, the factors which led to the location of an industry in a particular area may have changed so completely as to be unrecognisable; but, in spite of this, the industry remains. This is

EMPLOYMENT IN THE UNITED KINGDOM: 1974
(AS A PERCENTAGE OF TOTAL EMPLOYMENT)

OCCUPATION	% OF TOTAL EMPLOYMENT
Agriculture	1·9
Mining	1·6
Food, drink and Tobacco	3·3
Chemicals	1·9
Metal manufacture	2·4
Engineering	8·6
Shipbuilding	0·8
Vehicles	3·6
Textiles	2·5
Leather	0·2
Clothing	1·9
Bricks and Pottery	1·3
Timber	1·3
Paper and Printing	2·6
Other Manufacturing Industries	4·4
Construction	6·0
Gas, Electricity, Water	1·5
Transport	6·8
Distributive Trades	12·1
Business and Professional Services	19·0
Catering and Hotels	3·5
Government	2·6
Local Government	4·3
Other Services	6·0

known as *industrial* or *geographical inertia* and many industries in Britain show traces of it. There are many reasons for this.

1. Factories last a long time and it is expensive to change their location.

2. The people who work in the industry may be highly skilled and they may not want to move.

3. The closing down of a factory may cause high unemployment and many social problems in an area. This the government may find unacceptable and the management too expensive.

4. The industry may depend on other industries in an area and this makes movement from that area more difficult.

It is not surprising, therefore, that, in an old established industrial country such as Britain, the present distribution of industry shows traces of earlier patterns or that, in order to understand the development of modern industries, it is necessary to study their history.

In recent years the situation has been further complicated by government policy. Faced with the decline of many traditional industries such as textiles and shipbuilding and with the high unemployment often associated with this, the government has tried to encourage new industries to move into the areas most seriously affected. It has done this by giving grants to firms willing to make the move and by restricting development in the more prosperous areas of the South East and the Midlands. Such a policy has affected the distribution of industry in Britain but it has done little to halt the decline of certain industries and the rise of others.

Industrial Growth since 1945

'GROWTH' INDUSTRIES	% GROWTH SINCE 1945	'DECLINING' INDUSTRIES	% GROWTH SINCE 1945
Coal and oil products	+396	Food and drink	+90
Chemicals	+402	Metal production	+48
Engineering	+218	Shipbuilding	−13
Vehicles	+153	Metal goods	+63
Timber products	+148	Textiles	+41
Paper and publishing	+156	Leather, footwear, clothes	+117
Gas, electric	+278	Mining	−35
		Construction	+74

AVERAGE GROWTH FOR ALL INDUSTRY 126%

Energy: the changing pattern and its effect on industrial development

The availability of fuel and power is one of the main factors influencing the location of industry. The first major concentration of industry took place on the coalfields from the eighteenth century onwards. Coal had been used as a fuel in certain industries before this time but it ranked behind water power and wood as a source of energy. The eighteenth century, however, saw its introduction as a source of power in steam engines and as a fuel in many industries, notably the manufacture of iron. In both cases the use of coal was inefficient and it became usual for factories and furnaces to be located on coalfields where fuel could be obtained cheaply. This meant that it often became necessary to carry other materials to the

Walking dragline at work in an open-cast mine in the north east, removing over burden so as to expose the coal seam. Note i) the size of the machine in comparison with the bulldozer; ii) the 'feet' used for moving the machine wheels or tracks would sink under the weight; iii) the devastation caused and the need for reclamation after mining.

coalfields—a development which became relatively easy with the building of first the canals and later the railways. Improvements in transport, however, meant that coal could also be moved away from the coalfields to other industrial locations and, as industrial processes were improved to make more efficient use of coal, these alternatives became more attractive. For example, improvements in the steam engine reduced coal consumption by seventy-five per cent, while in the steel industry the proportion of coke to ore has been reduced from three to one to almost one to one.

Changes such as these obviously weakened the attraction of the coalfields and it is not surprising that, during the twentieth century, the development of more flexible forms of energy such as electricity and oil, and more flexible types of transport has encouraged the growth of industry in areas away from the coalfields. Nor is it surprising that the industrial development of the coalfields reflects this pattern, with a rapid rise to pre-eminence in the late eighteenth century, a long period of dominance during the nineteenth century, followed, in most cases, by a period of comparative decline. Exceptions to this pattern do exist but overall it is remarkably consistent.

Coal Mining

In Britain the rocks which make up the *coal measures* were laid down during Carboniferous times, some two hundred and eighty million years ago. They are sedimentary rocks and they form the upper part of a series of rocks which includes limestone, sandstones and mudstones, and coal. Examination of these different rocks indicates that they were deposited in conditions which were changing rapidly.

ROCKS OF THE CARBONIFEROUS PERIOD

TYPE OF ROCK	CONDITIONS UNDER WHICH DEPOSITION TOOK PLACE
The coal measures (youngest rocks)	Deltas were formed at the mouths of rivers. Tropical swamps flourished and the decaying vegetation formed the coal *seams*. Increased deposition by the rivers formed the rocks between the coal seams.
Sandstones, mudstones, Millstone grits	Rainfall increased and rivers deposited vast amounts of sediment in the sea.
Carboniferous limestone (oldest rocks)	Made up of the remains of creatures which lived in a clear, tropical sea. The lack of sediment indicates desert conditions.

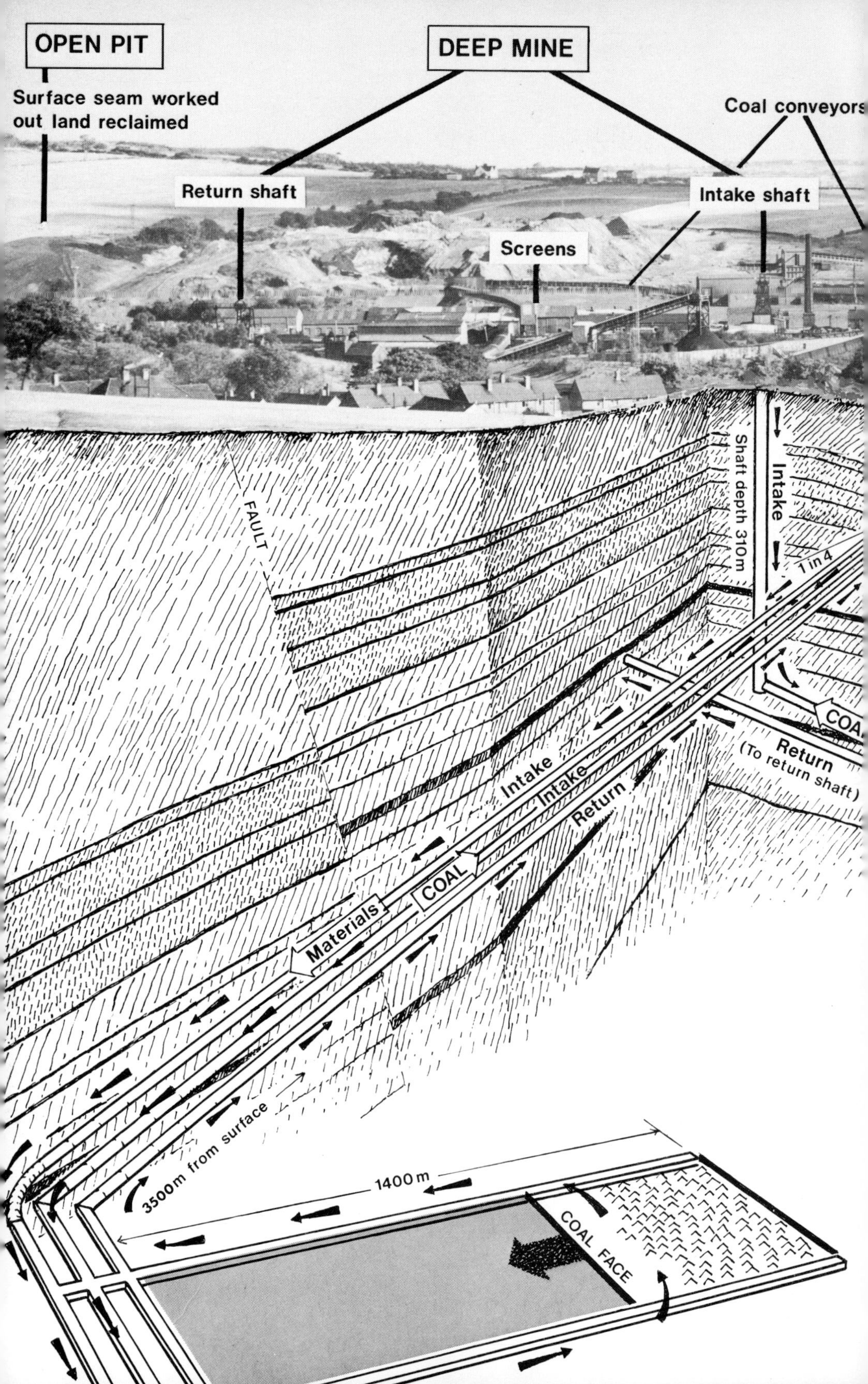

OPEN PIT
Surface seam worked out land reclaimed
DEEP MINE
Coal conveyors
Return shaft
Intake shaft
Screens
FAULT
Shaft depth 310m
Intake
1 in 4
COAL
Return
(To return shaft)
Intake
Intake
Return
COAL
Materials
3500m from surface
1400 m
COAL FACE

DRIFT MINE
Bunker for loading rail-way waggons
Intake & return
Railway
COAL + IRONSTONE Largely worked out
GREAT ROW
1 in 8
WINGHAY
COAL FACE
METHODS OF MINING
(Based on Silverdale Colliery North Staffordshire)
DEEP MINE
DRIFT MINE
VENTILATION
RETREATING FACE
COAL SEAM
COAL SEAM BEING WORKED
CAVED AREA
(Drawing not to scale)

Deposition took place over a large part of what is now the British Isles, with the result that, for its size, Britain possesses a larger area of coalfield than any other country in the world. Not all of this coal remains on the surface. In some places it has been buried under more recent rocks to form *concealed coalfields*; while, in others, the coal has been completely removed by erosion. It is these changes which have produced the present distribution of coalfields in Britain (see page 87).

Mining began in areas where the coal seams actually outcrop on the surface (the *exposed coalfields*), probably in pre-Roman times. It was not until the eighteenth century, however, that technology developed sufficiently for large scale mining to take place. By the end of the century many improvements had taken place and the three basic methods of mining used today had been introduced. These are:

1. Open cast or open pit methods, in which overlying rocks are removed and a seam is exposed on the surface. This method is cheap, when carried out on a large scale, but it does involve restoring the land when mining is finished.

2. Drift mining which involves driving a tunnel or roadway down to the seams.

3. Deep mining which makes use of vertical shafts to reach seams more than a thousand metres below the surface (see pages 84 and 85).

Because early miners exploited seams near to the surface, there are few places in Britain today where open cast coal mining can be carried out. As a result, virtually the whole of Britain's coal production is raised from deep mines and drifts.

The British coal industry reached its peak in the early years of the twentieth century and, since then, there has been a long period of decline, slow at first and then much more rapid. Many factors have contributed to this decline:

1. Overseas markets were lost during the First World War and coal exports never fully recovered afterwards.

2. During the depression of the 1930's industries were run down and the demand for coal was low.

3. Competition from new fuels, particularly oil increased the rate of decline after the Second World War.

The effects can be seen in the statistics quoted on page 87.

Using this information, complete the following exercise:

a) i) During which period did coal production increase most rapidly? Why did this happen?

ii) When did coal production reach its peak? What was the output in that year?

Coal mining

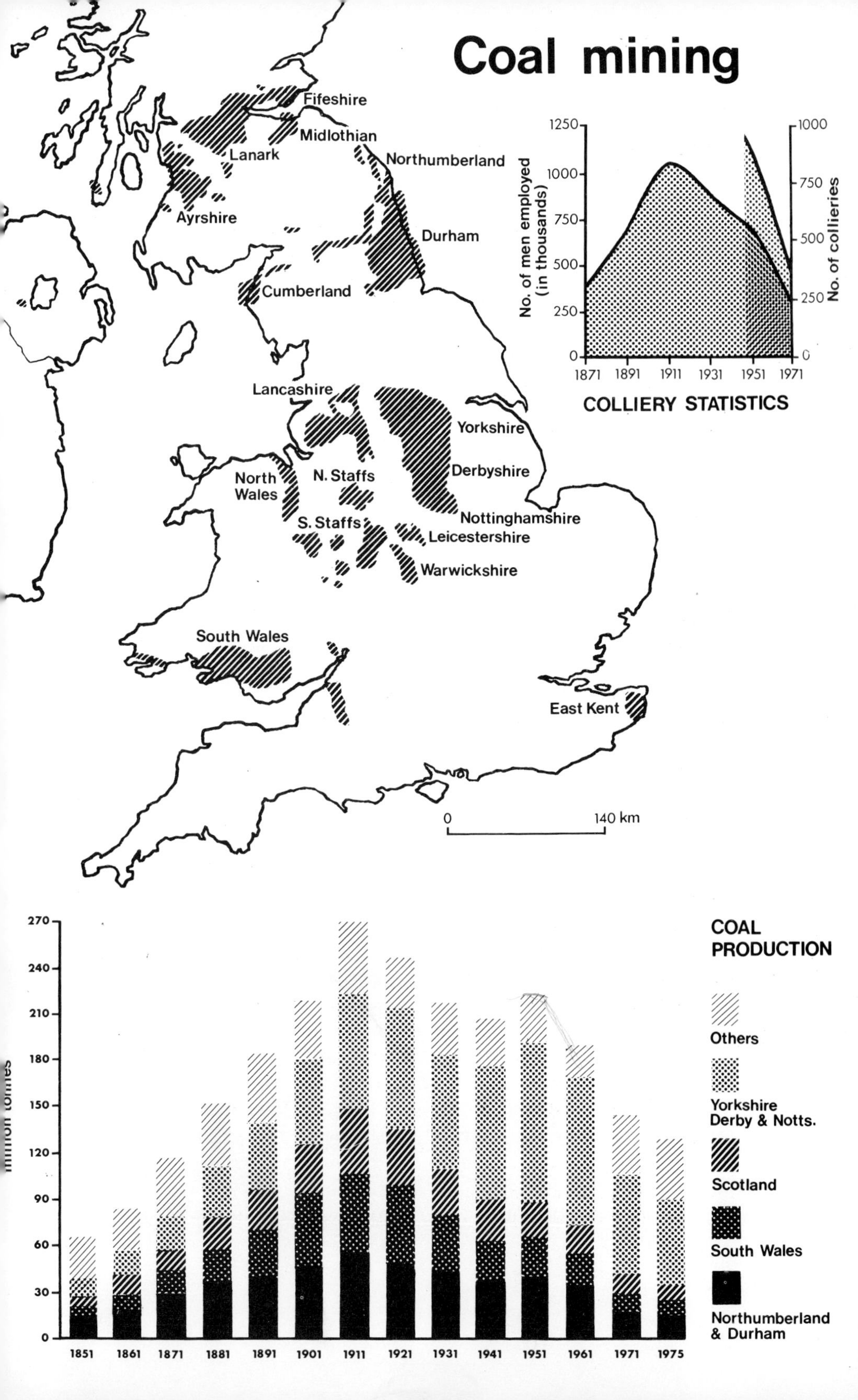

Fifeshire
Midlothian
Lanark
Northumberland
Ayrshire
Durham
Cumberland
Lancashire
Yorkshire
Derbyshire
North Wales
N. Staffs
S. Staffs
Nottinghamshire
Leicestershire
Warwickshire
South Wales
East Kent
0
140 km
1250
1000
750
500
250
0
No. of men employed (in thousands)
1000
750
500
250
0
No. of collieries
1871
1891
1911
1931
1951
1971
COLLIERY STATISTICS
270
240
210
180
150
120
90
60
30
0
million tonnes
1851
1861
1871
1881
1891
1901
1911
1921
1931
1941
1951
1961
1971
1975
COAL PRODUCTION
Others
Yorkshire Derby & Notts.
Scotland
South Wales
Northumberland & Durham

iii) During which period did production decline slowly? Why did this decline take place?
iv) When did the period of rapid decline begin? What was the main cause of this?

b) Write a brief account of recent changes in the coal industry, paying particular attention to output, labour force, size and number of collieries, the main centres of mining, markets for the coal.

MARKETS FOR COAL (AS A PERCENTAGE OF TOTAL PRODUCTION)

	1951	1971
Power stations	17	56
Coke ovens	10	15
Industry	19	12
Domestic users	18	11
Gasworks	12	0·1
Railways	6	0
Exports	3	1

These trends are rather misleading, however, and the future of the British coal industry is not as bleak as would first appear. Rapid increases in the price of oil (see page 119) and the appreciation that oil reserves are becoming seriously depleted has meant that coal has become more attractive as a source of energy and that production is likely to increase. Unfortunately this is not easy, for pits which have been closed can rarely be re-opened, and new development is very expensive. The latter course is the one being followed by the National Coal Board, however, and vast new reserves have been discovered in the Selby district of Yorkshire, the Vale of Belvoir and in north Oxfordshire, far beyond the present limits of the coalfields. Exploitation of the Selby field alone (by means of a vast new drift mine) could produce ten million tonnes of coal a year.

A rise in the demand for coal may also reverse another trend in the British industry; namely the decline of some old established mining areas and the concentration of mining in Yorkshire and Nottinghamshire, where the thick and relatively undisturbed seams can be worked more easily by modern automated methods. This trend has been apparent for most of the present century (see page 87) and mining could have virtually disappeared in some areas had it not been for the fact that they produce special types of coal which are in short supply.

This is an important consideration in the coal industry. Ordinary household or *bituminous* coal is plentiful and can be produced most cheaply in the Yorkshire area. Better quality coals such as *coking* coal (for use in the steel industry) and *anthracite* (a hard, clean, smokeless coal) are in short supply and mining continues in the areas where they occur, regardless of geological difficulties. It is this factor more than anything else which has allowed mining to continue on a large scale in Durham and South Wales.

In discussing the coalfields, however, it is important to remember that we are not simply considering one industry, large and significant though this may once have been. Instead we have to remember that, during the nineteenth century, the coalfields became the major industrial areas of Britain, each with its own complicated pattern of industrial development. This development still influences the location of industry in Britain today.

Although a variety of industries grew up on the coalfields, many of them were dependent upon one basic industry—the steel industry.

Industrial Study: The Steel Industry

Iron is one of the most common minerals in the earth's crust and it has been used, in metallic form, by man for more than two thousand years. In spite of this, its manufacture is far from simple. This is particularly true of the manufacture of the most useful alloy of iron—*steel*.

Using information given on pages 90 and 91, complete the following exercise by filling in the missing words.
Steel is an alloy of iron, containing . . . . carbon. If the metal contains less carbon it forms wrought iron which bends easily. With more carbon, the metal is brittle and is known as cast iron.
Three basic raw materials are used in the manufacture of steel.
They are: 1. . . . . 2. . . . . 3. . . . .
The first stage of the manufacturing process is to pass the raw materials (in granule form) under a heat source which fuses them together to form a cinder called . . . . This is then fed into a . . . . furnace which produces molten . . . . iron, which contains . . . . carbon. The carbon content of this iron is too high and reheating is necessary to burn off the surplus to form steel. This is carried out in a . . . . furnace which is charged with . . . . iron and . . . . metal. The steel, in the form of ingots, is taken to the mills where it is re-heated and rolled into bars, plates and strips.

For centuries the problems of controlling the carbon content of the iron remained unsolved and steel making was a hit and miss

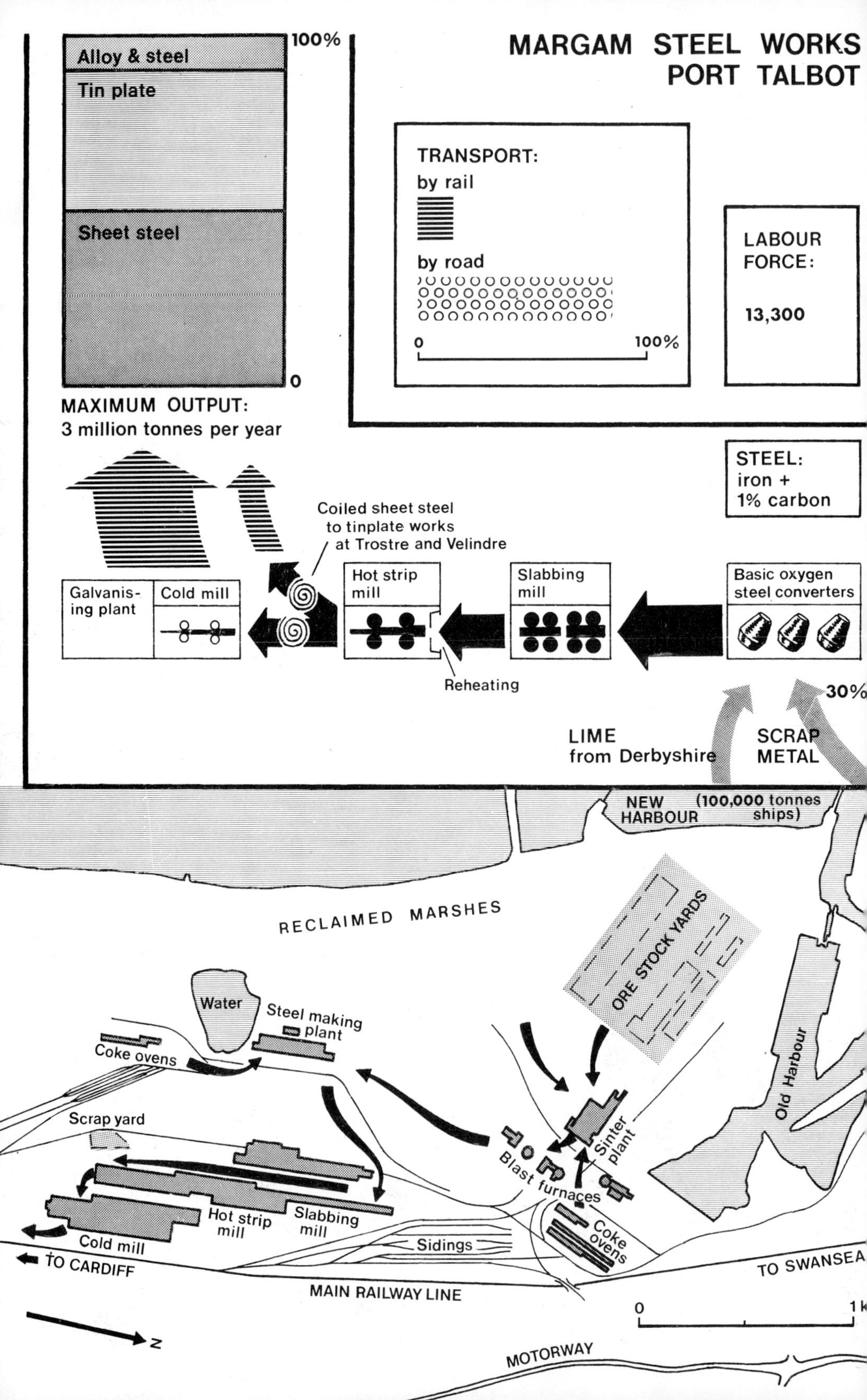

MARGAM STEEL WORKS
PORT TALBOT
100%
Alloy & steel
Tin plate
Sheet steel
0
MAXIMUM OUTPUT:
3 million tonnes per year
TRANSPORT:
by rail
by road
0
100%
LABOUR FORCE:
13,300
STEEL:
iron +
1% carbon
Coiled sheet steel
to tinplate works
at Trostre and Velindre
Galvanising plant
Cold mill
Hot strip mill
Slabbing mill
Basic oxygen steel converters
Reheating
30%
LIME
from Derbyshire
SCRAP
METAL
NEW
HARBOUR
(100,000 tonnes
ships)
RECLAIMED MARSHES
ORE STOCK YARDS
Water
Steel making
plant
Coke ovens
Scrap yard
Sinter
plant
Blast furnaces
Old Harbour
Hot strip
mill
Slabbing
mill
Cold mill
Sidings
Coke
ovens
TO CARDIFF
TO SWANSEA
MAIN RAILWAY LINE
0
N
MOTORWAY

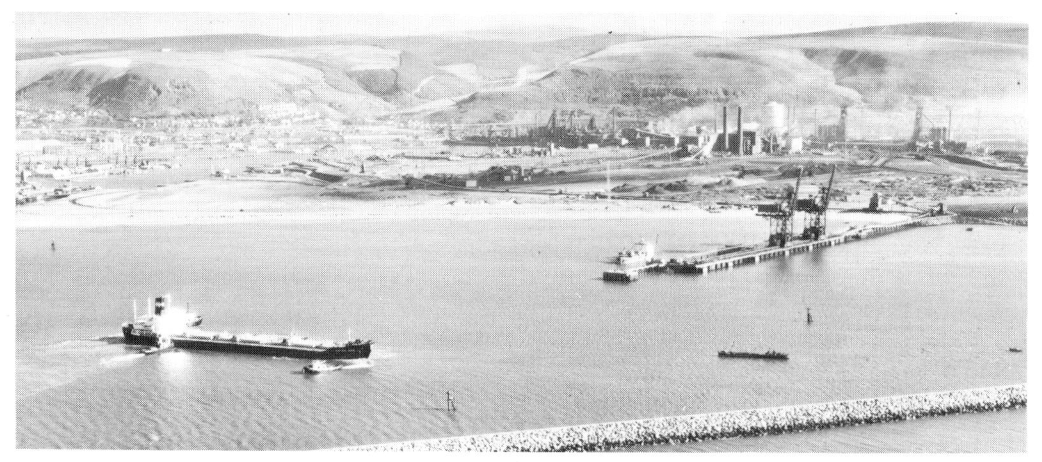

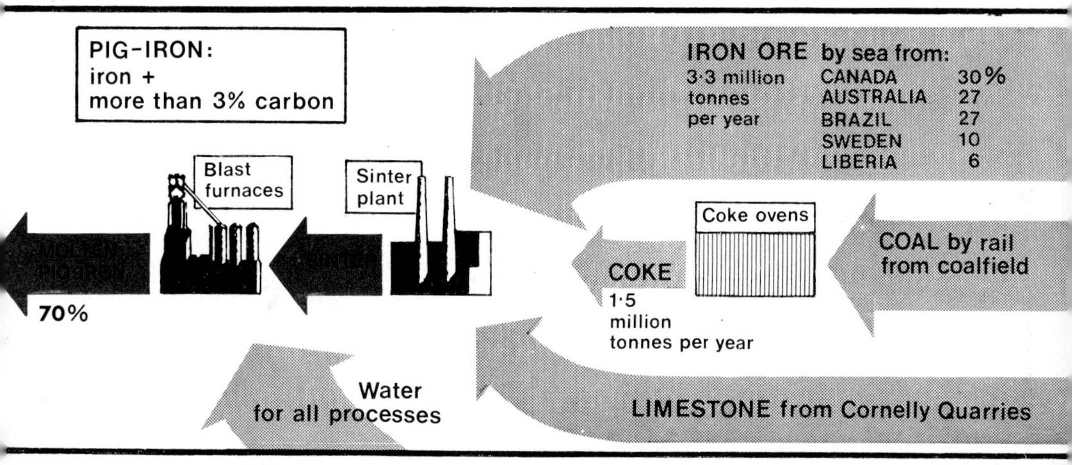

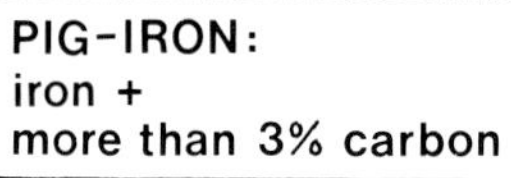

PIG-IRON:
iron +
more than 3% carbon
IRON ORE by sea from:
3·3 million
tonnes
per year
CANADA 30%
AUSTRALIA 27
BRAZIL 27
SWEDEN 10
LIBERIA 6
Blast
furnaces
Sinter
plant
Coke ovens
COAL by rail
from coalfield
COKE
1·5
million
tonnes per year
70%
Water
for all processes
LIMESTONE from Cornelly Quarries

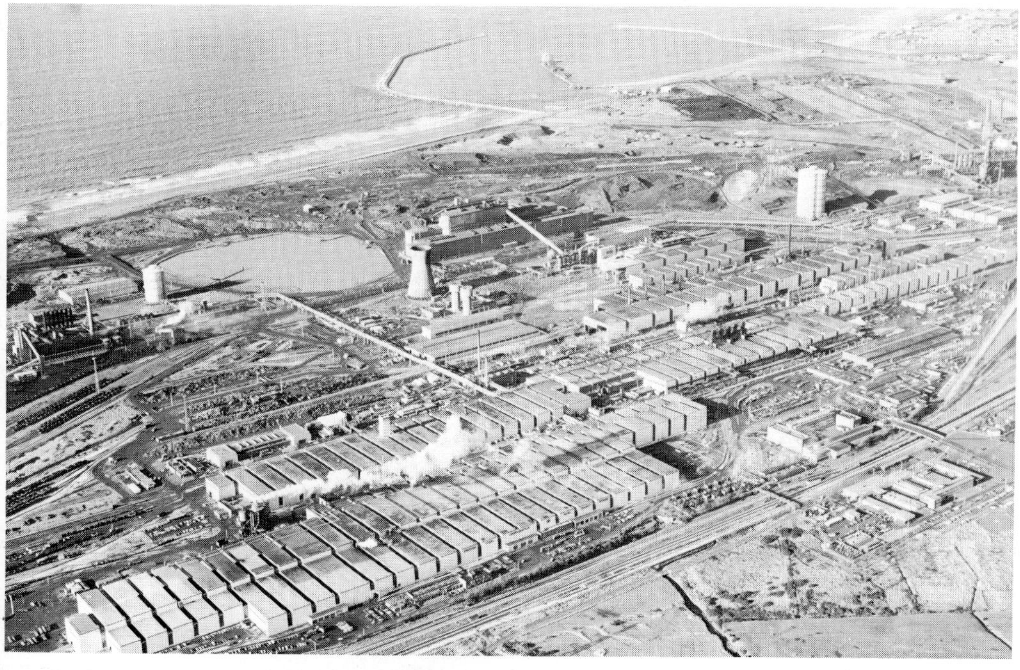

process. During the eighteenth century the Huntsman process allowed small quantities of steel to be made but it was not until the introduction of the Bessemer converter in 1865 that the large scale modern steel industry developed. Initially, this industry grew up on the coalfields but, during the twentieth century, changing circumstances have produced a very different pattern of location.

The Location of the Steel Industry

The following factors influence the choice of site for a steel works:

1. Availability of raw materials.
2. Water supply—millions of gallons are required for cooling purposes each day.
3. Availability of labour.
4. A large area of flat land, suitable for building.
5. Transport facilities.
6. Access to markets—preferably nearby.

Of these factors, the availability of raw materials is the most important. This is because a steel works consumes millions of tonnes of coal, iron ore and limestone, and, in the manufacturing process, the bulk of this material is considerably reduced. It is, therefore, cheaper to transport the finished product than the raw materials, and this encourages the location of works near to supplies of one or more of the raw materials. Other factors are taken into consideration but they tend to be of less importance.

Using the information given on pages 90 and 91, describe the factors which favoured the choice of Margam, Port Talbot as the site for a major steel works.

If it is difficult to identify the factors which influenced the siting of a single steel works, it is even more difficult to explain the distribution of the industry through Britain as a whole. This is largely because the present distribution pattern is the product of several earlier patterns, each of which reflected changes in the technology of steel making and in the sources of raw materials. The most important of these changes are summarised in the table.

It is clear from the table that three types of location have been favoured by the British steel industry during the last one hundred years.

1. *Coalfield locations* were favoured throughout the nineteenth century because coal was used in much larger quantities than ore in the making of steel. In addition, for much of the century, the coal measures were the main source of iron ore and the coalfields were,

	MATERIALS	PROPORTION USED TO MAKE ONE TONNE OF STEEL	SOURCES (MILLION TONNES)	
1913	Coal	2	Most coalfields produced some coking coal	
	Iron ore	1	Coalfield ore	1·5
		(Varies with quality of ore)	Jurassic ore	5·3
			Imported ore	7·4
	Main works located near coal supplies.	Production 7·7 mill. tonnes		
1974	Coal	0·6	Durham, South Wales, Yorkshire Imported supplies	
	Iron ore	1	Coalfield ore	0·0
		(average, but varies	Jurassic ore	6·8
		with quality of ore)	Imported ore	21·4
	Mains works located on the Jurassic ore deposits or on the coast.	Production 26·6 mill. tonnes		

therefore, the ideal location. Towards the end of the century, however, these ores began to run out and other sources had to be used. This strongly influenced the future location of steel works.

2. *Ore field locations* became more attractive as steel making methods improved and less coke was needed for the furnaces. As a result it became economic to transport coal and coke to the ore supplies, particularly to the large deposits of Jurassic ore in Lincolnshire and Northamptonshire. These deposits were of low grade ore (25–35% iron) which would be expensive to transport to the coalfields. (Why?) But they occur on the surface and are cheap to mine, using the open pit methods described on pages 82 and 86. It is not surprising, therefore that major steelworks were built at Scunthorpe and Corby to exploit these advantages.

3. *Coastal locations* had obvious advantages as soon as Britain began to import iron ore, and existing centres near the sea, such as Middlesbrough, began to develop at the expense of some inland centres. As our dependence on imported ores increased so new coastal locations were developed, particularly in South Wales. The

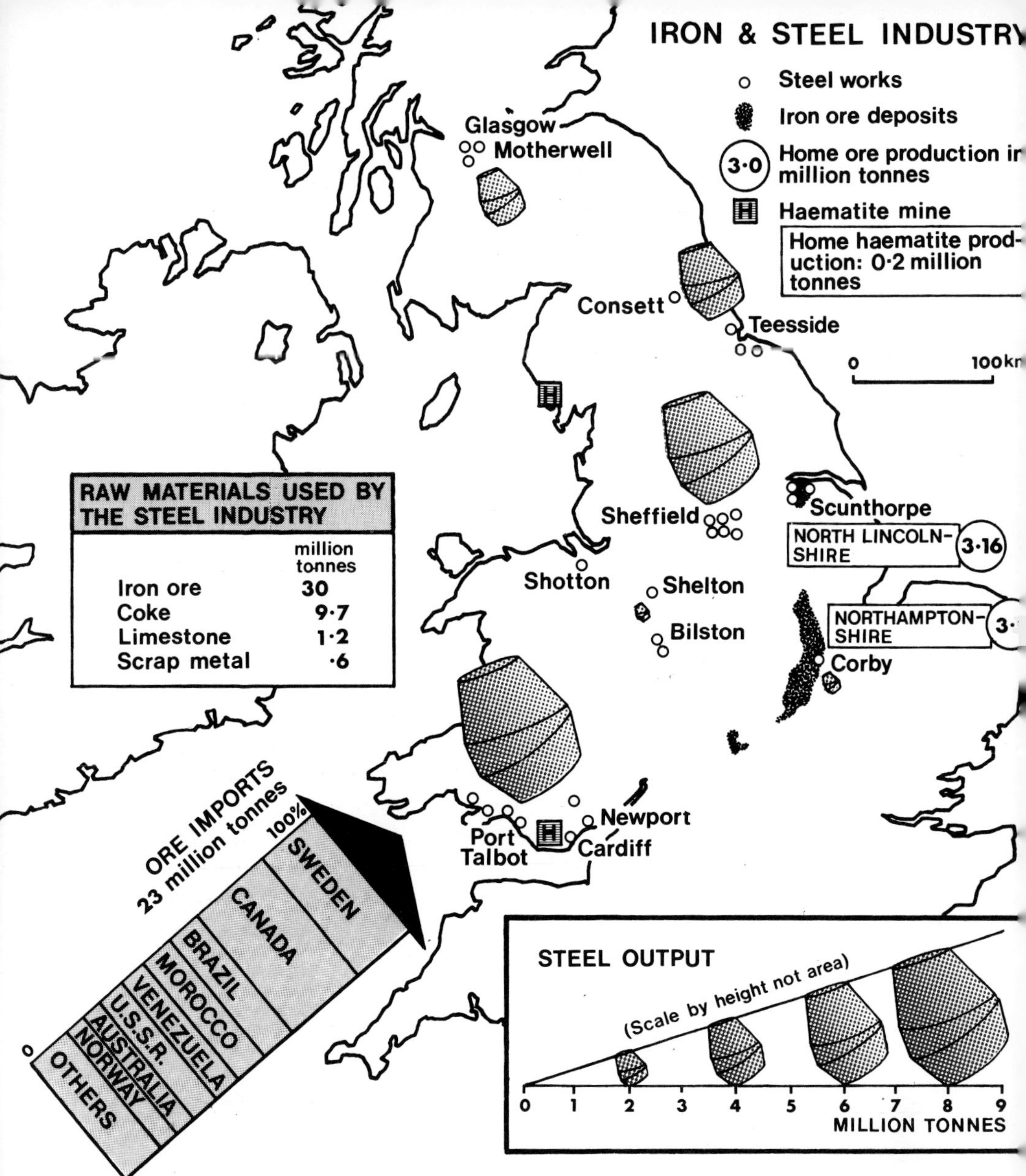

reasons are obvious. Imported ores have to be handled at ports and this marks a *break of bulk point* which has attracted the steel industry.

Because demand for steel has grown steadily and because the forces of geographical inertia are strong in the steel industry (see page 81), each new generation of steelworks has tended to be added to the existing pattern. This has produced the confused pattern of the distribution which exists today.

a) Study the steel producing areas shown opposite and, for each area:
   i) state whether its location is on a coalfield, on an ore field or coastal
   ii) identify the likely sources of raw materials. (Reference to the index will allow you to find further details of the areas).

b) It is proposed to build a large new steelworks in Britain and three possible locations have been discussed. They are:
   Scunthorpe, Teesside and Hunterston, on the Firth of Clyde.
   i) Which of these sites would you favour? Give reasons for your choice.
   ii) If the works is built, which of the centres mentioned above is likely to be run down?

Even with the building of new works, however, the decline of the inland centres is likely to be slow. Some, like Sheffield, have retained their importance by concentrating on high grade alloy steels. Others have installed electric arc furnaces and become dependent upon supplies of scrap metal rather than ore. And, in several cases, such as Consett, the government has prevented closure because of the unemployment which is likely to occur in a town almost wholly dependent upon the one industry.

The contraction of coal mining and the shift of the steel industry to the coast has seriously weakened the industrial base of the coal field areas. Furthermore, many of the other industries which brought prosperity to the coalfields during the nineteenth century have declined. It is not surprising, therefore, that unemployment on many coalfields has been high or that governments have tried to halt the decline. It is important to remember, however, that, although signs of industrial decline can be seen in all coalfield areas, the nature and extent of the decline varies considerably from place to place. This can be seen by studying the coalfield areas of Lancashire and Yorkshire; neighbouring coalfields with similar industrial pasts but very different futures.

Regional Study: The Lancashire Coalfield, a coalfield in decline

In 1900 the Lancashire Coalfield was the centre of Britain's major industrial region, almost a quarter of the country's population lived there. Today it is a depressed area with declining industries and a static or declining population. Here, therefore, in an extreme form, can be seen the problems of Britain's coalfields.

Coal Mining

Over a large area of the Lancashire coalfield the coal measures are exposed on the surface. This encouraged the early development of

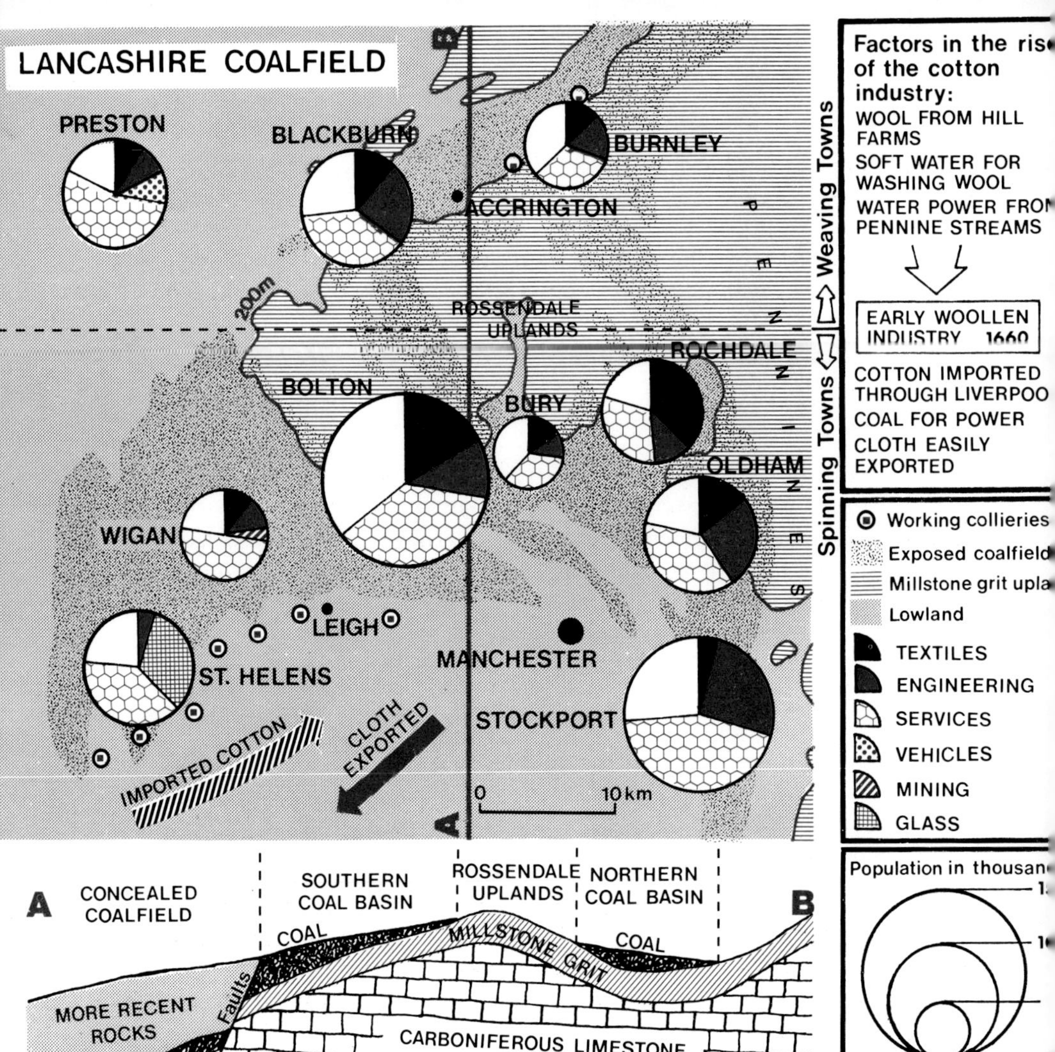

mining and paved the way for the enormous increase in production which took place during the nineteenth century. By 1900 output had reached twenty million tonnes but since then there has been a steady decline to the present figure of eight million tonnes. To a large extent, this reflects the general trends in the coal mining industry, but there are local problems which make the future of the Lancashire coalfield very uncertain.

1. Over much of the exposed coalfield, the best seams have been worked out and mining is no longer economic.

2. The steep dip of the rocks and extensive faulting means that, in the west, the coal measures are deeply buried under more recent

rocks, making mining impossible. The area of concealed coal is therefore very small and located mainly on the southern margins of the coalfield.

3. Folding and faulting have also made mining difficult and production costs are much higher than those for Yorkshire and the Midlands.

It is these local problems which determine the present pattern of mining in the area.

Referring to page 96, complete the following exercise:

a) i) Why is mining impossible in the central area of the coalfield?
   ii) In which area is mining concentrated today?
   iii) Give reasons for this development.

b) Wigan, Golborne and Burnley are three centres which are, or have been, important in the development of the mining industry. From the following list choose three statements, each of which summarises the situation in one of the centres.
   i) Situated on the exposed coalfield. Many active pits.
   ii) Situated on the concealed coalfield. Many active pits.
   iii) Situated on the exposed coalfield. All pits closed.
   iv) Situated on the concealed coalfield. Two pits active.
   v) Situated on the exposed coalfield. Two pits active.

c) Using information given on page 87, write a brief account of the changing pattern of mining in Lancashire when compared with the other British coalfields.

The decline in production and the concentration of mining in a few large pits has resulted in the loss of thirty thousand jobs since 1950. Serious as this may have been, it pales into insignificance when compared with the decline which has taken place in the major industry of the coalfield—the textile industry.

The Textile Industry

For a period of two hundred years, from the early eighteenth century onwards, Lancashire depended almost entirely upon one industry—the cotton industry. At its peak in 1914, the industry was the largest in Britain, comprising two thousand mills and a labour force of six hundred and twenty thousand. Today there are fewer than six hundred mills still open, employing less than one hundred thousand workers. This indicates collapse on a scale seen in no other industry in Britain and it is important to examine the reasons for it.

Prior to the introduction of cotton in the late seventeenth century, Lancashire was simply one of many relatively unimportant woollen producing districts in Britain. In common with almost all of these areas, the industry depended upon the availability of raw materials, particularly wool; supplies of soft water for washing and cleaning the wool; and fast flowing streams for direct water power. Such an industry, serving purely local markets, was a far cry from the industry of international importance which was to develop in the next hundred years. Many factors contributed to this change but among the most important were:

1. Cotton was a new fibre, recently introduced from India and it was not accepted in many of the other textile producing districts. In Lancashire the woollen industry was not so strongly organised and there was little united opposition to the introduction of cotton.

2. Liverpool was already becoming involved in the slave trade (see page 14) and was an obvious landing place for cargoes of cotton from the West Indies.

3. The growing port of Liverpool gave access to world, rather than local markets.

4. Once established, the industry proved to be a forcing house for a change. This can most clearly be seen in two developments:
   a) the invention of new machines which were to revolutionise the industry, making it into the first mass production, factory based industry
   b) the introduction of steam power as early as 1786—and, of course, Lancashire had a local coalfield as a source of fuel.

During the nineteenth century the cotton industry grew rapidly. No other textile producing region in Britain or abroad, could equal it for the efficiency of its organisation or the development of new techniques, and, as a result, it captured markets for cheap cloth throughout the world. Mills were established in towns throughout the coalfield and they leave a legacy which can be seen today.

Study page 99 and, using information given in this section, complete the following exercise:

a) List the factors which have contributed to the location of the cotton industry here.
b) Describe the possible advantages of the site of the mill labelled 4.
c) The numbers 1, 2, and 3 indicate different types of housing. Which of the following statements most accurately describes each type?
   - Terraced housing near to a mill
   - Semi-detached houses near to a mill
   - Terraced housing in the suburbs
   - Flats on cleared land
   - Flats on the outskirts of the town
d) Describe how relief has influenced the shape of the town.

Although spread widely throughout the coalfield region, the cotton industry varied considerably from area to area. In the south, around Manchester, there was a concentration on *spinning*, while in the north, *weaving* was more important. Such specialisation was only made possible by the development of an efficient system of communications between the two areas.

This industry reached its peak in the early years of the twentieth century, and, since then it has declined and changed out of all recognition. Many factors contributed to this decline; the most important being:

General view of Bolton, Lancashire.

1. The loss of export markets, particularly during the two world wars. This was of great importance to an industry which at its peak exported more than three quarters of its total production.

2. Industries were established in former market areas and these began to compete with Lancashire.

3. In comparison with such industries, the Lancashire mills were outdated both in terms of organisation and machinery. As a result pressures built up within the industry and, in order to survive, great changes had to be made. These have produced the present day textile industry in Lancashire; an industry which is a far cry from the mighty cotton industry which once dominated the area. When compared with this, the modern industry is:

1. Much smaller, with an output which is about one quarter of the figure for 1950.
2. More dependent on artificial fibres.
3. Controlled by large firms, particularly chemical firms which make the fibres.
4. Concentrated in factories which are larger and tend to contain all the processes (*vertical integration*).
5. Designed to produce a small range of high quality goods with which they can compete with foreign rivals.
6. No longer a major exporter and has lost even the home market for cheap textiles.

To sum up, a great international industry has been reduced to one of little more than local significance, and, in the process, more than half a million jobs have been lost.

The Lancashire Coalfield today

The collapse of two major industries—and the resulting decline in industries dependent upon them—has left the area economically very weak, and many attempts have been made to reverse the trend.

The first, and most obvious step was to slow down the rate of decline in existing industries. The concentration of mining in fewer large pits and the restructuring of the cotton industry, described above, were moves in this direction—moves which have met with some success in limiting the loss of jobs.

The second move was to attract new industries and this proved very difficult. In the battle to create new jobs, much emphasis has been laid on two advantages possessed by the region—easy access to the motorway network and the availability of cheap premises in the form of disused cotton mills. Few towns on the coalfield are far from the M6 and the completion of the M61 and M62 has created junctions which are possible growth points. The advantages of the new motorway network have still to be felt but the policy of offering old mills as cheap premises for new industries has already proved successful, in spite of their obvious disadvantages. Such premises have tended to prove most attractive to service industries, particularly those involved in distributing goods.

DISADVANTAGES	ADVANTAGES
1. Floor space divided into floors.	1. Large area.
2. Access to upper stories difficult.	2. Cheaper than building new premises.
3. Machinery difficult to install upstairs.	3. Near the town centre.
4. Road access often difficult.	4. Labour nearby.

The overall effect of these policies has been to diversify industry in towns which were over-dependent upon coal and textiles. In this, however, the degree of success has varied from place to place.

Study page 96 and complete the following exercise:

a) i) Name two towns which still depend heavily upon textiles.
   ii) In each case state whether it was a spinning or a weaving town.
   iii) Name the town which is most dependent upon engineering.
   iv) Name two towns with a varied industrial structure.
   v) In each case, name the manufacturing industries located there.

b) i) Arrange the towns in descending order of importance according to A. Population and B. The proportion of the population employed in service industries. Calculate the correlation coefficient.
   ii) Describe any relationship which exists between the two lists.

c) Referring to page 99, describe the advantages and disadvantages of the mill marked 5 as a site for new industries.

It is clear from the map that the larger centres have been most successful in attracting new industries, particularly service industries. The smaller towns and villages have remained dependent upon a narrow range of industries and have suffered severely during times of depression.

But mere size and range of industry is no guarantee of success. This can be seen in the case of *Manchester* which developed as the centre of the cotton industry. Although never a major manufacturing centre, Manchester became the business and commercial centre for the industry and the service centre for the entire region. In addition, important clothing and textile engineering industries developed. The opening of the Ship Canal in 1895 enabled the city

MANCHESTER: EMPLOYMENT

TYPE	% OF WORKFORCE
Engineering	10
Clothing	5
Paper and printing	5
Textiles	1
Other manufacturing	15
Construction	4
Shops and warehouses	15
Transport	7
Other service industries	40

to replace Liverpool as the main cotton port and to develop important port industries. Many of these industries were located on the Trafford Park industrial estate which was established on the banks of the canal and which at its peak, gave employment to a labour force of 50,000. The collapse of the cotton industry was reflected in a decline in the industries which serve it and Manchester suffered seriously from this. Furthermore, since many of the larger modern ships cannot use the canal, the port has tended to decline and, with it, many of the industries on the industrial estate. The extent of this change can be seen from the present structure of employment in Manchester.

One of the factors which has contributed to the continuing decline of the coalfield has been government policy for, in spite of its problems, the region has never been declared a development area, and has not, therefore, qualified for aid in attracting new industries. This is because unemployment has never been exceptionally high—largely because of migration from the area and because many of the unemployed were women who were not included in the unemployment figures. The effects of this lack of aid have been serious and the coalfield has tended to lose industry to nearby development areas, particularly Merseyside.

Regional Study: The Yorkshire–Nottingham Coalfield, a coalfield of growth

The Yorkshire–Nottingham coalfield is the largest of the British coalfields and economically one of the strongest. Although it has been affected by the general decline of the coalfield areas in Britain, at no time, has the decline reached the proportions seen in areas such as Lancashire.

Using information given on pages 87 and 104, complete the following exercise:

a) i) In 1901 the Yorkshire–Nottingham coalfield produced 7/12/25/30% of the coal produced in Britain.
   ii) In 1975 the coalfield produced 15/20/30/36/44% of total production.
   iii) During this period production has increased by 1/20/50 million tonnes.
   iv) Why has the share of British coal production increased more rapidly than the production of the coalfield itself?
   v) When did production reach its peak on the coafield? What was production in that year? Why has it declined?

b) Study the map of the Lancashire coalfield (page 96).
   i) Which coalfield has the largest area of concealed coal measures?
   ii) Which geological factors caused this?
   iii) A new extension of the concealed coalfield has been found in the north east. Name the town at its centre.

It is clear that coal production has in fact declined from its peak in 1951 but it is also obvious that this decline has been much slower than the decline of the British coal industry overall. As a result, the Yorkshire area now produces about forty per cent of the country's output. There are many reasons for the concentration of mining in this area; a concentration which has taken place at the expense of less fortunate areas.

1. Productivity is higher than in other coalfield areas. This is because the seams are thick and relatively undisturbed, which allows the efficient use of mechanical cutters and reduces production costs.

2. Reserves of accessible coal are greater than anywhere else in Britain. Once again geological conditions have favoured the area, for the coal measures dip very gently eastwards under more recent rocks and, as a result, the area of the concealed coalfield is very large. Its full extent is not yet known but new discoveries in the Selby and the Vale of Belvoir area indicate that it extends much farther north and south than was previously suspected.

3. The coalfield is also well positioned in relationship to possible markets. The demands of local industry, particularly the steel industry, and the Trent valley power stations (see page 113) provide a sound basis for the mining industry and it is likely that production will begin to increase again as energy costs rise and links with the Midlands and the South East are improved.

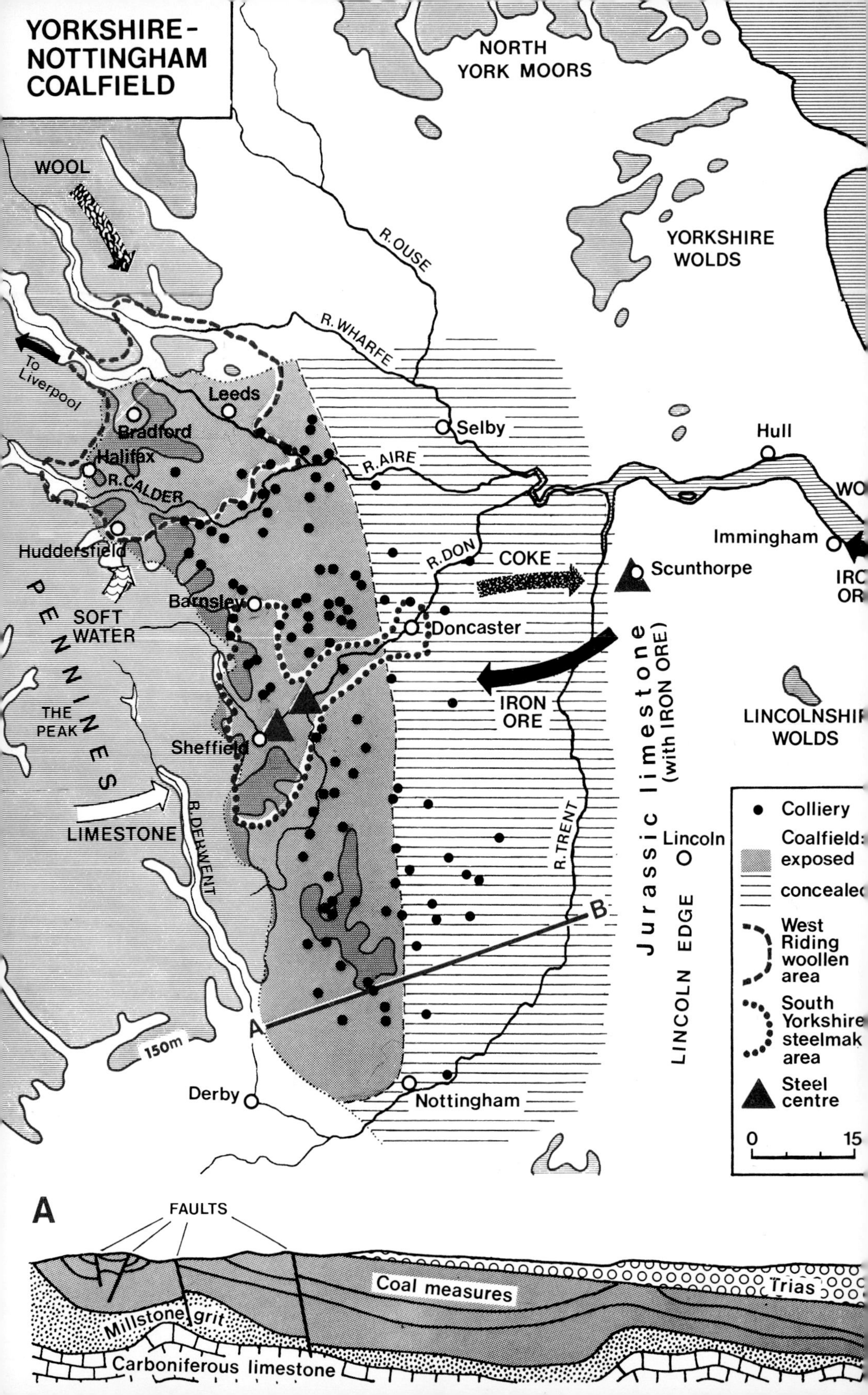
YORKSHIRE-
NOTTINGHAM
COALFIELD
NORTH
YORK MOORS
WOOL
R. OUSE
YORKSHIRE
WOLDS
R. WHARFE
To
Liverpool
Leeds
Selby
Bradford
Halifax
R. AIRE
Hull
R. CALDER
Huddersfield
Immingham
R. DON
COKE
Scunthorpe
PENNINES
Barnsley
SOFT
WATER
Doncaster
Jurassic limestone
(with IRON ORE)
THE
PEAK
Sheffield
IRON
ORE
LINCOLNSHIRE
WOLDS
LIMESTONE
R. DERWENT
R. TRENT
Lincoln
LINCOLN EDGE
Colliery
Coalfields:
exposed
concealed
West
Riding
woollen
area
South
Yorkshire
steelmak
area
Steel
centre
0
15
A
B
150m
Derby
Nottingham
A
FAULTS
Coal measures
Trias
Millstone grit
Carboniferous limestone

THE WEST RIDING WOOLLEN INDUSTRY

Origins:

Local supplies of:

1. Wool 2. Soft water for washing the wool 3. Water power.

Gave rise to a '*domestic*' industry similar to those found in other parts of Britain, but less important than the industries in East Anglia and the West Country.

Growth:

Rapid development during the eighteenth and nineteenth centuries.

1. Near to Lancashire where great changes were taking place in the cotton industry.
2. Local coal supplies encouraged the introduction of steam power.
3. Concentration on cheap cloth for which there was a growing market in the new industrial towns. (Contrast the West Country which continued to make high quality cloth).

During this period the industry changed to a *Factory* industry.

Structure:

1. Firms generally small.
2. Increased dependence on wool imported through Liverpool and Hull.
3. A high degree of specialisation, not between spinning and weaving, as in Lancashire, but between types of cloth eg:
   a) Heavy woollens in the Huddersfield area.
   b) Worsteds in the Bradford area.
   c) Cloths made from waste, eg Shoddy in the Dewsbury area.
4. Supplied a mass produced clothing industry which also found a growing market.

THE SOUTH YORKSHIRE STEEL INDUSTRY

Origins:

Local supplies of:

1. Iron ore from the coal measures. 3. Fireclay for lining Furnaces.
2. Limestone from the Pennines. 4. Charcoal.

Gave rise to a small scale industry scattered along the edge of the Pennines.

Growth:

1. Rapid development from the end of the eighteenth century onwards with the introduction of coke smelting. Local supplies of coking coal were available.
2. Early change to steel with the introduction of Huntsman's Crucible process in the Sheffield area towards the end of the century.
3. Sheffield emerged as Britain's major producer of high grade steel.
4. In the mid-19th Century the introduction of the Bessemer converter (the first method of mass producing cheap steel) emphasised the importance of the area as an iron and steel producer.

Specialisation:

1. Sheffield concentrated on special steels.
2. Heavy steel making centred on the Don Valley around Rotherham.

Later Developments:

1. Local ores were quickly exhausted.
2. The industry came to depend on high grade ores imported through Immingham and poorer quality Jurassic ores from Scunthorpe.
3. In turn, coking coal was sent to the steel industry at Scunthorpe.

Although Yorkshire has retained a strong position when compared with other coalfields, it is important to remember that there are marked variations within the area itself. For example, the eastward movement of mining onto the concealed coalfield has been accompanied by the closure of many pits on the exposed coalfield. This has not resulted in high levels of unemployment, however, because jobs were created in the new pits to the east and because the other basic industries of the coalfield have remained relatively strong. This emerges quite clearly from a study of the two major industries—the West Riding woollen industry and the South Yorkshire steel industry. Both display all the characteristics of traditional coalfield industries, with rapid growth during the nineteenth century followed by a long period of stagnation or decline during the twentieth century. Both, however, have proved much more resistant to decline than similar industries in the Lancashire area.

The West Riding Woollen Industry

At the end of the seventeenth century the West Riding, like Lancashire, was one of many woollen producing districts in Britain, struggling, often unsuccessfully to compete with producers in East Anglia and the West Country. One hundred years later the Yorkshire industry had outstripped all its rivals and a period of expansion was beginning which was to enable it to lead the world. There were many reasons for this development.

a) Using information given on pages 104 and 105, write an account of the origins and growth of the West Riding woollen industry, under the following headings:
   i) Origins. ii) Rise to importance. iii) The extent of the industry and specialisation within it.

b) Complete this account by making notes about the industry today, using the following information.

The traditional woollen industry reached its peak at the beginning of the twentieth century and since then there has been a slow but steady decline in output, labour force and number of mills. The reasons for this decline are already familiar:

1. Competition from foreign countries has increased.
2. Outdated machines and factories have not been replaced.
3. The industry was made up of many small firms which lacked the capital needed to make improvements.

In spite of these problems, however, the rate of decline has never approached that seen in Lancashire during the collapse of the cotton industry. To some extent, this can be attributed to the fact that the woollen industry was not so dependent upon export markets and that its main competitors were in developed countries where production costs were also high. But, in order to survive, many changes have had to take place in the industry itself, including the closure of uneconomic mills, the amalgamation of firms to create large units with sufficient capital to make improvements, and the concentration on a more limited range of products such as high quality cloth and ready-to-wear clothing. The success of this policy can be seen in the employment statistics given below.

EMPLOYMENT IN MAJOR INDUSTRIES IN VARIOUS CENTRES IN THE WEST RIDING (EXPRESSED AS A PERCENTAGE OF THE WORKFORCE)

	WEST RIDING REGION	LEEDS	KEIGHLEY	SHEFFIELD
Coal mining	5	less than one per cent		
Metal manufacture	5	3	4	15
Metal goods	4	2	3	16
Engineering	7	8	14	6
Textiles	8	2	23	0
Clothing	3	11	2	0
Distributive trades	12	16	10	13
Professional services	13	18	10	16
Other services	21	23	13	19

Using these figures, and those for Britain (page 80), complete the following exercise:

a) Calculate the location quotient for the textile industry in i) the West Riding, ii) Leeds, iii) Keighley.

b) i) Which town has the most varied industrial structure?
   ii) Name three industries, other than textiles, located in the town. Which of these industries could be linked with the textile industry?
   iii) Why are service industries more important in Leeds?

Attempts to attract new industries have had some success and, as in Lancashire, great stress has been laid on the availability of old mills as industrial premises and ease of access to the motorway network. This, together with the availability of jobs in other parts of the coalfield, has, at times resulted in shortages of labour in the textile industry which has been overcome by using immigrant workers and by bringing in workers from coal mining centres which can offer few opportunities for work for women.

Competition for new industry has been great and the West Riding has tended to be less successful than the Sheffield district of South Yorkshire.

The South Yorkshire Steel Industry

The Don Valley, from Sheffield to Doncaster, is one of the leading steel making centres in Britain, producing twelve per cent of the national output and, what is more important, more than sixty per cent of the total output of special steels.

a) Using information given on pages 104 and 105, complete the following exercise:
   i) Give four reasons for the early establishment of the iron industry in the Don Valley.
   ii) Name and give the dates of two inventions which encouraged the development of the steel industry in the area.
   iii) Name four raw materials used in the steel industry today and, in each case, name one or more sources of the material.

b) The steel industry is a good example of *geographical inertia*. Explain what is meant by this term and why the industry has remained in the area. (Pages 89 to 94 will help).

Two factors in particular have contributed to the continued prosperity of the steel industry in South Yorkshire. In the first place, the market for special steels such as toughened and stainless steel, has increased rapidly and this has benefited the Sheffield area, which has concentrated on such products since the late eighteenth century. And secondly, the opening up of the Lincolnshire iron ore fields around Scunthorpe, and the location of major steel works there has allowed links to develop between the two areas which have benefited both.

Two views of a steelmaking plant in the Sheffield area.

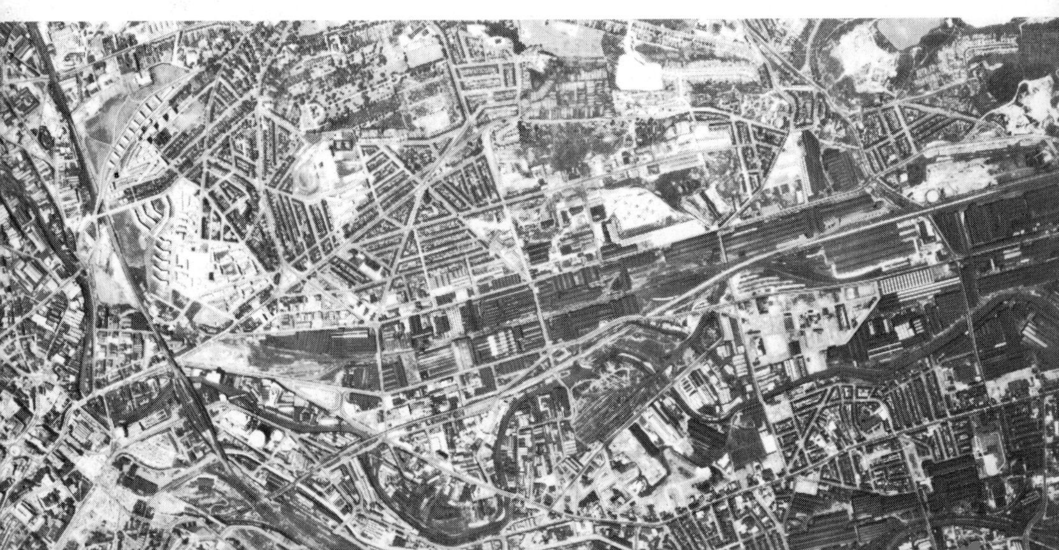

The future of steel making, particularly heavy steel making in the Rotherham area, does present many problems.

a) The factors which may influence the location of a steel works are given on page 92.
   i) Which factors favoured the choice of the site in the photograph below?
   ii) What are the disadvantages of the site?
b) Compare the lay-out of the works with that of the Margam plant (page 90). Which is the more efficient? Give reasons for your choice.
c) Describe the problems of making improvements to the works shown in the photograph.

The problem of difficult sites and the need to transport high grade imported ores from the ports means that the Don valley cannot compete with coastal locations in the mass production of heavy steel. In spite of this, investment in the steel industry has continued and production of special steels is expanding.

The advantages of the coalfield in terms of position have already been mentioned and accessibility has played an important part in attracting new industry to the area. For most of the twentieth century the Doncaster area has benefited most from this, largely because of its position as a rail centre near to the A1. Industries such as light engineering, agricultural engineering, rope and fibre making and clothing (often closely related to the nearby steel and textile industries) moved into the area and this, together with the eastward movement of mining made the town a growth point. Although the building of the motorways has not weakened the position of Doncaster it has produced new growth points which are likely to be developed and it is significant that, with the opening of the M62, the size of the market within a four hour road journey of the West Riding has almost doubled from twenty-two to thirty-seven millions.

Both regional studies indicate that the twentieth century has seen a reduction in the importance of the coalfields as centres of industry. Many factors have contributed to this decline but underlying them all is the development of alternative forms of energy, particularly electricity, oil and natural gas.

Electricity Supply

Although electricity was initially a coal based source of power, it was flexible enough to allow industry to break free of coalfield locations. At first it could be generated only on a small scale and was too expensive for widespread use in industry, but, during the twentieth century, power stations have become larger and more efficient and they have been linked together by means of a power *grid* which gives access to markets throughout the country. This flexibility is not reflected in the location of power stations, however, and it is important to remember that their location is influenced by factors just as binding as those influencing any other industry. Fortunately many different types of power station can be used to generate electricity, each influenced by different location factors. This has allowed the industry to become widely distributed.

POWER STATIONS
HYDRO
NUCLEAR
Coal fired
Oil fired
Gas fired
THERMAL
0
100km

In order to understand this pattern of distribution, therefore, it is necessary to appreciate the factors which have influenced it.

TYPE OF POWER STATION	% OF OUTPUT
Thermal: Coal fired	60
Oil fired	25
Gas fired	3
Nuclear powered	10
Hydro-electric	2
TOTAL OUTPUT	72,600 megawatts

Thermal Power Stations

In thermal power stations the turbines are driven by steam which is produced by heating water, using fuels of various kinds. In siting a thermal power station, therefore, the following factors have to be considered:

1. Availability of coal, oil or gas for fuel.
2. Availability of water, in small quantities to produce steam and in vast quantities for cooling.
3. Easy access to markets, since the transmission of power is expensive.
4. Sites suitable for the disposal of waste products, particularly ash.

The map on page 111 clearly reflects the importance of these factors and thermal power stations tend to be concentrated in certain well defined areas.

1. Most are situated in densely populated parts of the country, usually near to rivers or the coast. This is most apparent in the London area where the combination of a large population and considerable industrial activity has created an enormous demand for electricity. Many of the power stations have been built on the banks of the Thames which provides not only water but also a means of transporting coal and oil.
2. Other concentrations occur on the coalfields. This can be seen in Yorkshire, Lancashire, South Wales and Lanarkshire.
3. Similarly oil-fired stations have been built near to the major oil terminals and refining centres, eg Pembroke near to the Milford Haven terminal in South Wales.

One area—the *Trent Valley*—overshadows all others, however, and it is here that the interplay of the various location factors can most clearly be seen.

Study the maps and photographs on pages 114 and 115 and complete the following exercise:

a) i) How much electricity can the Trent Valley stations produce?
   ii) Draw a pie diagram to show this as a proportion of the total capacity of all British power stations.
   iii) Name the four largest power stations in the Trent Valley. What proportion of the Trent Valley output can they produce?

b) i) Why is so little water required to produce steam?
   ii) Large amounts of water are required for cooling. Why does this not cause an enormous lowering of the water level in the river?
   iii) What does the water cool?

c) i) From which coal field does West Burton obtain coal?
   ii) How is it transported to the power station?
   iii) Study the layout of the coal stock yards. How have they been designed to make the delivery of coal easier?
   iv) Name three other coalfields which serve the Trent Valley power stations.
   v) How does the power station dispose of its waste products?

d) Explain why the Trent Valley has developed as the main centre for the generation of electricity in Britain.

The dominance of the Trent Valley has continued in spite of the rapid increase in the number of oil-fired stations during the 1960's. The price rises imposed by the oil producers in the early 1970's together with fears about future world reserves of oil, have led to no more oil-fired stations being planned. This means a return to coal which will ensure the long term future of the Trent Valley.

Nuclear Power Stations

The same price increases and fears, when added to the difficulties of increasing coal production, make alternative sources of power attractive. In Britain most interest has centred on the use of nuclear power as a replacement for conventional fuels. This has many attractions:

1. A power station requires relatively small amounts of fuel.
2. New 'breeder' reactors can produce their own fuel.
3. As a result they can be built in areas with plenty of water since water for cooling is the only raw material required in bulk.
4. Stations are relatively cheap to run although building costs are high.

In spite of such advantages, however, development has been slow and nuclear power stations still only supply ten per cent of our electricity. This slow rate of progress is the result of a series of problems which have proved difficult to solve.

POWER SUPPLIES FROM THE TRENT VALLEY

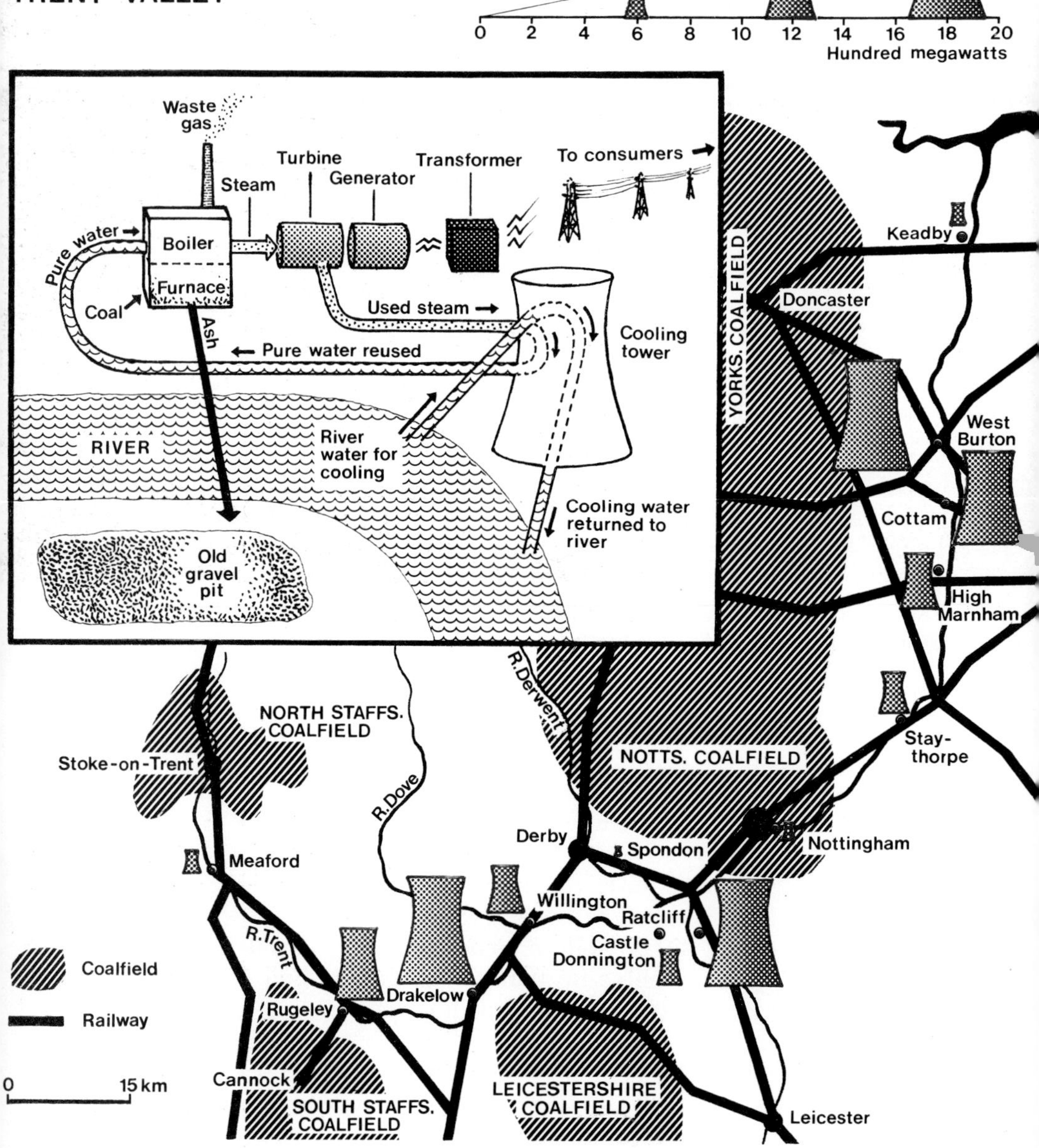

The West Burton power station in the Trent Valley. Note the 'merry go round' system of delivering coal to the stockyards.

1. The building of nuclear power stations has presented engineering problems which have caused serious delays.

2. People are afraid of nuclear energy and, as a result, early power stations were located in unpopulated areas. Sites nearer to centres of population are now being used but this has produced demands for increased safety precautions which, in turn, increase costs.

3. Transmission costs from remote sites are high.

4. The disposal of radio-active waste is difficult since it remains dangerous for thousands of years.

Until these problems are overcome, therefore, nuclear energy is unlikely to challenge conventional forms of energy.

a) Study the photograph of the power station, locate it on the map on page 111 and explain why the site was chosen.
b) Describe and explain the distribution of nuclear power stations in Britain.

Wylfa nuclear power station, Anglesey. The station houses Magnox reactors which generate 1,800 mW. Note coastal location and remoteness of site.

Hydro-electric Power Stations

Of all forms of power, hydro-electricity is by far the most attractive. Requiring no expensive fuels and using apparently inexhaustible supplies of water to drive turbines, power can be produced cheaply once the stations have been built. But, unlike other forms of power, development is virtually controlled by a number of physical factors which can be changed only with difficulty and at great expense. These include:

1. Large amounts of rainfall.
2. Heavy run-off.
3. Steep slopes which give a good 'head' of water to operate the turbines.
4. Facilities for storing water either in natural lakes or in man-made reservoirs. These can be used to regulate the flow through the power station.

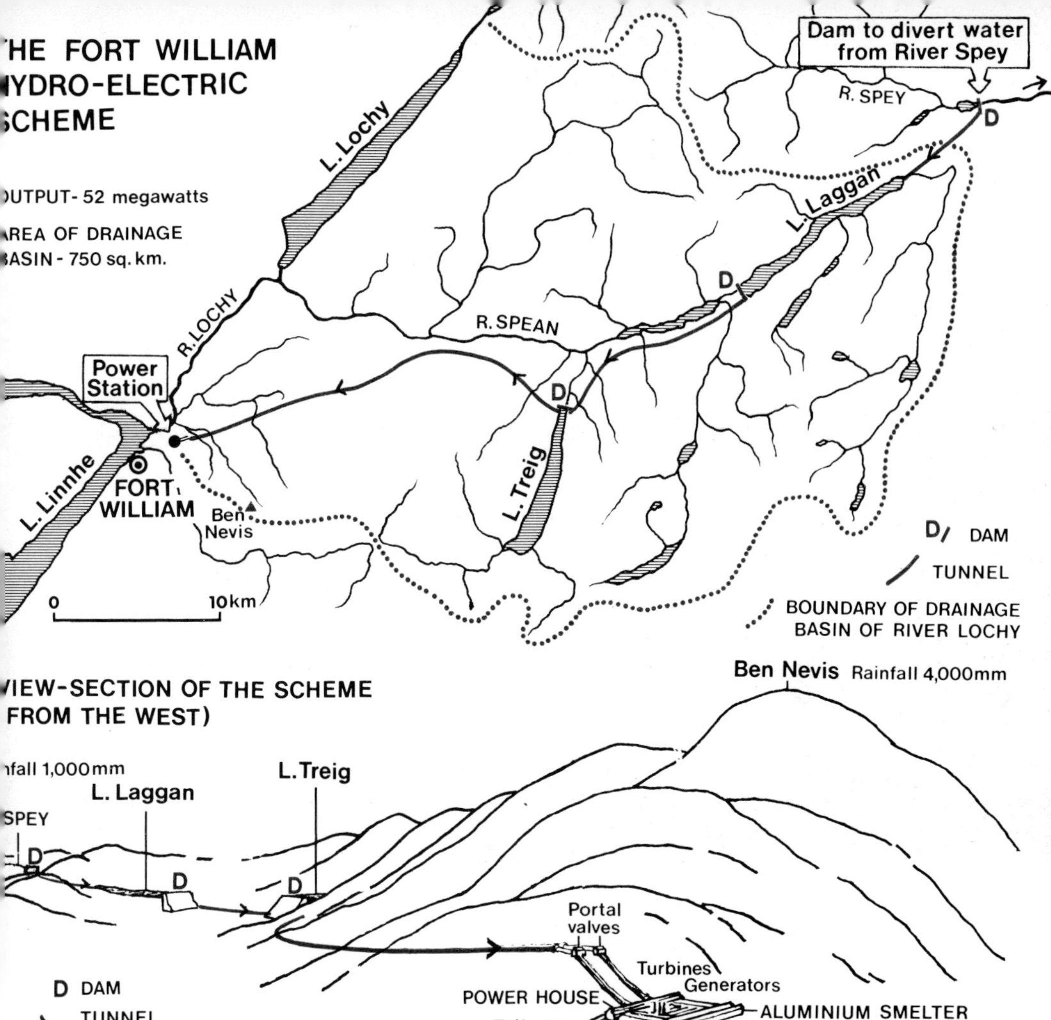

5. Large drainage basins which give a more regular flow of water in the rivers.

Unfortunately few places in Britain can meet all these requirements and there are few good sites for development.

Study the diagram above and complete the following exercise:

a) What are the advantages of the area as the site for a power station?

b) How has the size of the drainage basin been enlarged?

c) What is the output of the station?
Where is the electricity used?
How does it compare in size with the West Burton station?

d) Referring to page 111 describe and account for the distribution of hydro-electric stations in Britain.

The limits on development in Britain are obvious and it is not surprising that the largest of the Trent Valley stations produces more electricity than all of the hydro-electric stations in the Scottish Highlands added together. Furthermore, prospects for further development are slight since the best sites have been used and the remaining drainage basins in the mountains are either too small or lack rainfall. Only by developing sites lower down rivers where the increased volume of water may offset the reduced speed, could output be increased, and this is unlikely since it would involve flooding large areas of agricultural land.

The Problems of the Electricity Supply Industry

It is the function of the Central Electricity Generating Board to meet the changing demand for electricity in Britain. In order to do this efficiently, a complex network of cables, known as the National Grid, has been laid down to link together the major generating areas and the main centres of population. This enables power to be diverted to where it is needed at any given time—an important asset if, for example, the weather varies from one part of Britain to another.

Even with such a system problems occur. For example:

1. The prediction of future demand is difficult and mistakes can be costly.

2. Demand for electricity varies enormously during the day. Some of the peaks are predictable, eg at lunchtime and in the early evening. Others, such as those which occur at the end of a popular television programme, are less predictable. Predictable or not, such peaks have to be met and there must be spare generating capacity to do this. Such capacity is expensive since it remains unused for a large proportion of the time.

3. The building of a modern power station is a difficult engineering job and delays may occur which can disrupt forward planning.

4. Even more important, in these days of possible shortages, the thermal station is one of the least efficient ways of producing power. More than half of the heating value of the fuel is lost and this means that electricity is a wasteful and expensive form of energy.

Oil and Natural Gas

While the development of alternative—but often coal based—forms of energy, such as electricity and town gas, weakened the attraction of the coalfields, it was not until new fuels were introduced that their dominance as industrial areas was completely broken. And, of these new fuels, oil was by far the most important.

Prior to the First World War there was little demand for oil. After the war, however, the situation changed, slowly at first and then with ever increasing speed, as oil became the major source of energy in the world. The significance of this change in Britain can be seen quite clearly in the graph below. Such a growth reflected the superiority of oil over all its rivals. As a fuel it could be used more efficiently than coal and was much more flexible; it was also an invaluable raw material for industry, particularly the chemical industry; and it could be transported cheaply and easily by bulk tanker or along pipelines. This was of great importance to Britain which depended almost entirely upon imported oil.

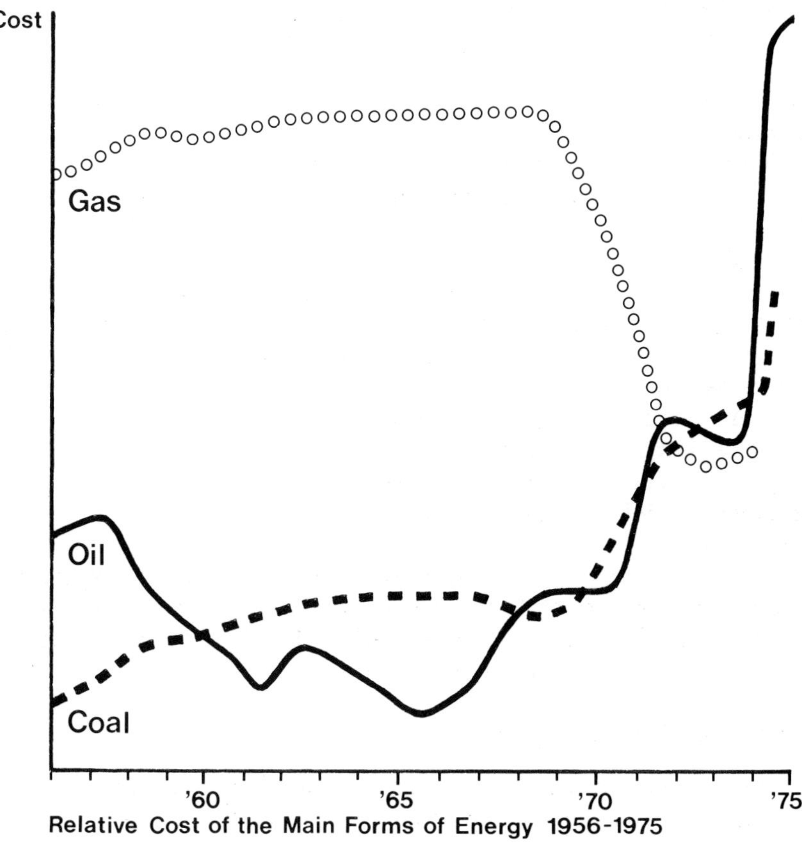

Relative Cost of the Main Forms of Energy 1956-1975

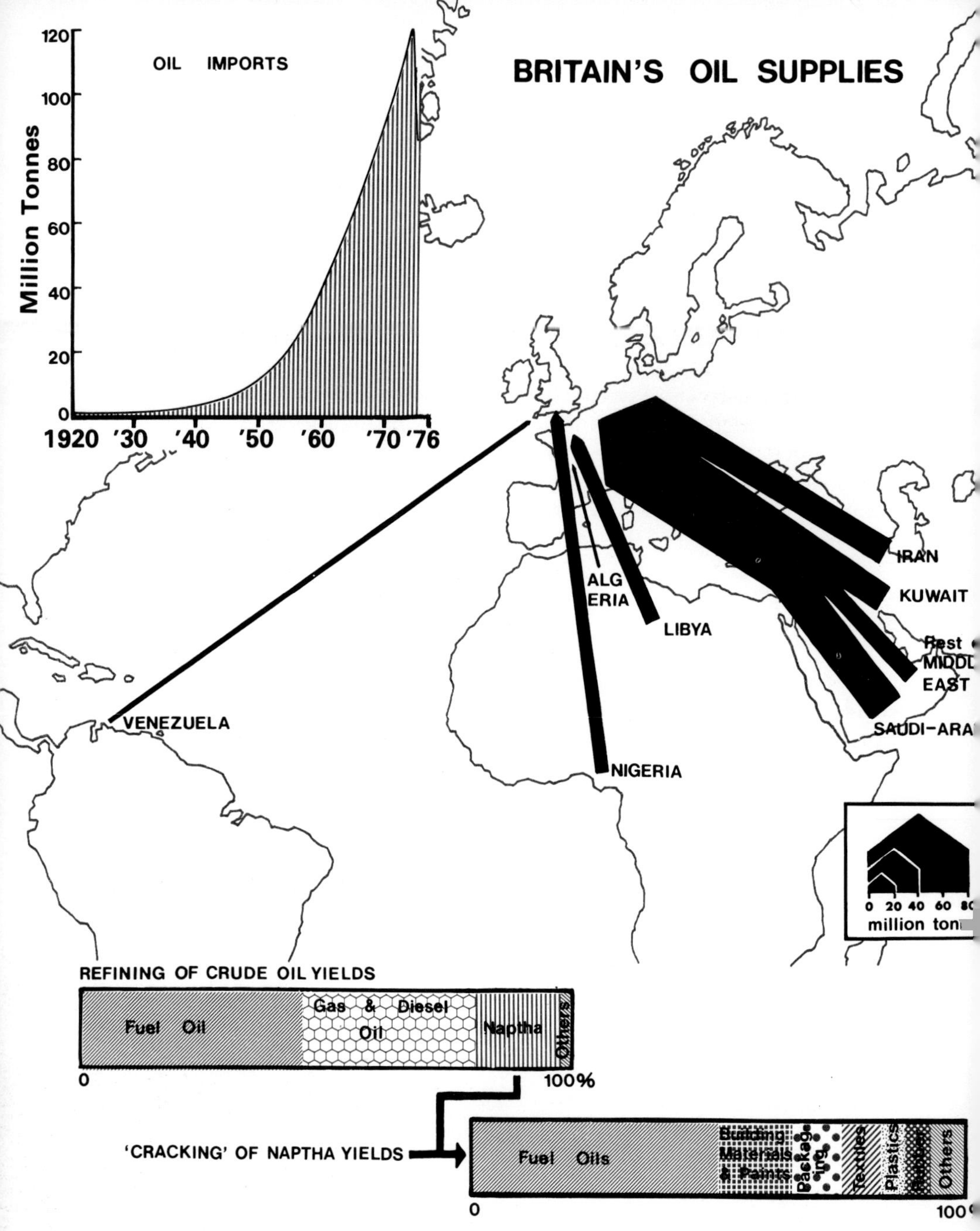

Using information given above, complete the following exercise:

a) i) How much oil did Britain import in 1950 and 1970?

ii) Name the main suppliers of oil to Britain, arranging them in order of importance.

b) What proportion of the oil is used:
  i) as a source of energy?
  ii) as a raw material for industry?

c) i) Name the canal through which most of Britain's oil used to pass.
  ii) With the closure of this canal in 1967, around which cape did the oil pass?

The oil is transported by tanker from the oil producing countries to Britain where it is refined. This is the usual state of affairs since it is cheaper to transport oil in a crude state rather than after refining when it has been broken down into many by-products, each of which requires separate treatment. As a result, the main refineries were built in the ports where the oil was landed (important 'break of bulk' points) and pipelines were constructed from the refineries to the markets. This apparently efficient system was put under great strain, however, with the closure of the Suez Canal in 1967, which enforced the use of the much longer Cape of Good Hope route. In order to reduce the cost of such a journey, the size of tankers was increased rapidly, from one hundred thousand tonnes in 1967 to half a million tonnes in 1975. These tankers were much too large for existing oil ports in Britain and terminals had to be built on deep water inlets, usually on the remote west coast. From these terminals, the oil is transported by pipeline or coastal tanker to the existing refineries.

Milford Haven oil terminal.

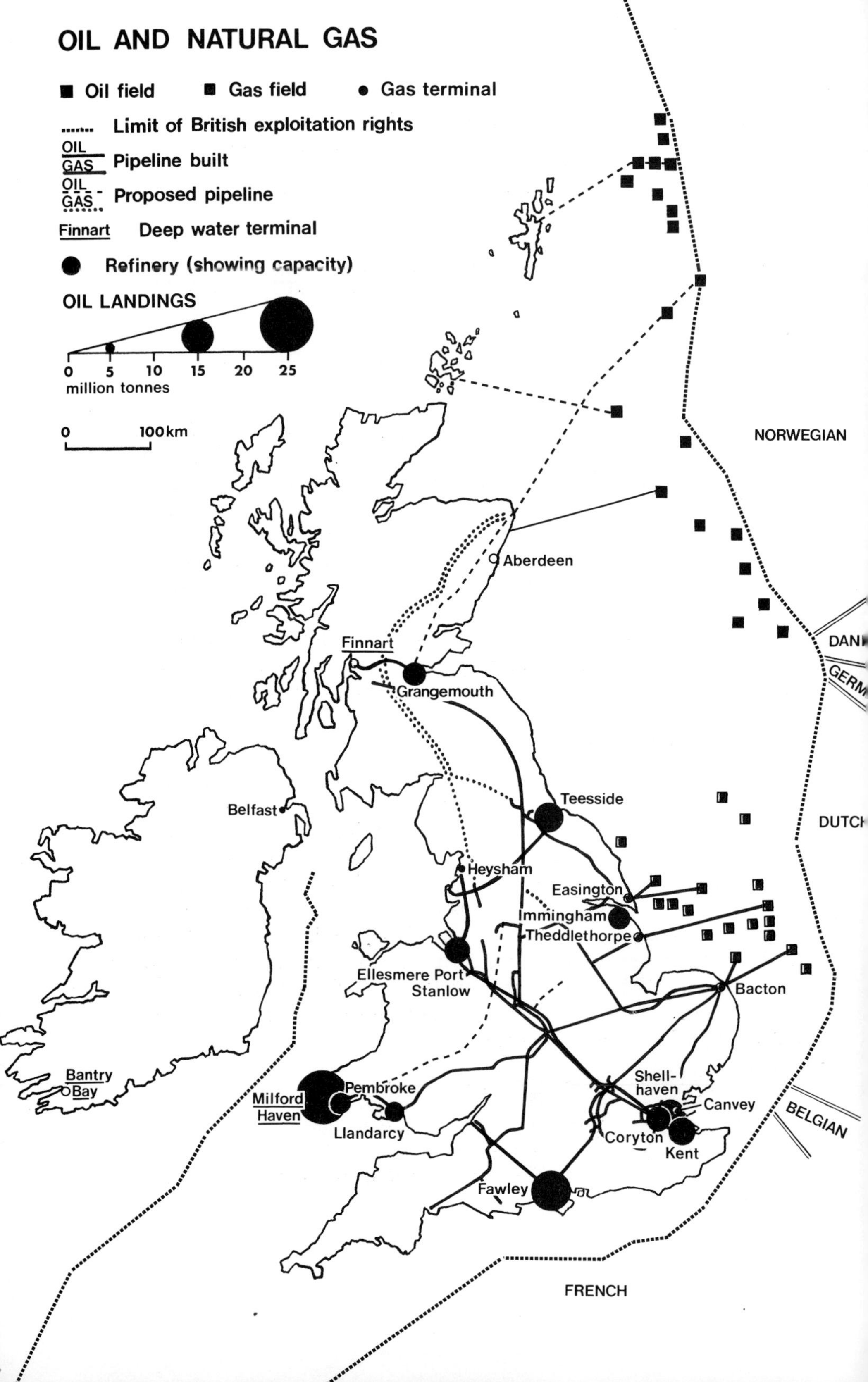

OIL AND NATURAL GAS
Oil field
Gas field
Gas terminal
Limit of British exploitation rights
OIL
GAS
Pipeline built
OIL
GAS
Proposed pipeline
Finnart
Deep water terminal
Refinery (showing capacity)
OIL LANDINGS
0 5 10 15 20 25
million tonnes
0 100km
NORWEGIAN
Aberdeen
Finnart
Grangemouth
DAN
GERM
Belfast
Teesside
DUTCH
Heysham
Easington
Immingham
Theddlethorpe
Ellesmere Port
Stanlow
Bacton
Bantry
Bay
Milford
Haven
Pembroke
Llandarcy
Shell-
haven
Canvey
Coryton
Kent
BELGIAN
Fawley
FRENCH

Using information given on page 122, complete the following exercise:

a) i) Name the main refining centres in Britain, arranging them in order of importance.
   ii) Name those which are likely to have developed since the introduction of super-tankers.
   iii) Name three deep water terminals.
   iv) Which of these has developed as a refining centre in its own right?
   v) Name two refining centres linked to these terminals by pipeline.

b) Study the pattern of oil pipelines in Britain.
   i) Try to explain the pattern (page 175 will help.)
   ii) How are oil products moved from the pipeline terminals to the actual users?

The effect of these changes on refining centres has varied. Some, like the Thames estuary sites and Teesside, are capable of handling all but the largest of modern tankers. Others, like Grangemouth and Llandarcy have been linked to deep water terminals. And even the most difficult of sites, such as Stanlow on the Mersey estuary has maintained its importance by using smaller tankers, until an unloading terminal can be constructed in deep water, off the coast of Anglesey.

Whether such a terminal will be necessary is doubtful, for the oil industry is facing yet another major change brought about by the discovery and exploitation of oil and gas deposits in the North Sea.

North Sea Oil and Gas

In the late 1950's one of the largest natural gas fields in the world was discovered at Gröningen in the Netherlands. This stimulated interest in the rocks underlying the North Sea, and, in the early 1960's major gas fields were discovered there, close to the east coast of Britain. Within a few years interest had shifted to the northern part of the North Sea where geological conditions appeared to favour the occurrence of oil. Exploration went ahead rapidly and many oil fields were discovered, particularly in the far north off the coast of the Shetland Islands. The first of these fields was brought into production in 1976 in spite of the difficulties of working in deep water, under difficult conditions.

Refer to the diagram on pages 124 and 125.

a) i) When was the oil formed in the North Sea?
   ii) In which type of rocks has it collected?
   iii) What prevents the oil from escaping?

b) Describe the problems likely to be faced when working in the North Sea.

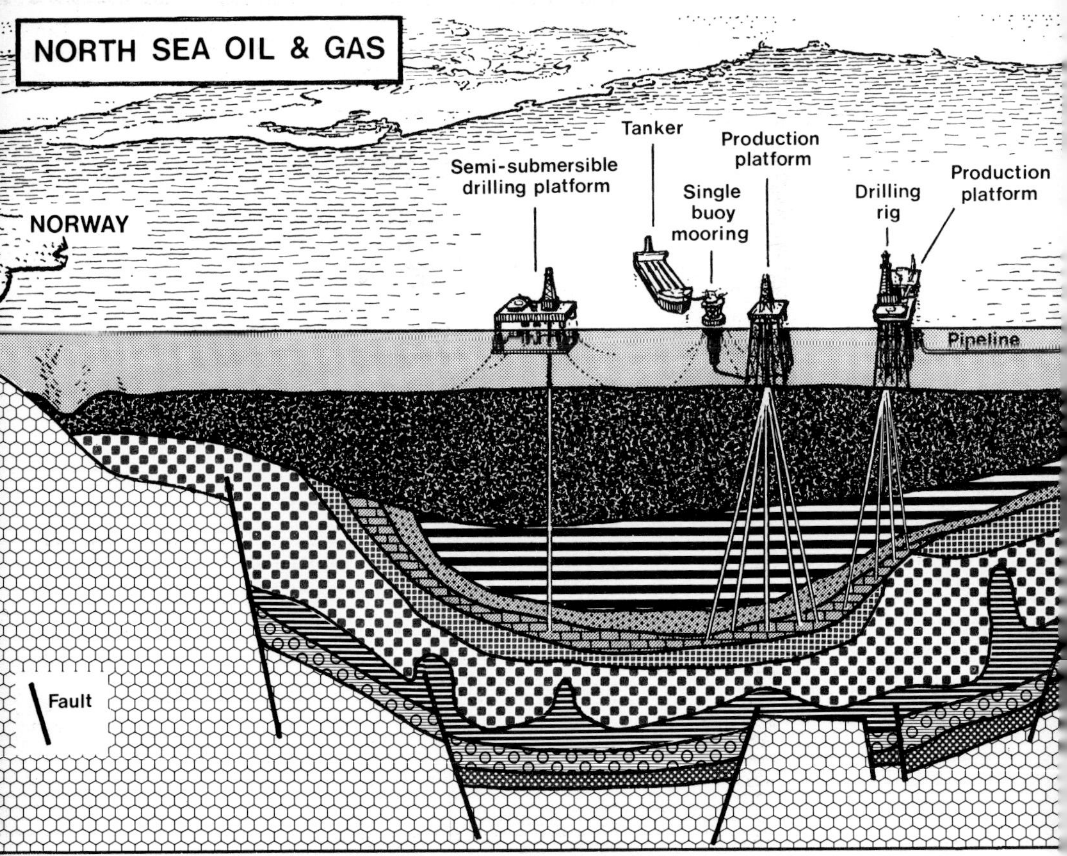

c) i) Describe how oil is obtained from the rocks.
ii) How is the oil transported to the land?
iii) Which refining centres are likely to benefit most from the development of the North Sea?

The development of North Sea oil will have profound effects on Britain.

1. The country will become virtually self-sufficient in oil and oil products. Imports will not stop entirely because it will still be necessary to import some heavy crude oil from the Middle East to mix with the lighter oils from the North Sea.

2. The distribution of the refining industry in Britain may be affected, with East Coast centres such as Teesside and Grangemouth benefiting. Large scale changes are unlikely, however, partly because of the effects of geographical inertia and partly because of the ease with which oil can be transported once pipelines have been built.

3. The costs involved in developing the North Sea are enormous and development has only been made possible by the price rises which have taken place during the early 1970's (See page 119).

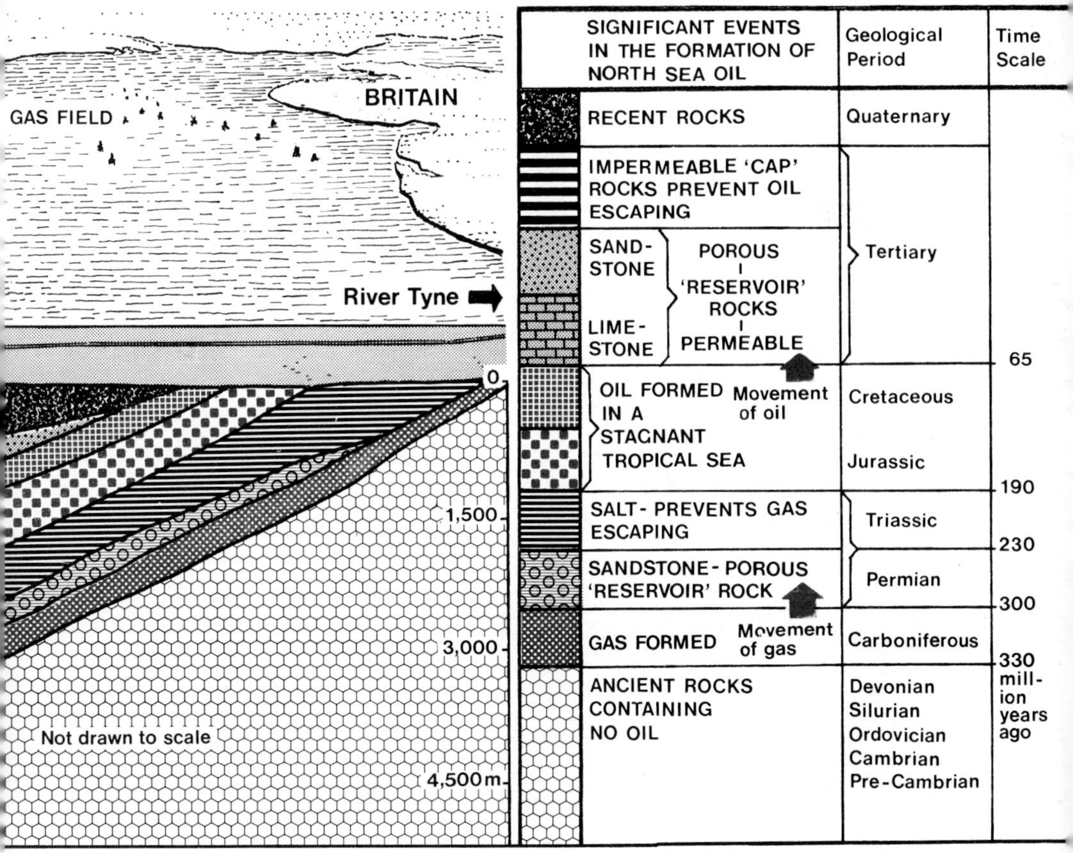
GAS FIELD
BRITAIN
River Tyne
0
1,500
3,000
4,500m
Not drawn to scale
SIGNIFICANT EVENTS IN THE FORMATION OF NORTH SEA OIL
Geological Period
Time Scale
RECENT ROCKS
Quaternary
IMPERMEABLE 'CAP' ROCKS PREVENT OIL ESCAPING
SAND-STONE
LIME-STONE
POROUS
'RESERVOIR' ROCKS
PERMEABLE
Tertiary
65
OIL FORMED IN A STAGNANT TROPICAL SEA
Movement of oil
Cretaceous
Jurassic
190
SALT- PREVENTS GAS ESCAPING
Triassic
230
SANDSTONE - POROUS 'RESERVOIR' ROCK
Permian
300
GAS FORMED
Movement of gas
Carboniferous
330 million years ago
ANCIENT ROCKS CONTAINING NO OIL
Devonian
Silurian
Ordovician
Cambrian
Pre-Cambrian

These changed oil from a cheap to an expensive form of energy and it has been suggested that alternatives should be found, leaving the remaining oil reserves for use as a raw material in industry. The extent to which this takes place in Britain will probably depend upon the extent of the North Sea oilfields and the success of exploration in the Celtic Sea.

USES OF OIL	%
Metal industries	24
Motor transport	24
Electricity generation	18
Heating	12
Chemical industry	7
Aviation	4
Others	11

Natural Gas

While the price of coal and oil has risen sharply since 1965, the price of gas has actually fallen, and gas is now the cheapest form of

Production platform in the Forties oilfield in the North Sea.

energy. Such a development seemed highly unlikely in 1964 when almost all gas was town gas, produced from coal. This was expensive and the market for it was declining rapidly. The situation was transformed by the discovery of large deposits of natural gas in the southern basin of the North Sea. Within ten years most parts of Britain had been linked by pipeline to the terminals at Bacton and Easington and appliances in households and factories throughout the country had been converted to use more efficient and stronger burning natural gas.

Furthermore, like oil, natural gas is a valuable raw material for the chemical industry and demand for it is likely to increase considerably. In order to meet these demands it is proposed to build pipelines to the North Sea oilfields to tap gas reserves which are at present being burnt off.

The significance of oil and natural gas as a raw material for the chemical industry cannot be overstated and it is important to examine the effects of the changes described in this chapter on that industry.

Industrial Study: The Chemical Industry

Of all the major manufacturing industries, the chemical industry has shown the most rapid and sustained growth in recent years, largely on account of the introduction of oil-based products. For much of the nineteenth century the industry was small, supplying the needs of other more important industries (eg dyes and bleaches to the textile industry), and making use of salt and by-products of the coking plants and steel works upon which Britain depended. During the twentieth century, however, the chemical industry has changed out of all recognition, both in the range of its products and in the scale of operations. In fact, it has become so vast that it is difficult to establish clear patterns within the industry, and, so complex are its products, that it is almost impossible to understand its links with other industries.

The modern industry depends upon two basic raw materials—salt and mineral oil—and the availability of these materials has strongly influenced its location.

Raw Materials

1. Salt is a traditional basic raw material of the chemical industry. It occurs in thick beds, deep underground. These deposits represent the remains of inland seas which dried up, under desert conditions, during the Triassic period, some two hundred million years ago.

The salt is obtained by pumping water, through bore-holes, into the beds and allowing solution to take place over a period of years. This produces brine, which is then pumped to the surface where it is either evaporated to produce salt crystals or piped in a liquid form for use in the chemical industry. Here most of the salt is converted into chlorine and sodium hydroxide—the basic materials for many branches of the industry.

Two major salt deposits occur in Britain—in the Weaver valley of Cheshire and in the Tees valley. Both have become important centres for the chemical industry.

2. Crude oil is a mixture of liquids and the refining process is designed to separate them. This is done initially in a fractioning tower, into which the oil, heated to 400°C, is passed as a vapour. Within the tower, the vapours rise and condense so that liquids of the same boiling point end up at the same level. This comparatively simple process breaks down the oil into a number of products, ranging from tars and heavy fuel oils, at the base of the tower, to petroleum, aviation fuel and gases at the top. Unfortunately the balance of these products does not coincide with the demand for

Llandarcy oil refinery, South Wales, which handles oil piped from the Milford Haven deepwater terminal.

them and, in recent years, a process has been developed to convert the heavier tars and oils into the lighter products which are in greater demand. It is this process, known as 'cracking', which produces most of the materials used in the petro-chemical industry.

Because of its strong links with the refineries, this branch of the industry has been attracted to the main refining centres on the coast.

Other raw materials, ranging from potash and sulphur to molasses, are used in the chemical industry but they are much less important than salt and oil. Furthermore, since many of these materials are imported, they have also tended to encourage the location of the industry near to ports.

The Location of Industry

Many factors play a part in the choice of a site for a chemical works. They include:

1. Availability of raw materials particularly oil and salt.
2. Access to port facilities for imported materials.
3. Good communications with suppliers of materials and markets.
4. Large areas of flat land for building.
5. Plentiful supply of water.
6. No large centres of population nearby, in case of accidents.
7. Possible government aid.

The importance of these factors can be seen in the development of the chemical works at Baglan Bay in South Wales.

British Petroleum built their works at Baglan Bay in 1963 and further extensions have made it one of the largest in Europe.

Using the information given, complete the following exercise:

a) i) Name four basic raw materials used at the plant.
   ii) For each raw material, state its source and explain how it is transported to the plant.
   iii) Locate the sources of raw materials and work out how far away from Baglan Bay they are.

b) Copy out the list of factors given above and underline those which might have influenced the choice of Baglan Bay as the site for the works.

c) Name the four most important end products of the works and, for each, name two major users.

Although apparently situated well away from the main market areas of Britain, the Baglan Bay plant does in fact have strong links with nearby industries, particularly the plastics industry at Barry, for which it provides most of the raw material.

One of the attractions of the site, apart from the availability of oil, was the fact that it was in a Development Region and government grants were available to help finance the project. This was very important since the cost approached one hundred and fifty million pounds. The aim of these grants was to bring new industries into declining areas and to provide new jobs for people employed in contracting industries. In the former aim, the Baglan Bay scheme has been successful, but, in the latter, success is less obvious since only two thousand permanent jobs were created in spite of the scale of investment. This tends to be typical of the chemical industry as a whole.

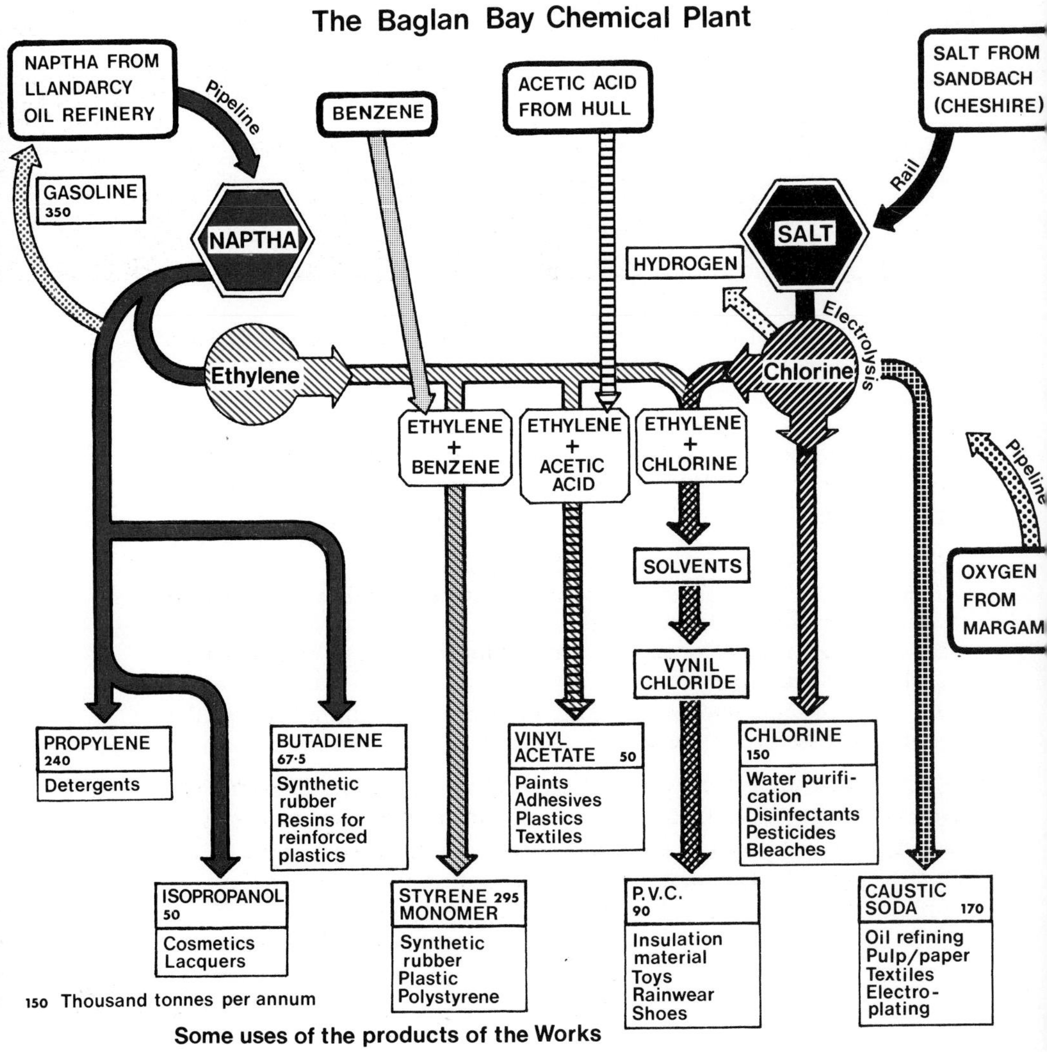

Baglan Bay chemical plant, South Wales.

In fact, many aspects of the Baglan Bay development are typical, and the factors which influenced its location are those which have influenced the distribution of the industry throughout Britain.

Study page 132 and complete the following exercise:

a) Name the three main centres of the industry (each takes its name from the river which flows through it).
b) Using the factors given above, account for the development of the industry in these areas.
c) Which of these factors have been most important in influencing the distribution of the industry as a whole?

This pattern of distribution, centred on the great deep water estuaries has developed over a long period of time and it is unlikely to change greatly in the near future. There is, however, likely to be some change of emphasis with the development of North Sea oil supplies and the east coast centres are again likely to benefit.

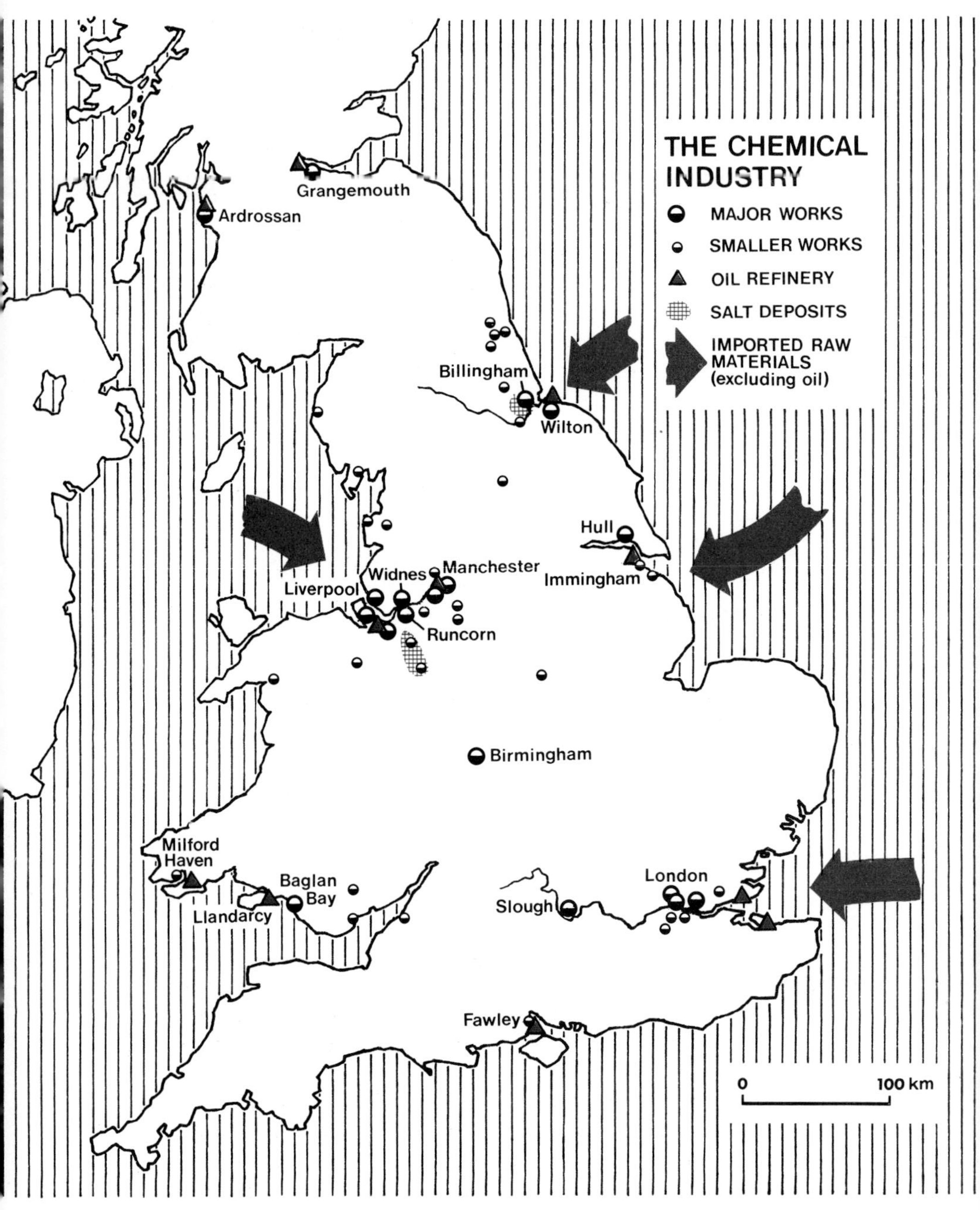
THE CHEMICAL INDUSTRY
MAJOR WORKS
SMALLER WORKS
OIL REFINERY
SALT DEPOSITS
IMPORTED RAW MATERIALS (excluding oil)
Grangemouth
Ardrossan
Billingham
Wilton
Hull
Immingham
Manchester
Widnes
Liverpool
Runcorn
Birmingham
Milford Haven
Baglan Bay
Llandarcy
London
Slough
Fawley
0
100 km

Communication: and the location of industry

The changes in the pattern of energy supplies to industry coincided with changes in communication, which have also had important effects on the location and distribution industries, and the development of industrial areas. Basically these changes are:

1. Before the Industrial Revolution, industries depended upon a very poor system of horse-powered transport.
2. The Industrial Revolution and the expansion of industry on the coalfields in the nineteenth century coincided with the development of canals and, even more important, railways.
3. Recent industrial changes have taken place against a background of improved road transport.

The extent of these recent changes can be seen from the following statistics:

TYPE OF TRANSPORT	% OF TOTAL CARRIED 1954	1975
a) FREIGHT		
Road	22·0	58·5
Rail	73·0	19·8
Coastal ships	3·0	19·4
Canals	1·0	0·2
Piplines	1·0	2·0
Air	0·01	0·1
b) PASSENGERS		
Road (Private)	38·9	76·9
(Public)	41·0	14·9
Rail	20·0	7·7
Air	0·1	0·5

The purpose of any system of communications is to move people, goods or even information from place to place, and developments take place in the system to improve the efficiency of these movements. Each system of communications forms a *network* in which *places* are joined together by lines or *links*. The efficiency of the network can be measured in terms of the ease with which movement can take place between the places in the network. This is known as *connectivity*—a concept which is of great importance to industry, since places with a high degree of connectivity tend to be favoured

A canal in industrial Staffordshire.

locations for industries. Unfortunately measuring the efficiency of a communications network is not easy. (Two simple methods which will be used in this book are given on page 309). Some of the problems stem from the fact that movements within the system can be measured in at least three different ways.

1. The usual measurement of a journey is its distance in kilometres. Such a measurement can, however, be misleading, particularly when considering the needs of industry.

2. Time—journey time is perhaps more significant, particularly with regard to the movement of passengers, and this is not always the same as distance.

3. Cost—for the majority of industries this is by far the most important consideration and, because the cost per kilometre of moving freight tends to decrease as the length of journey increases, the effects on industrial location can be very complicated.

If the efficiency of a transport system is measured in terms of time and money, any delay becomes a serious problem. Most delays occur where goods have to be handled (*breaks of bulk*), for example in ports or at railway termini, and it is significant that many of the recent improvements in transport have been designed to reduce handling and to eliminate such breaks. Against this, however, it is important to remember that industry tends to be attracted to places where breaks of bulk take place.

Communications, therefore, play an important, if complicated, part in the location of industry. Before 1750 most industry depended upon horse drawn transport and a totally inadequate system of poor quality roads. This made the movement of raw materials and finished products both difficult and expensive; factors which tended to restrict the size of industries and of the areas which they could serve. As a result industries became scattered throughout the length and breadth of the country and few of them ever grew to serve a national market. In fact, the development of large scale industry and its concentration in certain limited areas, such as the coalfields, became possible only with the development of new forms of transport, particularly the canals and railways.

Canals

For centuries the most effective way of moving freight was by boat and, by 1750, coastal and river traffic had become very important. Unfortunately Britain has few navigable rivers and, during the eighteenth century, plans were made to build a canal network which would link the major rivers. Construction started during the second half of the century and, within the space of seventy years, most parts of lowland Britain were served by the network. The effects on industry were immediate. Freight rates were reduced by more than fifty per cent and it at last became possible to assemble the raw materials require by large scale industry and to distribute finished products to markets throughout Britain and the world.

SHORTEST DISTANCE BY CANAL BETWEEN MAJOR CENTRES (KM)

	London	Bristol	Birmingham	Liverpool
London				
Bristol	283			
Birmingham	235	150		
Liverpool	390	285	145	
Hull	449	–	262	230

1. Average speed for the movement of freight by horse drawn barge — 4 km per hour.
2. Average cost of movement in 1840: By Road — 3p per tonne 1 km
   By Canal — .75p per tonne 1 km   By Rail — .5p per tonne 1 km

At the same time, however, there were, within the system, problems which were to render it incapable of competing with later forms of transport such as the railways.

Using the information given on pages 134 to 136, complete the following exercise:

a) i) Describe the vessel used on the canal.
   ii) Such boats are called 'narrow boats'. Why are they necessary?
   iii) How wide is the canal?

b) i) What is the feature shown on the photograph?
   ii) What is its purpose?
   iii) What effect do such features have on journey times?

c) i) Why are canals more suitable for freight transport rather than passenger transport?
   ii) The shortest distance by canal between five major cities is given in the table. Calculate the index of directness of the links and, using an atlas explain why they vary.
   iii) Why has it proved difficult to adapt the canal network to the requirements of modern industry?

The canals are, in fact, too narrow and too shallow, and, with the building of the railways, too slow and too much affected by relief. One or two of the later canals were larger but they tended to fall into disuse with the rest of the system and, like the older canals, are used only by leisure craft. Not until the end of the nineteenth century was a waterway built which was comparable with the large continental canals. This was the Manchester Ship Canal which was designed to take ocean-going vessels, and which is still important today.

Railways

An even more important influence on the development of British industry during the nineteenth century was the building of the railways. The impact of this form of transport can best be seen by studying the journey times shown in the table below.

TYPICAL TRAVEL TIMES 1840–1975

LONDON TO:	CAMBRIDGE	READING	BRIGHTON	TUNBRIDGE WELLS
By Road: 1840	6 hrs	$4\frac{1}{2}$ hrs	$5\frac{3}{4}$ hrs	4 hrs
By Rail: 1840	1 hr 50 m	1 hr 15 m	1 hr 40 m	2 hr 15 m
By Rail: 1900	1 hr 19 m	45 m	1 hr 20 m	55 m
By Rail: 1975	1 hr 03 m	37 m	58 m	52 m

Select three of the journeys and, for each:

a) Calculate the time saved during the periods 1830–1900 and 1900–1975.

b) Let the time in 1900 represent 100 and calculate the times in 1830 and 1975 as a proportion of this.

This clearly shows that, while the canals made the movement of freight economically possible, it was the railways which revolutionised the speed at which such movements could take place. Later improvements in transport did not produce such striking results.

The building of the rail network took place rapidly after 1830 and, by the end of the century, almost every town of any size had its rail link. This involved overcoming many problems.

Study the map on page 139 and the photograph below and complete the following exercise:

a) i) Locate on the map, three areas which have always been badly served by the railways.

Railway track following obvious topographical features.

ii) Which of the following factors may have prevented the building of railways there? Too many roads/High land/Too wet/Not enough room because of buildings.

iii) How has the problem of high land been overcome by the builders of the railway shown in the photograph, and how has an even gradient been maintained?

b) Using any Ordnance Survey map, compare the routes taken by the rail and main road link between any two places.

i) Describe the differences between the two routes.

ii) Calculate the index of directness for the two links. Why do they differ?

c) Which of the following statements best describes Britain's rail network? An even spaced network/an irregular network centred on Birmingham/a rectangular grid/a radial network centred on London.

The railways reached their peak in the early years of the present century and since then they have faced increasing competition from road transport. This has resulted in a decline in the importance of the railways (see the table on page 133) and the closure of many lines. Some of these closures were the inevitable result of the way in which the railways were built. For example during the boom years rival companies built several competing lines between the major cities and, as traffic declined, some of these lines became seriously under-used. After nationalisation such lines were closed. Most, however, were the products of the nature of rail transport itself and a series of problems arose which are summarised in the following table.

MOVEMENT OF FREIGHT BY RAIL BETWEEN LONDON AND GLASGOW (DISTANCE 630 KM)

STAGE	TIME TAKEN	
• Transport to station (by road)	2 hr	
Handling at terminal		30 min
Journey	8 hr	
Handling at terminal		30 min
• Transport to destination (by road)	2 hr	
Total	13 hr	

• This will vary according to whether the customer collects and delivers the consignment.

1. Goods cannot be moved from door to door.

2. A second form of transport (usually road) is needed to take goods and passengers to stations.

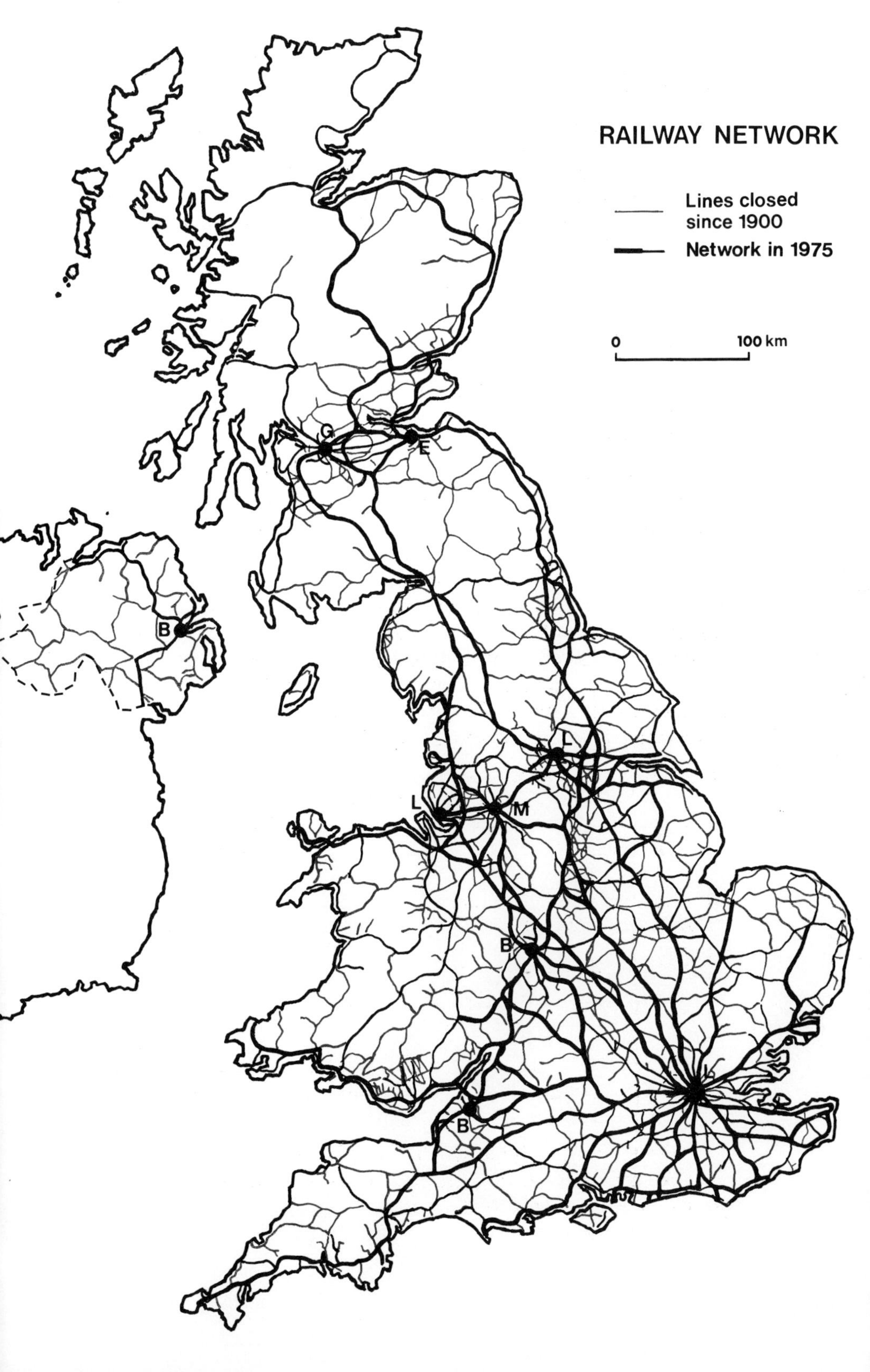

RAILWAY NETWORK
Lines closed since 1900
Network in 1975
0
100 km
G
E
B
L
L
M
B
B

3. As a result, the handling of goods, which is expensive and time consuming, makes up a large proportion of the journey time.

4. This, in turn, means that direct road transport is usually more efficient for short journeys and that the railways can only compete over long distances or with bulk commodities such as coal and iron ore.

5. It also means that a large labour force is required and this is expensive.

In an effort to halt the decline of the railways and to relieve pressure on the road system, many changes have been introduced, most of them aimed at improving the flexibility and reducing the costs of the railways. In addition to the closure of many branch lines these measures include:

1. The electrification of most main lines.
2. Increasing speeds on these lines by improving the track, signalling and trains.
3. The reduction of handling by automation and the introduction of container traffic.
4. The reduction of the labour force.

Freightliner container being transferred from road to rail.

Such measures have met with limited success, however, and the railways continue in Britain only because of large government subsidies.

Roads

The revival of road transport came with the development of the internal combustion engine. Prior to this, horse-drawn vehicles could not compete with the railways either in speed or in the amount of freight which could be moved. With the development of a more effective source of power, however, the natural advantages of road transport quickly became apparent.

MOVEMENT OF FREIGHT BY LORRY BETWEEN LONDON AND GLASGOW

STAGE	TIME TAKEN
Loading	1 hr
Journey	10 hr
Unloading	1 hr
Total	12 hr

Refer to the table above.

a) Represent the information given in the table in the form of a diagram similar to that shown on page 19.
b) List the advantages of road transport over rail transport.

The advantages of door to door transport, reduced handling and greater flexibility resulted in the rapid growth of road transport at the expense of other forms of transport (see the table on page 133). Such growth has, however, created a whole range of problems which must be set against the undoubted advantages of road transport.

Problems

Many of these problems stem from the weakness of the road network itself. Most of Britain's roads have developed over a long period of time and the system is largely unplanned. Built, in the first place to link settlements, such roads form long distance links which are not the most *direct* or the most *efficient*. Furthermore, since they were built to serve settlements, they were usually built through them. As a result, increased traffic has tended to produce *congestion*, particularly in towns where several roads may converge. Congestion is made worse by the fact that the majority of roads were designed to handle horse drawn traffic and are inadequate for modern motor vehicles, especially modern lorries.

The problems are greatest in our town centres where long delays can occur and where the traffic can cause damage to buildings as well as an unacceptably high accident rate. Improvements to the road system are obviously needed but the rapid growth in traffic (see table below) has meant that road building and improvement has been unable to keep pace and the problems of overcrowded roads and congested town centres remain.

YEAR	VEHICLES ON ROADS (MILLS)
1947	3·4
1952	4·9
1957	7·5
1962	10·6
1970	15·0
1975	17·2

The M6: a typical motorway. Note i) the avoidance of contact with existing road; ii) the avoidance, where possible, of built up areas.

Solutions

The obvious solution to such problems is to improve existing roads or to build new ones. This policy has been carried out in Britain in two ways.

1. For a long time it was thought that traffic problems could be overcome by straightening and widening roads and, if necessary by building *ring roads* to divert traffic around town centres. Such a

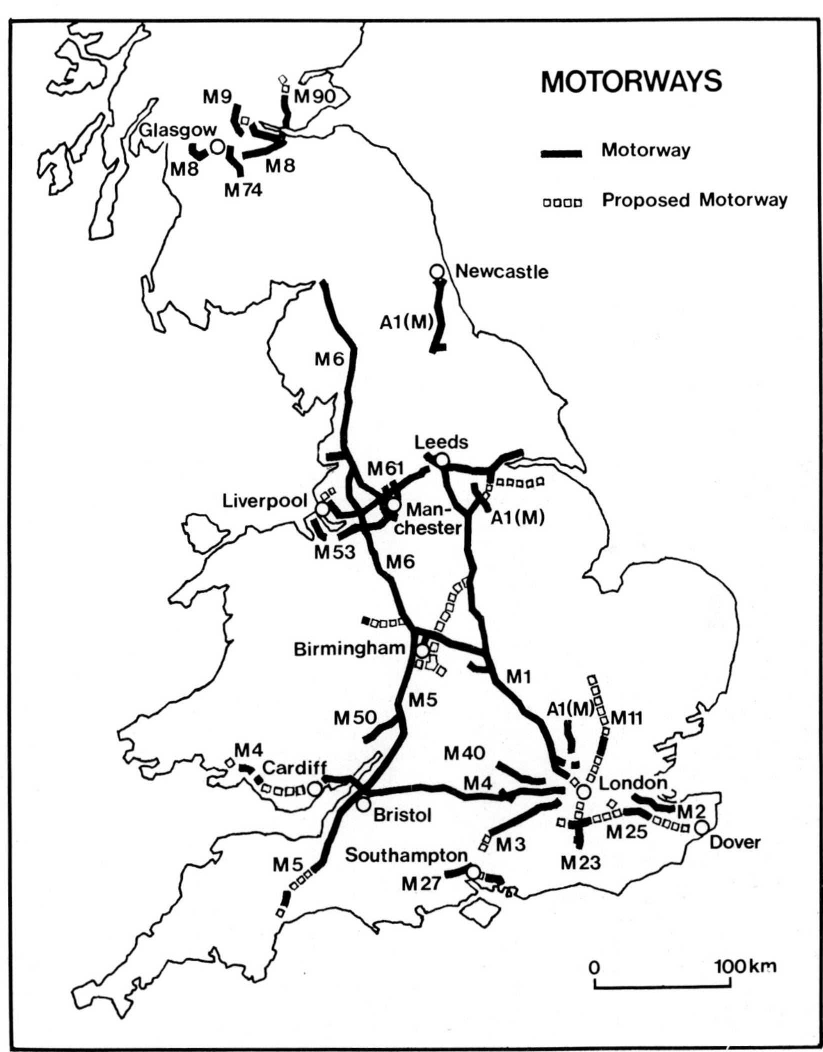

policy proved difficult and expensive, particularly in urban areas where land values are high and when many houses have to be cleared. Furthermore, piecemeal improvements tended to cause even greater congestion on unimproved stretches of road.

2. As a result, in the early 1950's the new policy was adopted, of building fast, direct links between the centres of population. These roads, known as *motorways*, form the basis of our modern transport system, and they have had important effects on the economic development of the country.

a) Travel times between centres have been reduced by up to a half.

b) Congestion in some town centres has been relieved, although not to the extent anticipated.

ROAD DISTANCES BETWEEN MAJOR CENTRES (KM)

	London	Bristol	Birmingham	Liverpool	Manchester	Leeds	Newcastle	Glasgow
London		*175* 165	*175* 165	*315* 290	*295* 265	*305* 275	*435* 400	*630* 565
Bristol	**190** 165		*140* 125	*255* 220	*255* 230	*310* 270	*455* 400	*585* 510
Birmingham	**190** 165	**145** 125		*145* 125	*125* 115	*175* 150	*320* 280	*300* 410
Liverpool	**330** 290	**255** 220	**160** 125		*55* 50	*115* 105	*245* 200	*340* 290
Manchester	**330** 265	**255** 230	**160** 115	**60** 50		*65* 55	*205* 170	*340* 300
Leeds	**335** 275	**350** 270	**220** 150	**115** 105	**70** 55		*145* 135	*335* 300
Newcastle	**500** 400	**515** 400	**390** 280	**280** 200	**235** 170	**165** 135		*230* 200
Glasgow	**655** 565	**620** 510	**485** 410	**385** 290	**385** 300	**440** 300	– 200	

Bold – using MOTORWAYS (av. speed: passenger – 110 kph)
freight – 80 kph

Italic – using ORDINARY roads (av. speed: passenger – 55 kph)
freight – 45 kph

Text – straight line distance

c) The network has resulted in changes in the accessibility of places; a factor which is likely to influence the future location of industry (see photograph on page 142).

a) Describe the ways in which motorways differ from other types of road, paying particular attention to size, directness, access, rules of use.

b) From the following list, draw up two groups of statements; the first to illustrate the advantages of motorways, the second the disadvantages.

1. Need a lot of land.
2. High speeds possible.
3. Safer because few bends and no oncoming traffic.
4. Expensive to build.
5. Difficult to build in urban
6. Boring to use.
7. Few hold-ups likely.
8. Cause congestion where the motorway ends.
9. Difficult to cross.
10. Access points infrequent.

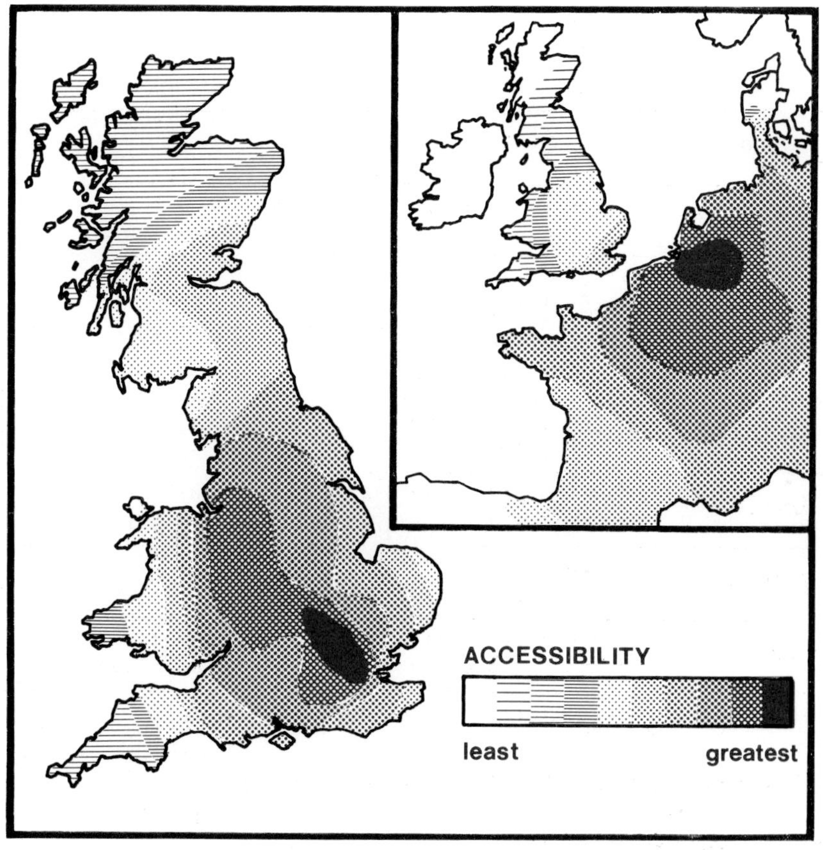

c) Refer to page 144 and complete the following exercise:
   i) Calculate the index of directness of the links shown, using a) motorways, b) trunk roads (see page 309).
   ii) Which centres have benefited most from the building of the motorways?

d) i) Calculate the accessibility of the places mentioned (see page 310).
   ii) Plot the cities on a blank outline map of Britain and against each write its accessibility number.
   iii) Join together places with an accessibility number less than 30 so as to form a continuous line. Compare this with the pattern shown on the map on page 145.
   iv) Places other than cities may become attractive to industry because of motorway accessibility. Identify two such places and mark them on the map. Calculate their accessibility and compare it with the accessibility of the cities.

The effect of this changing pattern of accessibility on the location of industry has been striking, particularly during the present century, when there has been a shift of industry from the coalfields to locations which are well served by road transport. Here raw materials can be assembled and, what is perhaps more important, the main markets and ports are within easy reach. As we have seen however, transport systems cannot be viewed simply in terms of distance and, when time and cost are included the concept of accessibility becomes very complicated. Furthermore the effects of accessibility have to some extent been countered by government policies which have been designed to divert industry away from the most accessible areas of the South East and the Midlands. In spite of this, the effects can be clearly seen in the case of certain industries; for example, the motor industry.

Industrial Study: The Motor Industry

The motor vehicle industry is the largest single manufacturing industry in Britain, and the most important exporting industry. It is also one of the most complex of modern industries, showing in its structure and distribution, the effects of many different factors.

The industry developed in the last decade of the nineteenth century as an extension of the engineering industry. As such, it tended to grow up on the coalfields where the engineering industry was already highly developed; where there were supplies of locally produced metals; and where the labour force was already skilled. Of these areas, South Staffordshire, Lancashire and Lanarkshire were most important. Throughout this early period, the industry was on a small scale, producing a large range of hand-made cars, in very small numbers.

The modern industry is completely different, and the changes in its structure have produced marked changes in its distribution.

1. After the First World War *mass production* methods were introduced from the United States and these are now used in all factories.

2. This has involved the establishment of vast mechanised *assembly lines* which are supplied with *components* from factories scattered over a wide area.

3. Large amounts of capital were needed for such developments and this has led to small firms amalgamating and to many firms being taken over by the main American motor manufacturers.

4. Vehicles are now produced in vast numbers.

5. The range of models is much smaller and parts have been standardised to allow the more efficient use of mass production methods.

6. Skilled labour has been replaced by semi-skilled labour.

Part of the Leyland assembly line at Longbridge.

7. A large proportion of the output is exported.

Such an industry, with its need to assemble the two thousand components which make up a modern motor car; and to distribute the millions of vehicles produced each year, is obviously very different from the early motor industry. And equally obviously, it tended to be attracted to different locations.

Location of the Motor Industry

Some of the early car making centres have remained important. Of these, the Midland centres are the most significant and the factories in the Birmingham/Coventry area have continued to grow. This is largely due to their strong links with the component industries on the nearby South Staffordshire coalfield, and to their accessibility and central position, which has allowed parts to be brought in from as far afield, as South Wales.

The importance of accessibility and transport can also be traced in the later development of the industry in the South Midlands and London areas for, as the movement of components became easier, the industry was attracted to these densely populated areas, and large assembly plants were built in Oxford, Luton and Dagenham. These were all located in the area of greatest access to markets (see page 145) and within easy reach of major ports.

As a result of these developments, by the late 1950's the motor industry was concentrated in a fairly narrow corridor stretching from London to Birmingham. The advantages of this location in terms of transport were obvious and they were increased by the building of the early motorways.

It is surprising, therefore, that the major developments since then have taken place on Merseyside and in Scotland, areas which are far from the main component industries and which are much less accessible. Both are development areas, however, and the movement of the motor industry there has been the direct result of government pressure. The building of assembly plants at Halewood, Ellesmere Port and Linwood has encouraged the development of some component industries nearby but most of the parts are still transported from London and the Midlands—an expensive operation made possible only by the improvement of road and rail links between the regions.

Using information given on page 149, complete the following exercise:

a) i) Name the main car assembly plants.

ii) Name four major component industries.

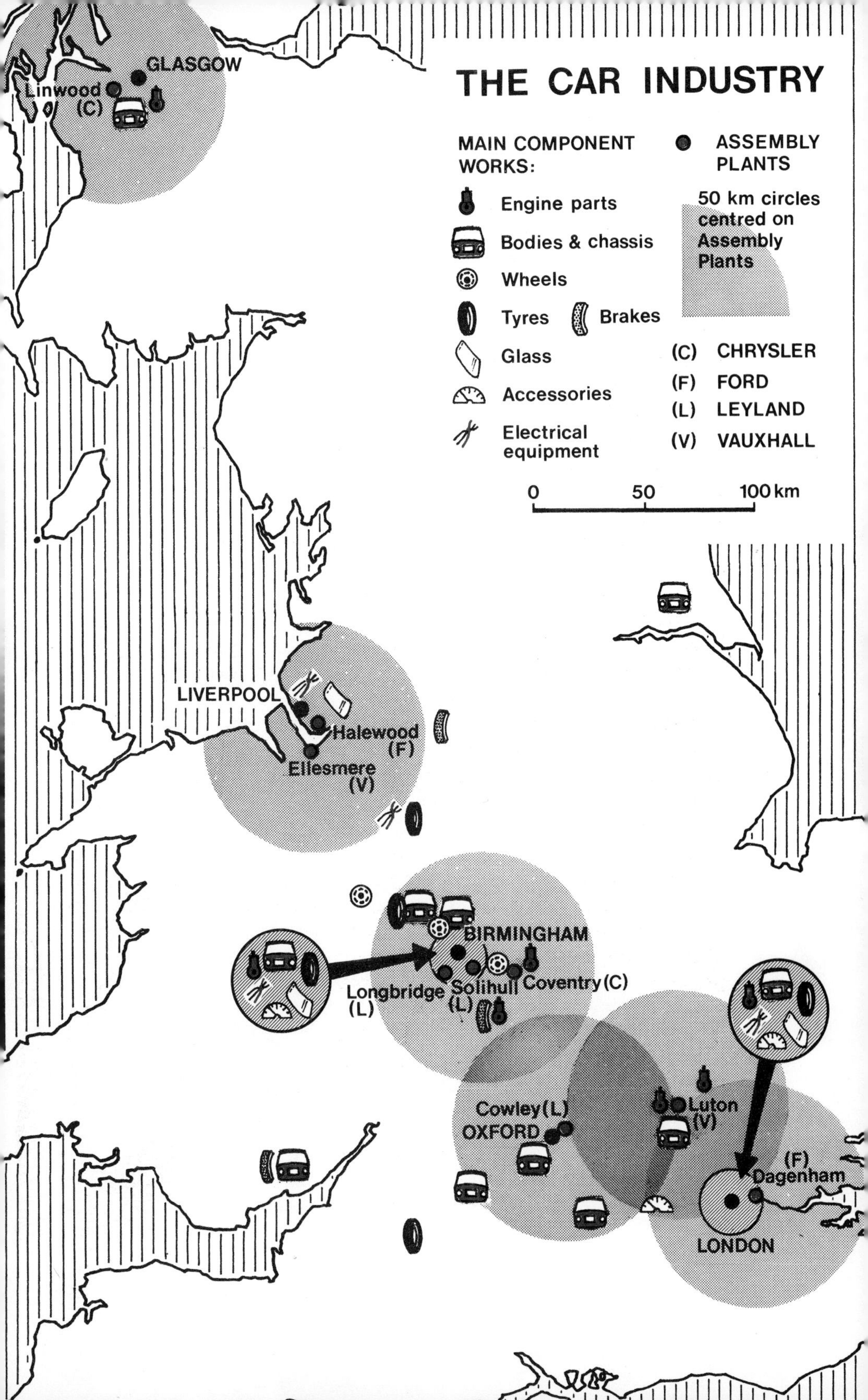

THE CAR INDUSTRY
MAIN COMPONENT WORKS:
Engine parts
Bodies & chassis
Wheels
Tyres
Brakes
Glass
Accessories
Electrical equipment
ASSEMBLY PLANTS
50 km circles centred on Assembly Plants
(C) CHRYSLER
(F) FORD
(L) LEYLAND
(V) VAUXHALL
0
50
100 km
GLASGOW
Linwood (C)
LIVERPOOL
Halewood (F)
Ellesmere (V)
BIRMINGHAM
Longbridge (L)
Solihull (L)
Coventry (C)
Cowley (L)
OXFORD
Luton (V)
(F) Dagenham
LONDON

iii) For each of the assembly plants name one major source of supply of each component.

b) Select three contrasting car manufacturing centres and, for each, describe the main advantages and disadvantages of its location.

Problems of the Industry

Since the late 1960's the motor industry has been beset by a number of serious problems:

1. Competition from other countries such as West Germany and Japan has increased.

2. In comparison with these countries, the industry in Britain is out of date and unable to compete. This is reflected in a much lower level of productivity.

3. The industry depends on more than two thousand different parts reaching an assembly line. Disruptions in the supply of one of these parts can bring large sections of the industry to a halt and cause large numbers of workers to be laid off. Such disruptions have occurred frequently and labour relations are often poor. In an attempt to overcome the problem the major manufacturers have taken over many of the component industries.

In spite of its problems the motor industry remains of vital importance to the British economy, consuming one fifth of the nation's steel production, sixty per cent of the rubber production and more than a quarter of the output of non-ferrous metals such as lead and zinc. It is also a major earner of foreign currency and some sections of the industry such as trucks and commercial vehicles remain among the most important in the world.

The complexity of the factors influencing the development of British industry are apparent in the case of the motor industry but they can be seen perhaps more clearly in a study of a major industrial region.

Regional Study: The West Midlands

For the past two hundred years the West Midlands has been one of Britain's leading industrial regions. During this time its industries have changed considerably to keep pace with a changing world. In spite of this, however, traces of the old patterns still show through to give a very complicated picture, which summarises the development of British industry as a whole. These include:

1. Traditional industries, often based on agricultural raw materials, which grew up during medieval times. Typical of such in-

dustries are brewing at Burton and leather working at Stafford and Walsall.

2. The coalfield industries which emerged during the seventeenth century and which overshadowed, if not replaced, the traditional industries. Two coalfields in particular dominated this type of development; the South Staffordshire coalfield with its metal based industries and the North Staffordshire coalfield with its pottery industry.

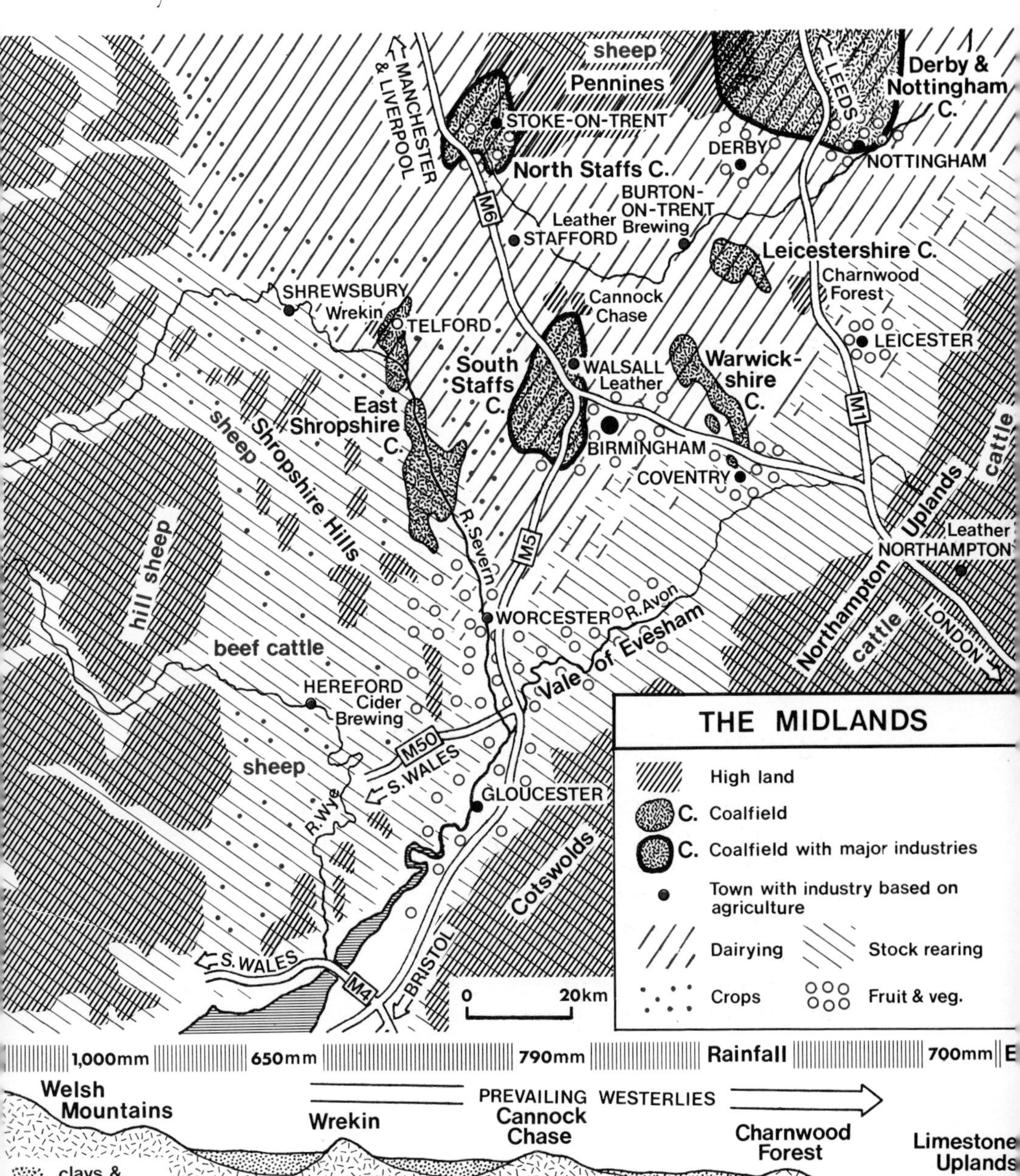

3. The industries which depend upon the central position of the Midlands and upon an efficient transport system. Such industries first appeared with the building of the canals and railways and they became dominant during the twentieth century, with the improvement in road transport.

Study page 151 and complete the following exercise:

a) Write a brief description of the agriculture of the Midlands, paying particular attention to the influence of relief and rainfall.
b) Name three industries which depend on agricultural raw materials. In each case name one centre important for that industry.
c) Name five coalfields in the Midlands.
d) Name the two most important and in each case name one industry associated with the coalfield.

Complicated as this picture may appear, the reality of industrial development in the Midlands is even more difficult to understand. This is largely due to the presence in the region of the Birmingham/Black Country conurbation, which has influenced development everywhere.

Birmingham and the Black Country

The conurbation, which comprises Birmingham and the towns of the South Staffordshire coalfield, ranks second only to London in terms of its size and concentration of industry. And, like London, for the last two hundred years it has been among the most prosperous of the British regions—a situation made possible by the variety of its industry and by its ability to adapt to changing circumstances.

During the nineteenth century the area emerged as a centre of heavy industry and the coalfield towns became most important. *Coalmining* developed there during medieval times but it was not until the eighteenth century that large scale production started. Thereafter, however, growth was rapid, particularly in the southern part of the coalfield which contained the famous Thick Seam, up to ten metres thick and lying near to the surface. Exploitation of this seam was easy and costs were low but it also devastated large areas of land, leaving it derelict and in need of reclamation. It was also worked out quickly and this led to the early closure of pits, a decline in production and the movement of mining northwards.

Using information given on the map opposite and the table on page 154, complete the following exercise:

a) i) When did coal production reach its peak?
   ii) During which period did production expand most rapidly?

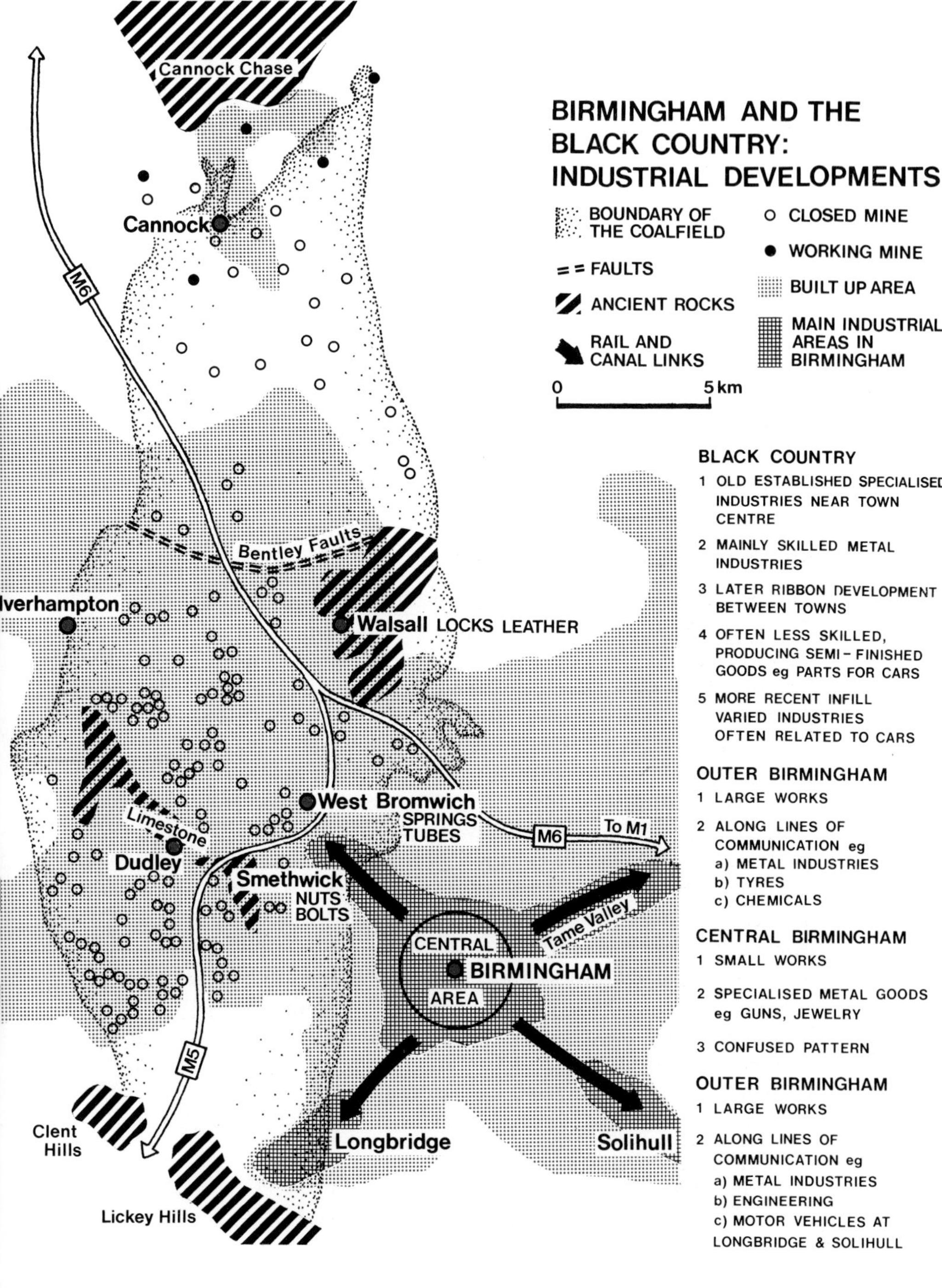
BIRMINGHAM AND THE BLACK COUNTRY: INDUSTRIAL DEVELOPMENTS
BOUNDARY OF THE COALFIELD
CLOSED MINE
WORKING MINE
FAULTS
BUILT UP AREA
ANCIENT ROCKS
RAIL AND CANAL LINKS
MAIN INDUSTRIAL AREAS IN BIRMINGHAM
0
5 km
Cannock Chase
Cannock
M6
Bentley Faults
verhampton
Walsall LOCKS LEATHER
Limestone
Dudley
West Bromwich
SPRINGS
TUBES
M6
To M1
Smethwick
NUTS
BOLTS
Tame Valley
CENTRAL
BIRMINGHAM
AREA
M5
Clent Hills
Lickey Hills
Longbridge
Solihull
BLACK COUNTRY
1 OLD ESTABLISHED SPECIALISED INDUSTRIES NEAR TOWN CENTRE
2 MAINLY SKILLED METAL INDUSTRIES
3 LATER RIBBON DEVELOPMENT BETWEEN TOWNS
4 OFTEN LESS SKILLED, PRODUCING SEMI - FINISHED GOODS eg PARTS FOR CARS
5 MORE RECENT INFILL VARIED INDUSTRIES OFTEN RELATED TO CARS
OUTER BIRMINGHAM
1 LARGE WORKS
2 ALONG LINES OF COMMUNICATION eg
a) METAL INDUSTRIES
b) TYRES
c) CHEMICALS
CENTRAL BIRMINGHAM
1 SMALL WORKS
2 SPECIALISED METAL GOODS eg GUNS, JEWELRY
3 CONFUSED PATTERN
OUTER BIRMINGHAM
1 LARGE WORKS
2 ALONG LINES OF COMMUNICATION eg
a) METAL INDUSTRIES
b) ENGINEERING
c) MOTOR VEHICLES AT LONGBRIDGE & SOLIHULL

b) i) In which part of the coalfield was mining concentrated in the early years?
ii) What evidence is there that mining has declined in this area?
iii) Where is mining concentrated today? Name the main centre.
iv) Which natural features have limited the spread of mining northwards?

c) i) Name two other minerals found in the coalfield areas
ii) Which industry did they give rise to?

SOUTH STAFFS: MINERAL PRODUCTION (000 TONNES)

	COAL	IRON ORE	IRON
1800	—	—	50
1850	4,500	950	754
1870	10,000	450	670
1900	8,500	51	800
1970	3,000	0	

The modern industry is, therefore, located at a considerable distance from the old industrial centres and mining is no longer important in the conurbation itself.

Much of the early prosperity depended upon one industry—the *iron industry.* From small beginnings this industry grew rapidly and, by 1850, the area had one hundred and forty furnaces, producing more than a quarter of Britain's iron output. It was the smoke from these furnaces which was to give the area its name—the Black Country. Today the region produces less than five per cent of the country's steel output and this small industry depends upon scrap metal and steel brought in from the major producing areas. Local raw materials have long been exhausted. In fact, when the industry was at its peak in 1870, local supplies of limestone from the Dudley area and iron ore from the coal measures were inadequate, and it was found necessary to import materials from North Staffordshire, Northamptonshire and South Wales. This situation has become more and more pronounced, producing a classic example of industrial inertia.

Although the basic iron and steel industry has declined, the industries which depend on it have not, and *engineering* and the *finishing of metals* are the leading industries of the area. The origins of these industries can be traced back to the small scale metal industries which existed in the area long before the nineteenth century. These industries produced a wide range of specialised goods and they

General view of Tipton in the Black country showing i) the mixture of housing and industry in the older parts of the town; ii) the importance of the canals (a) and railways (b) in the development of the town; iii) the large areas of land still available for development; the problem of derelict land (c); the use of reclaimed land (d).

expanded rapidly during the boom years, making use of the vast quantities of locally produced iron. In addition, non-ferrous metals were brought in from South Wales, for finishing and conversion into alloys. Although most of the towns involved in these industries were situated on the coalfield, the leading centre was *Birmingham*, some ten kilometres to the east. When local iron production declined, the metal using industries survived by using imported metals and they can still be traced in the industrial patterns of most of the towns. This survival was made possible by:

1. Industries concentrating on products which required only small amounts of metal and which were valuable in relationship to their size. This offset the high costs of importing metals.

2. Industries making use of the skilled labour force available in the area.

3. Industries making use of the central position of the region. In the early days this had been a disadvantage because road transport was bad and links with the ports were poor. The building of the canals changed this and Birmingham became the centre of the canal system. This allowed raw materials to be assembled and finished products to be distributed economically. The building of the railways merely emphasised the importance of this central position.

4. Industries changing to meet changing circumstances. This can be seen most clearly in the case of the motor industry. Many of the metal industries changed to produce components and large assembly plants were built nearby, in places of easy access. (See page 149.) Furthermore, other industries have developed to serve the

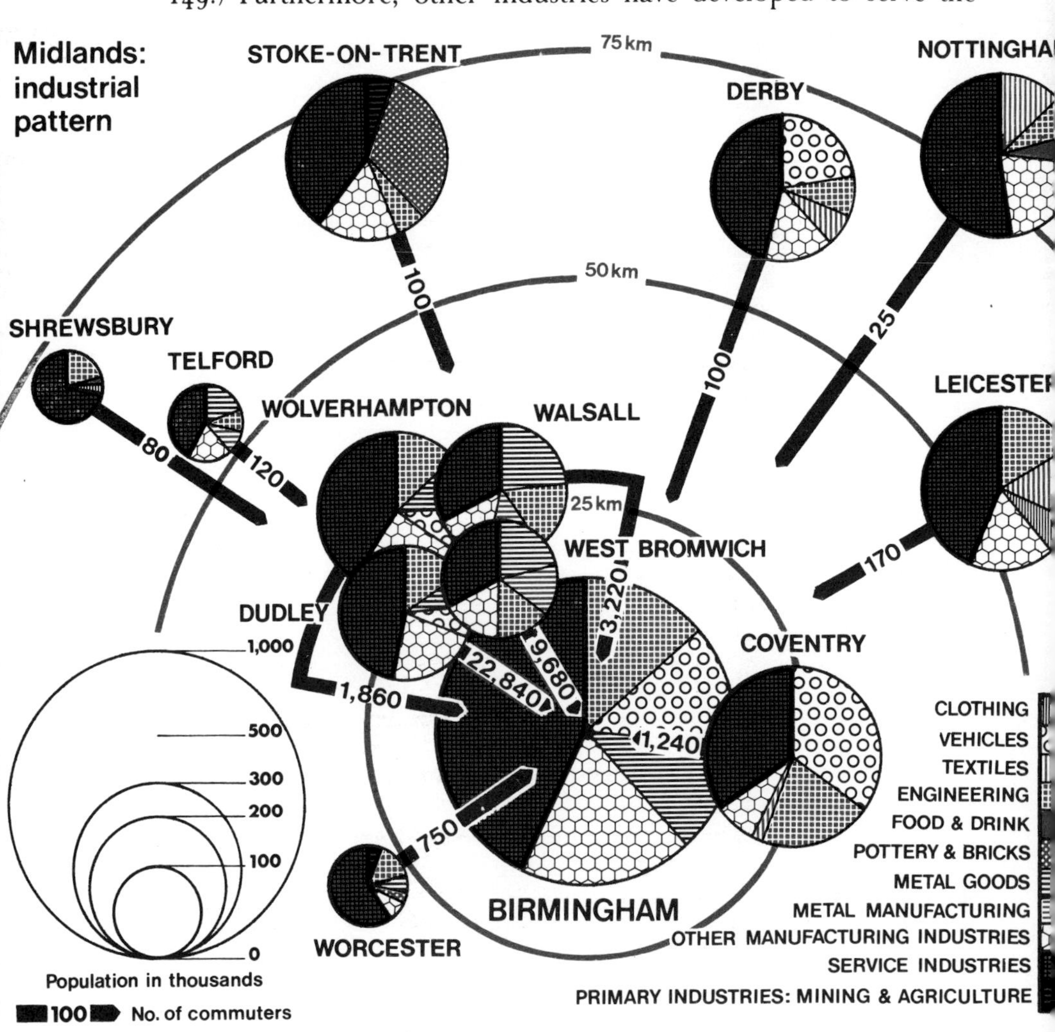

assembly lines, eg tyres, windscreens, plastics and paints, and this has produced a complicated group of industries which are closely linked one with another.

The continued development of the metal industries coincided with the growth of new industries, often located along the main lines of communication. Among these industries are those concerned with the production of consumer goods, and service industries involved in the distribution of goods; both attracted by the fact that nearly seventy per cent of the population of Britain is within three hours driving time of Birmingham.

EMPLOYMENT IN MAJOR INDUSTRIES
IN THE WEST MIDLANDS
(AS A PERCENTAGE OF
THE TOTAL WORK FORCE)

	WEST MIDLANDS REGION
Coal mining	less than 1%
Metal manufacture	5
Metal goods	7
Engineering	11
Vehicles	8
Distributive trades	10
Professional services	12

Complete the following exercise:

a) i) For each of the following—the West Midlands/Birmingham/West Bromwich—work out the proportion of the labour force employed in a) manufacturing industry, b) service industries, c) metal manufacture, d) engineering, e) vehicles. (Also see page 79).
   ii) For each calculate the location quotient for both manufacturing and service industries.
   iii) What does this tell you about the structure of industry in the region?

b) Name four old established metal working industries. For each:
   i) Name one centre (see page 153).
   ii) Name the raw materials likely to be used and explain where they are obtained from.

c) Describe the pattern of industry in Birmingham and explain how it has developed (see page 153).

d) i) Name three motorways which serve the West Midlands.
   ii) Using the map on page 143, calculate the index of accessibility of Birmingham. Compare it with that for Glasgow.

iii) Which of the following industries are likely to be attracted by the accessibility of the region—those requiring coal/those needing to assemble a large variety of materials/those distributing goods to a large and widespread market/those which reduce bulk raw materials?

With such a strong and varied industrial base the region has prospered and the growth of population and industry has been greater than in any other area outside London. This has produced problems of lack of space, congestion, shortages of labour and higher production costs. Even more important, it has led to the government encouraging industry to move out of the conurbation and into the development areas.

Industrial Development outside the Conurbation

Because of the strong links which exist between industries in the area, many firms were not prepared to move far away. When they were prevented from expanding in the Birmingham area, therefore, they moved short distances into the towns on the edge of the conurbation. These towns already possessed industries, often of a traditional kind, but this development resulted in the spread of industries of the type found in the conurbation, eg light engineering and metal working. Equally important, Birmingham draws a large proportion of its labour force from the surrounding towns and, with the improvement in communications, commuting has developed on a large scale. This too has caused the influence of the conurbation to be felt throughout the region. The pattern which has emerged from these developments is shown on page 156.

Study the map and complete the following exercise:

a) For each of the following centres:

Birmingham a regional centre	Walsall—a manufacturing town
Shrewsbury—a county town	Telford—a new town

describe its industrial structure. Explain why they vary.

b) Describe how the pattern of industry changes to the east of a line drawn through the centre of Birmingham.

c) Describe the pattern of commuting in the Midlands, paying particular attention to the area involved, the rate at which the influence of Birmingham declines and the impact on individual towns.

In some cases the influence of Birmingham and the Black Country is very strong. Telford and Redditch, for example, which were planned as new towns to house overspill from the conurbation (see page 227) are now struggling to attract new industries which will save them from becoming simply dormitory towns for Birmingham. In other cases, towns were far enough away and sufficiently large to resist the pull of the city. This can be seen in the case of Stoke-on-Trent which still depends to a great extent on the old established pottery industry. (For a more detailed account see page 229.) Generally, however, the surrounding towns have benefited from the movement of industry out of Birmingham for it has enabled them to share in the prosperity of the city.

For a long period the extent of this industrial strength and prosperity was best seen in *Coventry*—the centre of the British motor industry. During medieval times Coventry was an important market centre and cathedral town. The silk industry was introduced in the seventeenth century and this remained its main industry for the next two hundred years. During the nineteenth century its location near to the Warwickshire coalfield encouraged the development of a small metal working and engineering industry. All of these industries were on a small scale, however, and it was not until the twentieth century that rapid growth took place. The silk industry was strengthened by the introduction of artificial fibres and the engineering industry concentrated first on bicycles and later on motor vehicles. During the early years of the century, the first car assembly lines were built, machine tool factories grew up to supply them and a host of component industries appeared, making Coventry the boom town of the Midlands. Wartime bombing did not halt this growth and, after the war, a new city centre was built and access to the communications network improved. Prosperity seemed endless until the oil price rises in the early 1970's produced a slump in the motor industry and, within the space of two years, the boom came to an end.

Coventry's problems were seen to a lesser extent throughout the West Midlands which is heavily dependent upon the motor industry, and it remains to be seen whether industry there can once again adapt to changed circumstances.

The Estuaries of Growth: the development of industry on the coast

The final element in the pattern of distribution of industry in Britain has been the growth of large scale industry on or near to the coast. As we have already seen (page 7 onwards), ports have traditionally been centres of industry but, with the increased dependence of modern industry on imported bulk materials, industrial development on the coast has been rapid, and such locations now rival and even overshadow, the older inland centres. Much of this expansion has been concentrated on the so called 'Estuaries of Growth', ie the Mersey, Thames, Tees, Humber and, to a lesser extent, the Severn and Southampton Water, and it is expected that future industrial growth will be rapid here.

These estuaries share certain advantages in terms of attracting industry; among them:

1. Deep water moorings, capable of handling large modern ships.
2. Access to imported raw materials.
3. Space for development on the lowland beside the estuary.
4. Often existing ports with established industries.
5. Hinterlands which contain existing industries.

The extent of development has, therefore, tended to depend upon other factors, especially:

1. Remoteness from the main markets in Britain.
2. The difficulty of establishing links across the estuary which often acts as a barrier to communications. Humberside has suffered particularly in this respect.
3. Government policy which has favoured the estuaries in the Development Regions by offering aid to firms moving there. Merseyside has benefited most from this policy, while other areas, such as Humberside, have suffered.

The extent of the differences produced can be seen from a study of two contrasting areas such as Merseyside and Humberside.

Regional Study: Merseyside

As can be seen from the following statistics, Merseyside has become a major industrial growth point, growing much faster than the rest of the North West Region and faster than Britain as a whole. Much of this growth has taken place in recent years but there was already in the area an industrial centre—Liverpool—which was of major importance but which faced serious problems brought about by the decline of old established industries.

AREA	CHANGE IN NUMBERS EMPLOYED IN INDUSTRY: 1951–1971
Merseyside	+11%
North West England	+ 1%

Liverpool and the Early Industrial Development of the Estuary

In the space of two hundred years—from 1650 to 1850—Liverpool grew from a small fishing village to become Britain's leading port. Many factors contributed to this development, including:

1. The Mersey estuary provided a sheltered, deep water harbour.
2. The narrow entrance to the estuary meant that tidal currents were strong and this kept the channels free of silt.
3. The west coast situation was ideal when world, as opposed to European, trade began to grow.
4. The port served the Lancashire cotton industry which was the fastest growing industry in the world.

For much of this early period, Liverpool, like Bristol, depended on the infamous 'Triangular Trade'; shipping manufactured goods out to West Africa, slaves from West Africa to America, and sugar and cotton from America to Britain. And, even after the abolition of the slave trade, trading links with these areas remained strong.

The Manchester ship canal.

MERSEYSIDE

Built-up area

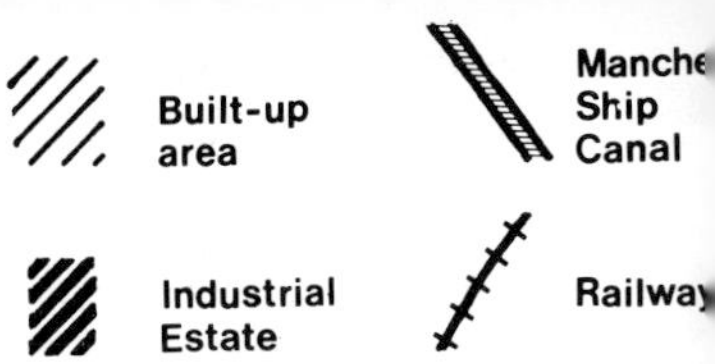

Manche
Ship
Canal

Industrial Estate

Railway

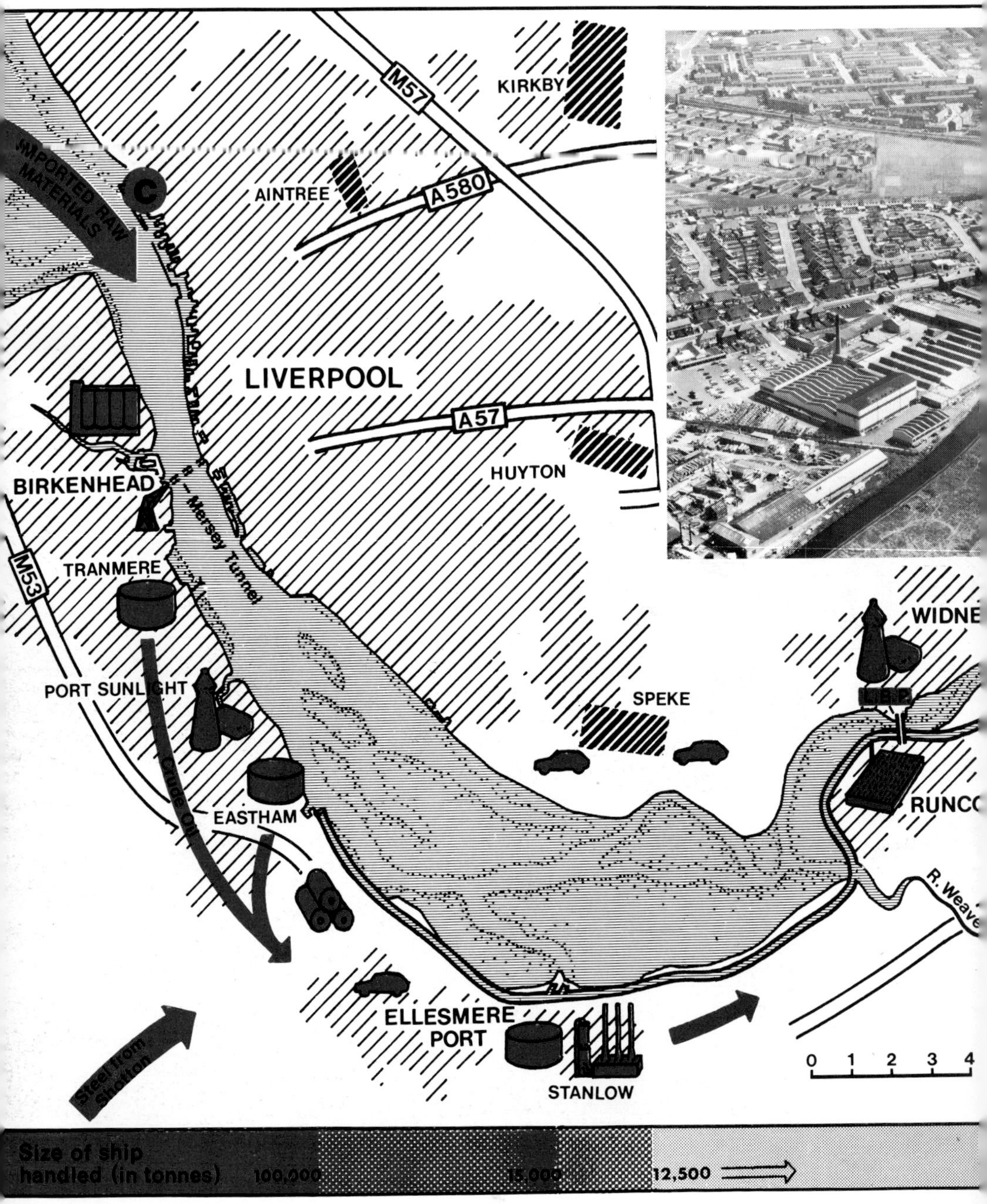

M57
KIRKBY
IMPORTED RAW MATERIALS
C
AINTREE
A580
LIVERPOOL
A57
HUYTON
BIRKENHEAD
Mersey Tunnel
M53
TRANMERE
PORT SUNLIGHT
Crude Oil
WIDNE
SPEKE
L.E.P.
RUNCO
EASTHAM
R. Weave
ELLESMERE PORT
STANLOW
Steel from Shotton
0 1 2 3 4

Size of ship handled (in tonnes) 100,000 15,000 12,500 ⟹

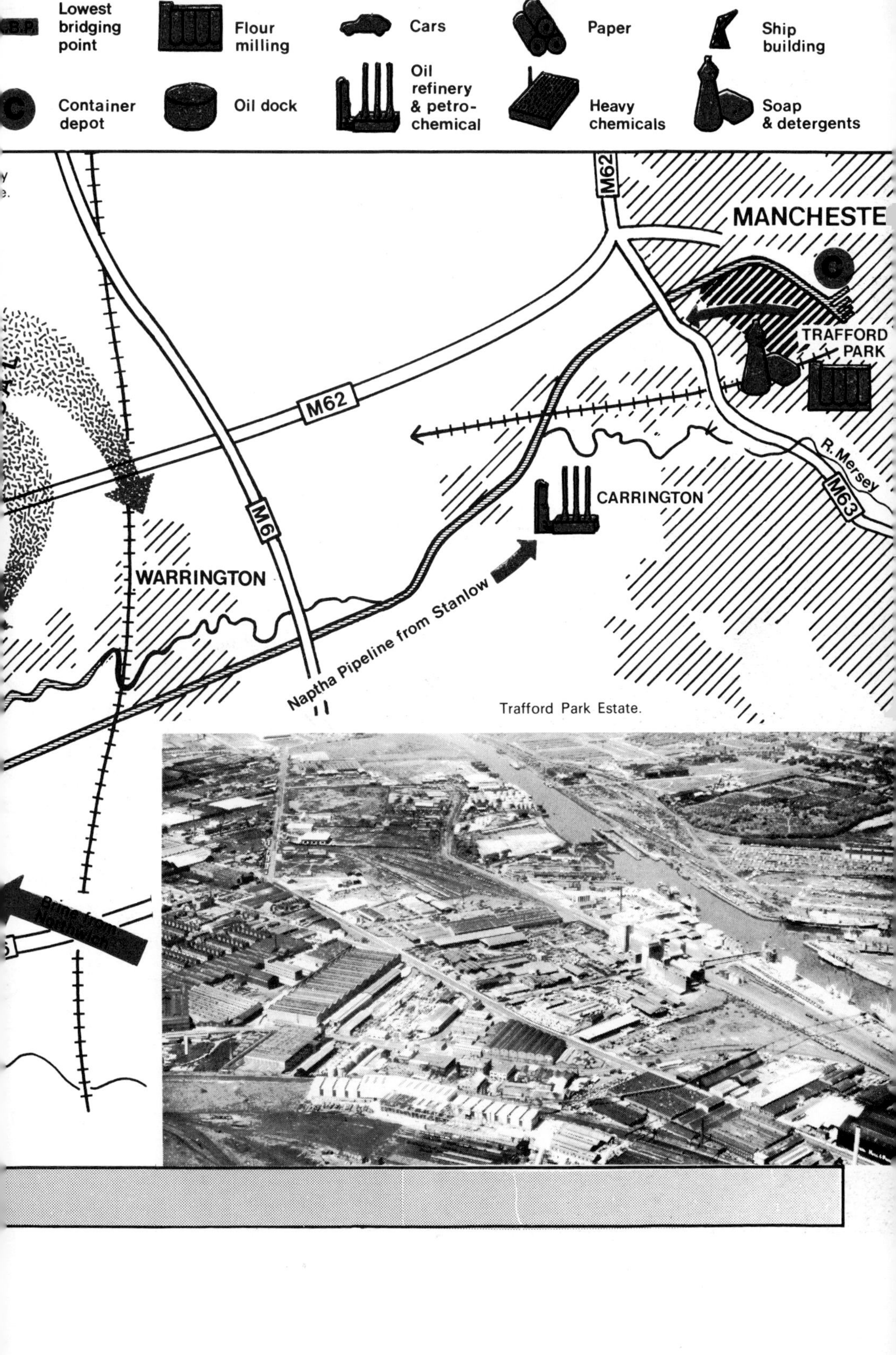

Trafford Park Estate.

Profitable as this trade was, however, it was the enormous growth of the cotton industry in the nineteenth century which established Liverpool as Britain's leading port. During this period a complex system of enclosed docks was built on both shores of the estuary and the port took on its present form. Since this peak in the nineteenth century the importance of Liverpool has tended to decline. The building of the Manchester Ship Canal at the end of the century diverted the cotton trade to Manchester and, even more important, the collapse of the cotton industry during the twentieth century left Liverpool with a hinterland which was much poorer than those of most of its rivals. In spite of this, however, Liverpool still ranks second only to London among the ports of Britain.

Using information given on pages 8, 9, 12 and 162, write an account of the trade of Liverpool, paying particular attention to the size of the port, the nature of its trade, the areas with which it trades and how it differs from its main rivals.

It was also during the period of growth that Liverpool emerged as an important industrial centre. Most of its industries were typical 'port industries', particularly those involved with the building and servicing of ships and the processing of imported raw materials such as grain, timber and, later, oil. Such industries provided a poor base for growth and industrial decline followed the decline of the port.

To some extent, however, the decline of Liverpool was offset by the rise of industries along the shores of the estuary and on the banks of the Ship Canal. Unfortunately, these industries, attracted to the area by the availability of imported raw materials, and because ports are major break of bulk points, were very similar to those already established in Liverpool and few of them proved to be growth industries. There were some exceptions to this general rule. The chemical industry which had grown up in the Widnes–Warrington area, using brine from Cheshire, expanded rapidly with the development of oil based chemicals, and the soap and detergent industries, which were closely related to it, also grew quickly. Such industries could not, however, halt the industrial decline of the area and unemployment remained high. It was this which prompted the government to grant the area Development Area status, in an attempt to increase the number of jobs on Merseyside.

Using information given on pages 162 and 163, complete the following exercise:

a) i) Name two traditional port industries located on Merseyside.
   ii) In each case name an important centre.
   iii) Why are such industries attracted to the area?

EMPLOYMENT ON MERSEYSIDE

INDUSTRY	NUMBER EMPLOYED (000's)	% CHANGE SINCE 1950
Primary industry	8	−44
Manufacturing industry	323	+8
Food and drink	51	+0·6
Chemicals	45	−5·4
Engineering	70	+20
Ships	17	+13
Motor vehicles	30	+136
Clothing	10	−20
Paper	18	+29
Service industries	385	+21
Transport	82	−22
Distribution	93	+19
All employment	779	+11

iv) Referring to the above table, work out the location quotient for each of the industries (see page 80 for the statistics for Britain).

b) i) Name three centres of the chemical industry.
ii) What are the advantages of locations on or near to the Ship Canal?
iii) What are the disadvantages in terms of future development?
iv) The chemical industry is a growth industry here. Why has employment declined?

c) i) What are service industries?
ii) Why are they so important in ports?

d) Which industries have grown most rapidly since 1950?

Recent Industrial Development

As in most Development Areas, government policy aimed at:

1. stimulating declining industries
2. attracting new industries.

Although there have been successes in both of these fields, there have also been problems, many of which relate to difficulties in the port itself. In fact, the Mersey estuary contains a large number of small ports rather than a single port, and most of them are out of date and too shallow to handle modern vessels. The oil ports at the entrance to the Ship Canal clearly illustrate the problems. Built to handle ships of less than 20,000 tonnes, they have been developed to handle first 60,000 tonne and then 100,000 tonne tankers. Further deepening of the channel has proved expensive and is still not likely to allow the largest modern tankers to use the ports. As a result an off-shore terminal has been proposed to keep the industry on Merseyside. Many out of date systems have been closed and development has been concentrated at the seaward end of the estuary where new docks with specialised handling facilities have been built and the Seaforth and Gladstone container terminal has been opened. Even here, however, development has been slowed by poor labour relations. Given such problems, together with the weakness of the hinterland, it is not surprising that the expansion of existing industries has failed to provide sufficient employment on Merseyside or that the area has come to depend quite heavily on attracting new industries.

Many firms, particularly those in the light engineering and food processing industries, moved onto *industrial estates* which were set up on the outskirts of Liverpool. But the greatest success has been in persuading Ford and Vauxhall to build car assembly lines at

The Ford car assembly line at Halewood. Note the area of land required for such a development.

Halewood and Ellesmere Port. There were good reasons for the car industry to move into Merseyside. The steelworks at Shotton, on the Dee estuary produced steel strip for bodies and components; there was a plentiful supply of labour; the Mersey ports were convenient for export; and the M6 and the main railway line provided good links with the parent companies in London and the South Midlands. Even more important, however, was the fact that the government brought pressure to bear on the companies to make the move, and gave financial aid.

The success of the policy can be seen in the employment statistics on page 165. The dangers are that industries have been diverted from areas such as the Lancashire coalfield which are equally in need of new industry.

Regional Study: Humberside

The Humber shares many of the advantages of the Mersey estuary but the pattern of development has been very different.

1. It has a deep water channel, capable of taking large modern vessels.

2. There is ample space on both sides of the estuary for further industrial development.

3. It is better situated, in terms of energy supplies, than any other estuary, being near to the Yorkshire coalfield and the North Sea gas terminal at Easington.

4. There is, in Hull, an existing port which could provide a centre for growth.

In spite of these advantages, however, progress has been slow and Humberside has tended to lag behind both in terms of the development of trade and in the spread of industry along the shores of the estuary. The extent of this failure is reflected in the development of the regional centre—Hull.

Kingston upon Hull is among the oldest of Britain's major ports. Well situated to handle trade with Europe, it grew steadily for a period of seven hundred years until the nineteenth century. Then, with the building of the railways, new docks were built and the port expanded rapidly, both in terms of general trade and as a centre for Britain's new deep sea fishing fleet. These early enclosed docks were situated on the sheltered waters of the River Hull, away from the main tidal flow of the estuary. They are now too small to handle modern vessels and have been closed down, and trade has shifted to newer docks on the estuary itself where there are eleven kilometres of waterfront; much of it still to be developed.

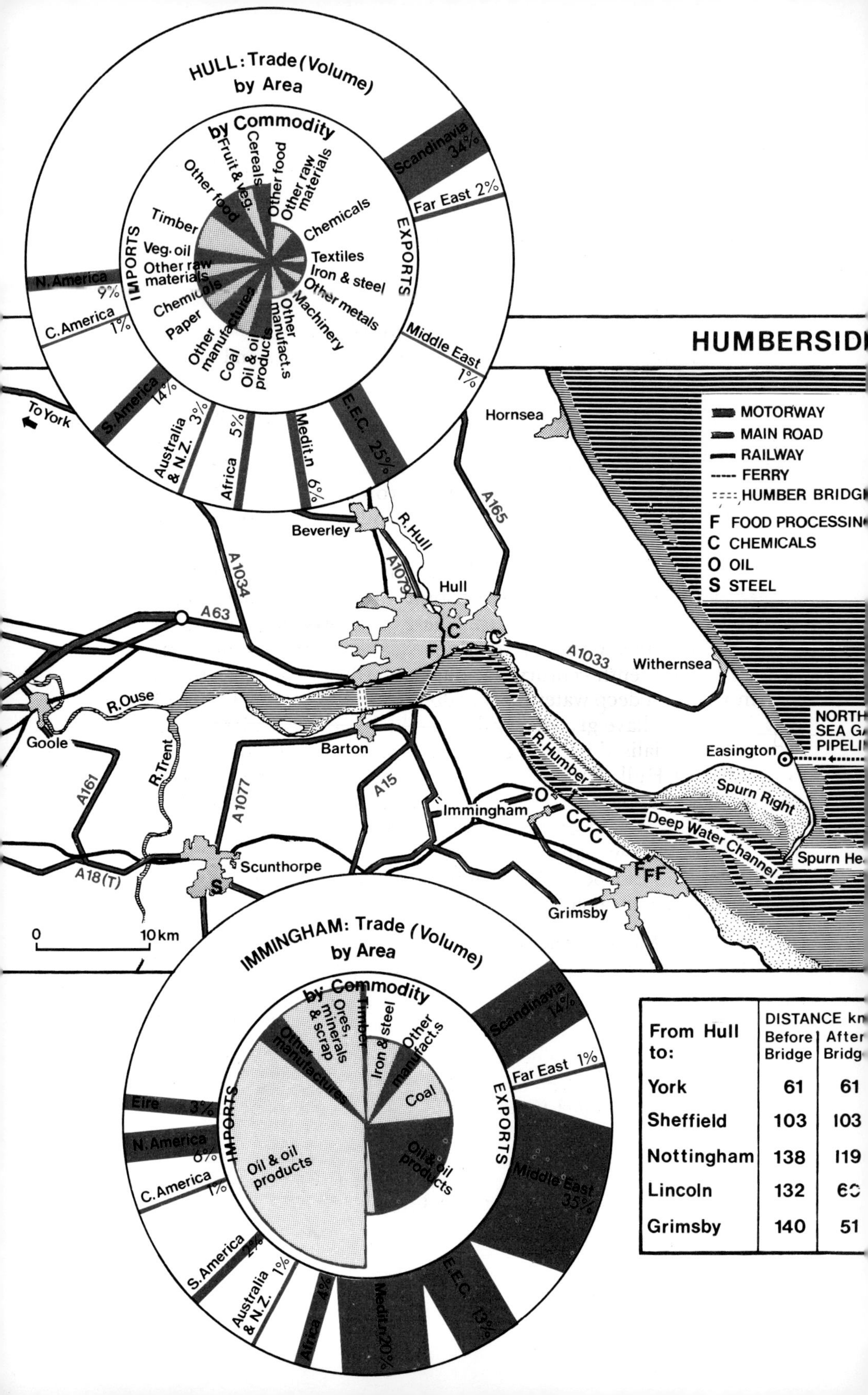

From Hull to:	DISTANCE k Before Bridge	After Bridg
York	61	61
Sheffield	103	103
Nottingham	138	119
Lincoln	132	[illegible]
Grimsby	140	51

Study page 168 and complete the following exercise:

a) i) Explain why Hull was the best site for a port on the Humber estuary.
   ii) Why were the early docks built on the River Hull?
   iii) Why is future port development likely to take place on the seaward side of Hull?

b) i) Write a brief account of the pattern of trade of Hull.
   ii) Describe how it differs from that of Liverpool.

c) i) Name one traditional port industry found in Hull.
   ii) Name one other industry located in the Hull area and explain why it has developed there.

Industrial growth has also tended to be slow and Hull has not emerged as a major industrial centre on a scale comparable with many other British ports. Furthermore, the development which has taken place has tended to be on a very narrow base, with traditional port industries dominating and with little influx of modern growth industries.

One reason for this is the fact that much of the recent industrial development has taken place on the south shore of the estuary, particularly at *Grimsby* and *Immingham.* These two ports were a product of the railway age. Grimsby developed as a leading fishing port during the nineteenth century (see page 77), while Immingham was built in 1912 as a deep water railway dock to handle the coal trade of Yorkshire. Both have grown rapidly since 1945, however, largely on account of specialised handling facilities which have been built there rather than at Hull. For example, Immingham has an oil terminal capable of handling 200,000 tonne tankers and is well situated to take North Sea oil and gas. Furthermore, its coal trade remains important and it now handles a large proportion of the iron ore supplies of Scunthorpe and the Don Valley which are sent there in the returning coal waggons.

Study pages 168 and 170 and complete the following exercise:

a) i) Give one physical reason why Immingham was chosen as the site for a new port in 1912.
   ii) How has the oil terminal been developed so as to handle the largest modern tankers?

b) i) Write a brief account of the pattern of trade of Immingham and explain how it differs from that of Hull.
   ii) Although Immingham handles a greater tonnage than Hull, the value of its trade is less. Why?

c) The photograph shows a range of dock facilities labelled a–d. They include 1. General Cargo. 2. Coal ore. 3. Container. 4. Roll on/roll off. 5. Oil. 6. Grain. 7. Timber. Identify each facility labelled.

Immingham dock.

The diversion of this bulk trade to Immingham has resulted in the development there of industries such as oil refining and chemicals. Grimsby has shared in some of this development but its major industries are food processing and freezing, using the fish landed in the port and the agricultural produce of the Lincolnshire farmlands. Most of this development is restricted to a narrow coastal strip and the interior is relatively undeveloped with the exception of the steel industry at Scunthorpe and the chemical industry which has been attracted to the isolated and sparsely populated lowlands of the Trent Valley (Why?).

Isolation is, in fact, one of the factors which have contributed to the slow rate of development of the region as a whole. For, unlike the Mersey, the Humber estuary has presented a formidable barrier to communications which is yet to be overcome. Apart from ferries, Goole is still the lowest crossing point of the river and, as a result, the two major ports and industrial areas of Humberside are isolated from each other and the hinterland of each is considerably reduced.

The building of new motorways and the completion of the Humber bridge should enable the estuary to unite, rather than divide, its two shores, and allow it to develop as a major industrial area. Some idea of the benefits to be gained by building the bridge can be obtained by calculating the index of directness of the links shown on the table on page 168 before and after its completion.

Equally important is the fact that Humberside was not declared a Development Area and, without government aid on the scale granted to these areas, it has found it difficult to attract new industries.

Such problems could, however, be a thing of the past for Humberside is in a position to benefit from our rapidly growing trade with Europe and from the landing of North Sea oil and gas.

Artist's impression of the Humber Bridge.

Population

The geography of population involves the study of many factors, the most important of which are 1) the growth of population, 2) its structure and, 3) its distribution.

1. Growth

The growth of population in Britain reflects, to a large extent, the economic changes already described in this book. Prior to 1700, Britain was a predominantly agricultural country and population growth was limited by the productivity of the land. As a result, numbers increased slowly, and the population of England and Wales grew from about two million in 1086, when the Domesday Book was compiled, to five million in 1700. During the eighteenth and nineteenth centuries Britain became the leading industrial and trading nation in the world and population growth was rapid, reaching ten million in 1801 and thirty seven million in 1901. In spite of improvements in farming, such growth outstripped the productivity of the land and Britain became dependent upon imported foodstuffs.

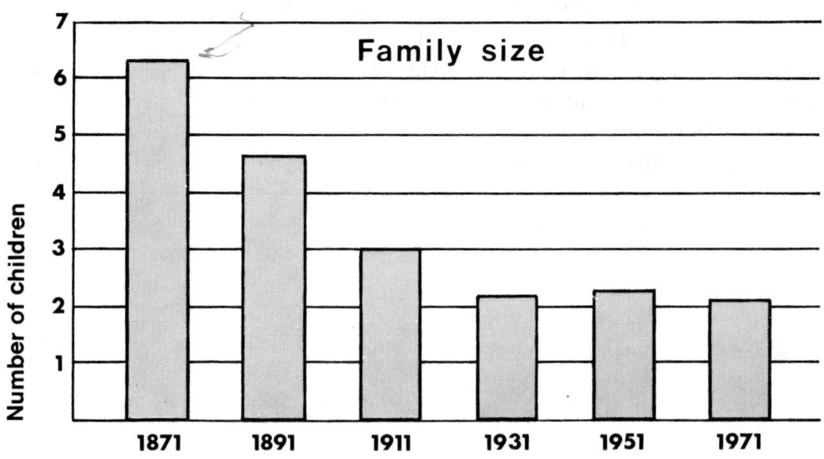

Population increase on this scale was stimulated by a decline in the *death rate*, and, in particular, the death rate among children. This was made possible by a variety of factors, including improvements in medical science, improved diet and improved hygiene. The full extent of population growth during the nineteenth century does not emerge from these figures, for this was a period of high emigration and millions of people left Britain to settle in North America and Australia.

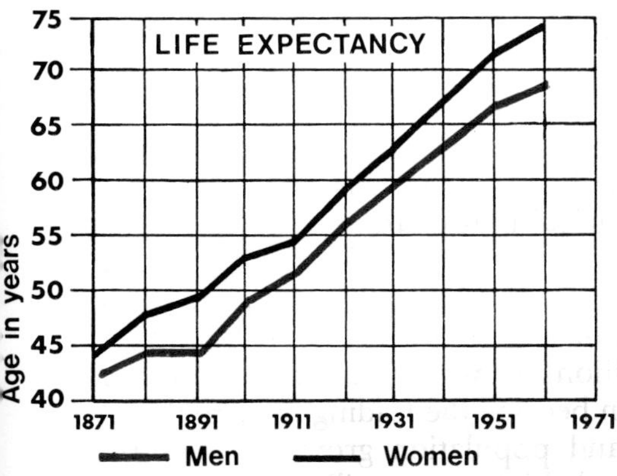

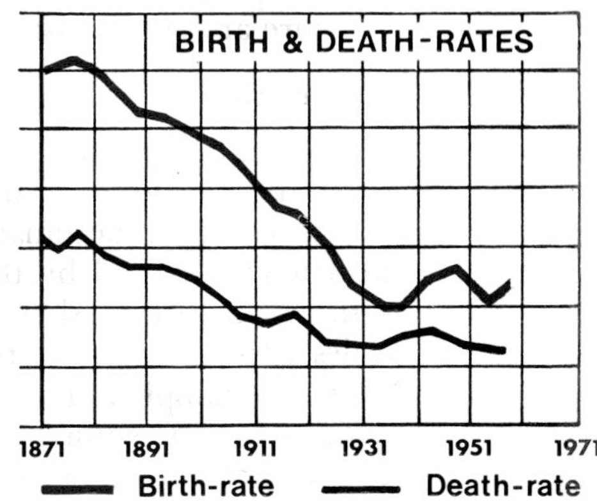

After 1900 the rate of growth started to slow down in spite of a continuing decline in the death rate and a decline in the rate of emigration. In fact, in recent years emigration has been balanced by an influx of immigrants from areas such as India, Pakistan and the West Indies which are, or were, members of the Commonwealth. Reasons for this trend are difficult to establish but the most important is obviously the decline in the *birth rate* which has been rapid since the turn of the century. There are no simple causes which explain this decline and it seems to have been the result of a decision by parents to limit the size of families. This has become so marked that there are now signs that the population has ceased to grow and may, in fact, be declining.

2. Structure

The changes described above have also had important effects on the structure of population in Britain. This can be seen from the graph showing the age structure. One hundred years ago the shape of this graph was an almost perfect pyramid, reflecting a high birth rate and a death rate high enough to reduce the population at a regular rate. Today the pattern is completely different.

Using information given in the diagram below, complete the following exercise:

a) Describe how the present day structure of population differs from that of the nineteenth century.

b) Which of the following factors has contributed to the change:

i) a decrease in the number of young people, ii) a general decrease in population, iii) a general increase in population, iv) an increase in the number of people over fifty, v) a decrease in the number of old people?

c) Using information given in the chapter, account for the changes which have taken place. Pay particular attention to the effects of the changes in birth and death rates, life expectancy and family size.

Once again the effects of the decline in both birth and death rates can be seen, with an ever-increasing proportion of old people in the population and a decreasing proportion of children.

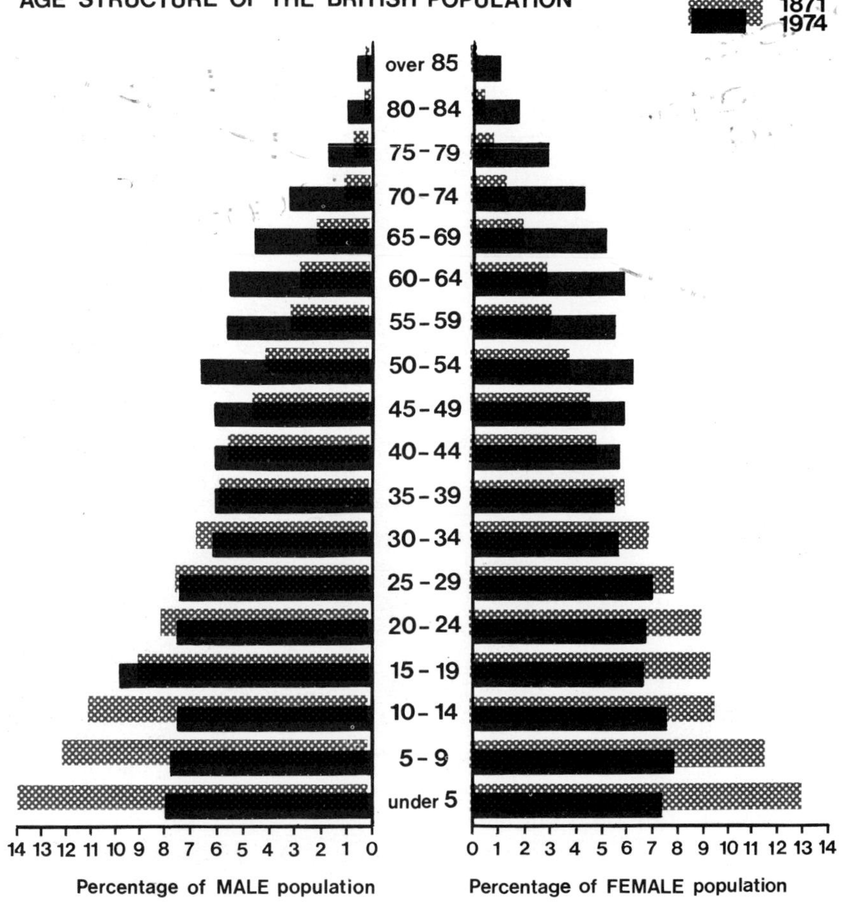

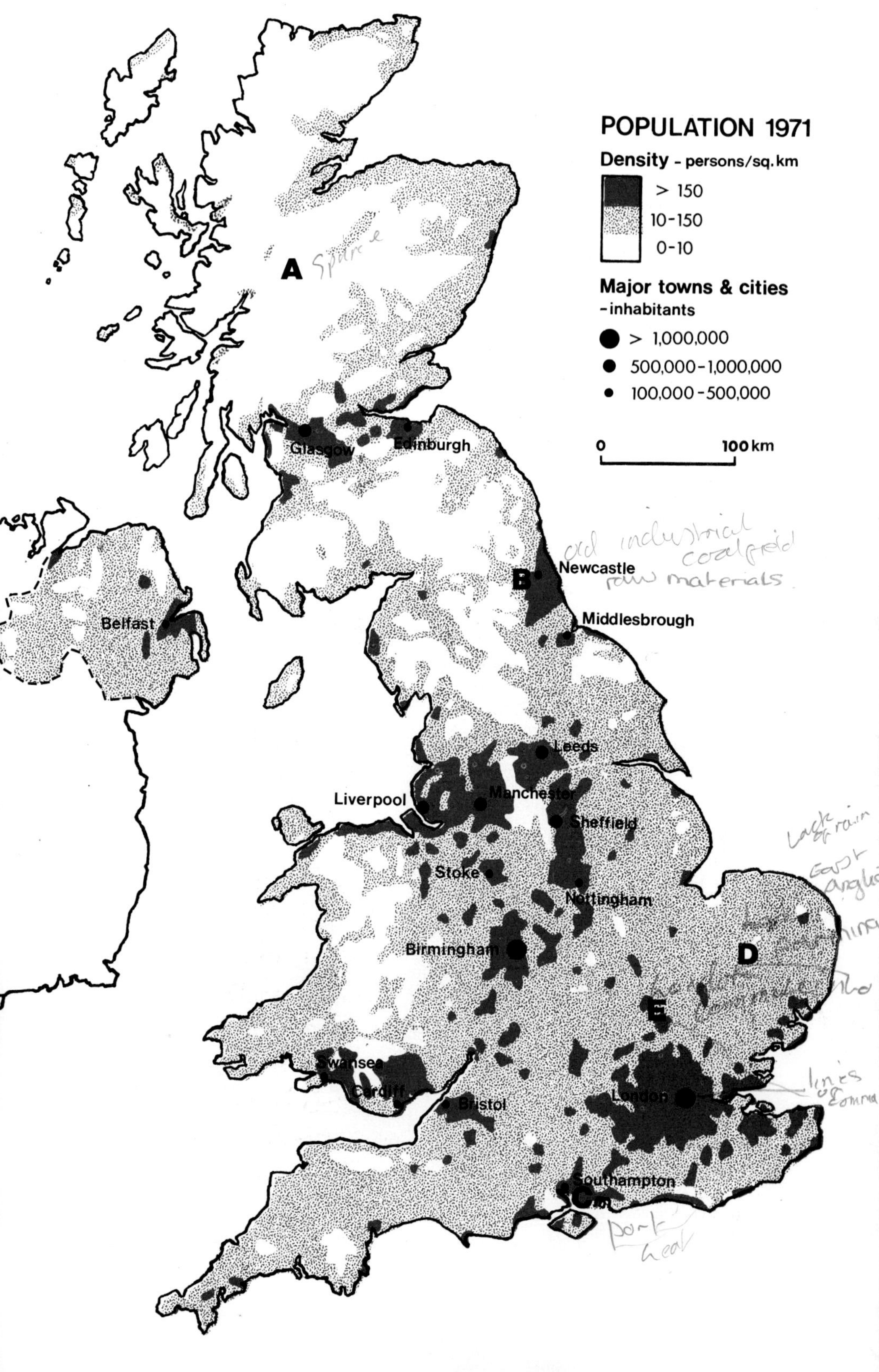
POPULATION 1971
Density - persons/sq.km
> 150
10-150
0-10
Major towns & cities
-inhabitants
> 1,000,000
500,000-1,000,000
100,000-500,000
0
100 km
A
B
C
D
E
Glasgow
Edinburgh
Newcastle
Middlesbrough
Belfast
Leeds
Liverpool
Manchester
Sheffield
Stoke
Nottingham
Birmingham
Swansea
Cardiff
Bristol
London
Southampton

3. Distribution

Many factors have contributed to the pattern of distribution of population in Britain today. Among the most important of these are:

1. Physical factors such as the relief of the land, the fertility of the soil and the amount of rain. These factors can be divided into two groups—those which have encouraged people to settle in an area, eg fertile soils (*positive factors*) and those which have discouraged settlement, eg high land and thin soils (*negative factors*).

2. The availability of resources, particularly fuel and raw materials for industry.

3. Economic factors, such as effective communications and access to markets.

A close study of the distribution pattern gives some indication of the importance of these factors.

Study page 175 and complete the following exercise:

a) Locate the areas labelled A–E and, in each case:

i) State whether it is an area of high or low population density.

ii) Give the average density of population.

iii) State which of the following factors has contributed to the pattern of population (more than one factor may be important in each case).

- high land
- fertile soils
- near lines of communications
- remoteness
- heat
- local raw materials
- deep water port
- heavy rainfall
- lack of rain
- cold

Reference to the maps on pages 26, 29, 30 and 143 will help.

iv) State which of the following groups of factors is most important:

- positive physical factors
- negative physical factors
- raw materials
- economic factors

b) Write a brief description of the distribution of population in Britain and the general factors which have influenced it.

It is important to remember that the distribution of population in Britain has changed considerably over the years, and is continuing to change. Among the most important of these changes have been:

1. During the eighteenth and nineteenth centuries, the movement of people from rural areas to the new industrial towns, which were then developing on the coalfields.

2. At the same time the movement of people from upland areas to the same industrial towns.

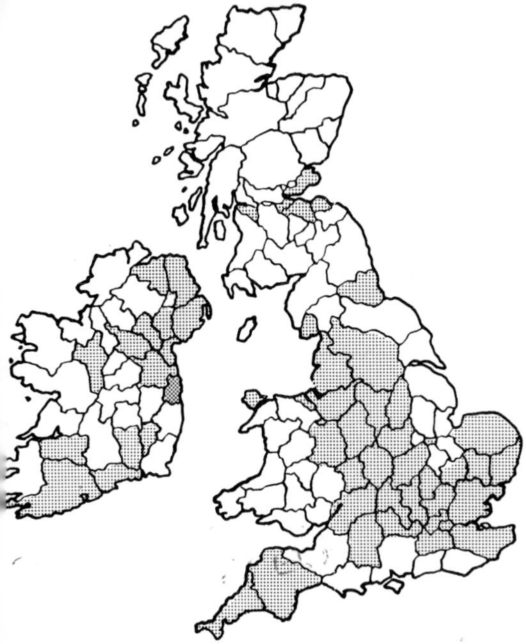

POPULATION GROWTH IN THE BRITISH ISLES

1751: 12·1 million

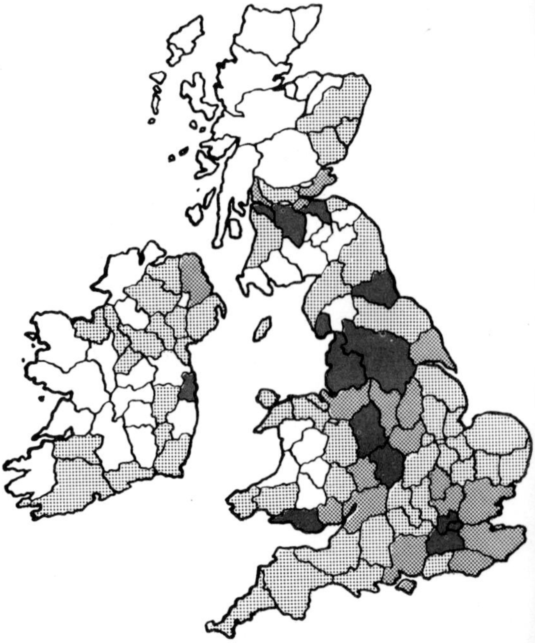

1911: 45 million

1971: 53·8 million

less than 40

40 – 149

150 – 400

more than 400

persons per square kilometre

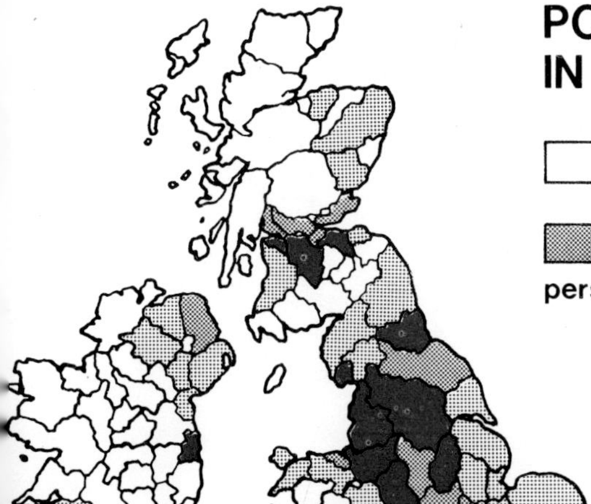

England

Ireland

Wales

Scotland

population in millions

0
10
20
30
40
50
60
70

1650 1700 1750 1800 1850 1900 1950 2000

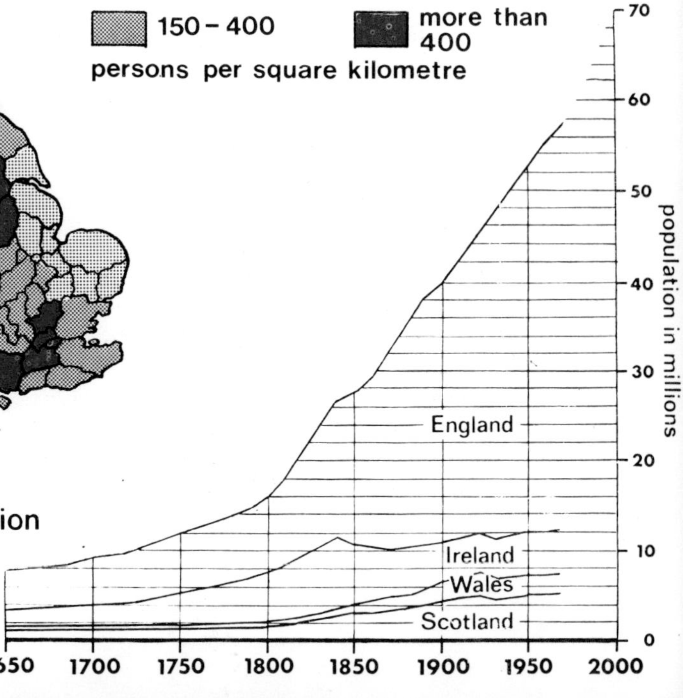

3. During the twentieth century, there has been a shift of population from the coalfields to the new industrial centres in South East England, the Midlands and on the coast.

4. Also during the twentieth century there has been a movement of population away from the centres of the great cities towards new residential areas on their outskirts. (See page 224.)

5. The influx of immigrants from Commonwealth countries since 1945. These people have been attracted to the large cities where work was available, and they have tended to move into the city centres where the population was declining and where cheap housing was available. The results of these developments can be seen in the following statistics.

POPULATION CHANGE IN THE STANDARD REGIONS AND CONURBATIONS

STANDARD REGION	AREA (000 sq km)	POPULATION (000's) 1911	1931	1961	1975	% CHANGE SINCE 1961
North	19	2,815	3,038	3,250	3,126	− 4
Yorkshire and Humberside	14	3,877	4,285	4,635	4,894	+ 5·5
East Midlands	12	2,263	2,531	3,100	3,728	+20
East Anglia	12	1,192	1,232	1,470	1,780	+21
South East	27	11,744	13,539	16,127	16,936	+ 5
South West	23	2,687	2,794	3,411	4,233	+24
West Midlands	13	3,277	3,743	4,758	5,178	+ 8
North West	8	5,796	6,197	6,567	6,577	+ 0·1
Wales	20	2,421	2,593	2,644	2,765	+ 4·5
Scotland	77	4,760	4,843	5,179	5,206	+ 0·5

CONURBATION	POPULATION (000's) 1911	1931	1961	1971	% CHANGE SINCE 1961
Greater London	7,160	8,110	7,992	7,452	−6·8
West Midlands	1,651	1,951	2,378	2,372	−0·3
West Yorkshire	1,590	1,655	1,704	1,728	+1·4
South-East Lancashire	2,328	2,427	2,428	2,393	−1·5
Tyneside	761	827	856	805	−6
Clydeside	–	1,690	1,802	1,728	−4·2
Merseyside	1,157	1,347	1,384	1,267	−8·6

Study the table opposite and complete the following exercise:

a) i) Draw graphs to show the population trends in

South East	East Anglia
Greater London	North West

ii) Using information given in the chapter, describe the factors which may have influenced these trend lines.

b) Draw a map to show the outline of the Standard Planning Regions (see page 79). On it plot the density of population of each region using the following scale (in persons per sq Km).

0 — 149	300 — 449
150 — 299	450 and over

c) i) On a second map, plot the percentage change in population of the regions since 1961. Use the following scale:

Less than 0	5–10
0–5	Over 10%

ii) Describe the distribution of population in Britain and explain how it has changed since 1961.

Although these trends are important, they are too general to provide a full understanding of the changing distribution of population in Britain. In order to do this it is necessary to identify the trends within various regions since each region reflects to a different degree the changes outlined above.

Regional Study: Wales

General Population Trends

The growth and movement of population in Wales clearly illustrates the general pattern described above. Wales is a predominantly mountainous country, with large areas of land over six hundred metres in height, and, for the past two hundred years, these upland areas have lost population. Much of the migration from the mountains has been to coalfield areas, particularly in South Wales, but migration to other parts of Britain and overseas has been important, and there has been a smaller movement to nearby lowland and coastal areas surrounding the mountains. The effects of these movements can be clearly traced in the statistics given on page 177.

Wales

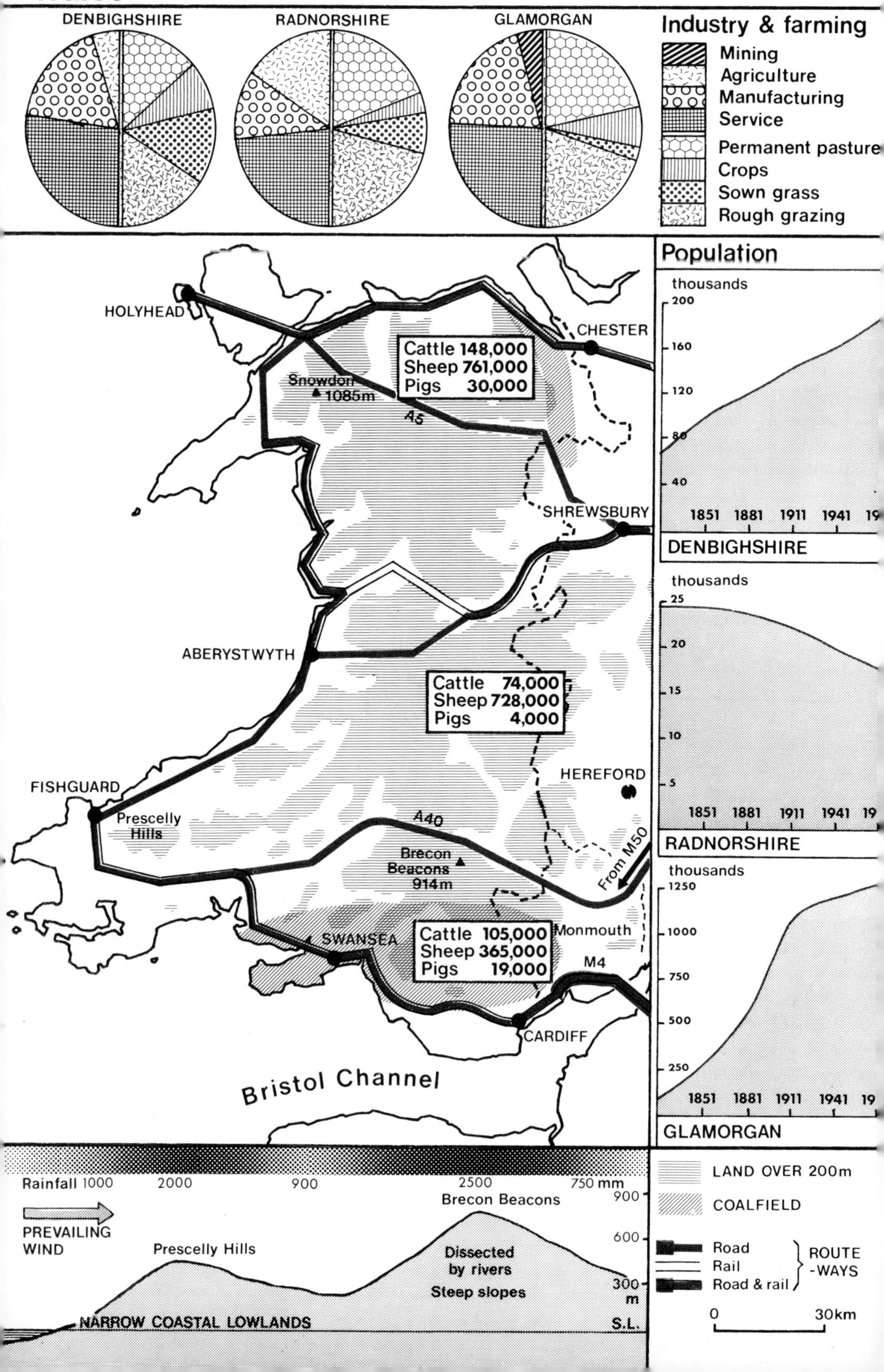

DENBIGHSHIRE

RADNORSHIRE

GLAMORGAN

Industry & farming

- Mining
- Agriculture
- Manufacturing
- Service
- Permanent pasture
- Crops
- Sown grass
- Rough grazing

HOLYHEAD

CHESTER

Cattle **148,000**
Sheep **761,000**
Pigs **30,000**

Snowdon 1085m

A5

SHREWSBURY

ABERYSTWYTH

Cattle **74,000**
Sheep **728,000**
Pigs **4,000**

HEREFORD

FISHGUARD

Prescelly Hills

A40

Brecon Beacons 914m

From M50

SWANSEA

Cattle **105,000**
Sheep **365,000**
Pigs **19,000**

Monmouth

M4

CARDIFF

Bristol Channel

Population

thousands

200, 160, 120, 80, 40

1851 1881 1911 1941 19

DENBIGHSHIRE

thousands

25, 20, 15, 10, 5

1851 1881 1911 1941 19

RADNORSHIRE

thousands

1250, 1000, 750, 500, 250

1851 1881 1911 1941 19

GLAMORGAN

Rainfall 1000 2000 900 2500 750 mm

PREVAILING WIND

Prescelly Hills

Brecon Beacons

Dissected by rivers

Steep slopes

NARROW COASTAL LOWLANDS

900, 600, 300 m, S.L.

LAND OVER 200m

COALFIELD

ROUTE-WAYS: Road, Rail, Road & rail

0 30km

Using the information given opposite, complete the following exercise:

a) Refer to the graph on page 177, showing population growth in Britain, and write a brief account of population trends in Wales.

b) The three counties illustrated on the map include one which is located in the mountains, one which is located on the coalfield, and one which contains both mountain country and coastal lowland. Identify each.

c) In which county has the population:
   i) declined most rapidly
   ii) increased most rapidly
   iii) increased rapidly at first and then less rapidly
   iv) declined and then increased.

d) The following list contains factors which may have influenced the migration of people from mountain areas. Select the ones which apply to Wales.
   i) The environment is harsh with steep slopes/dense forests/thin soils/swamps/permanent ice/heavy rainfall/exposed situations.
   ii) The hill country is too poor for agriculture with the exception of large scale cereal production/dairy cattle in the valleys/market gardening/sheep rearing.
   iii) Earthquakes are common.
   iv) Disease is widespread.
   v) The area is remote and communications are difficult.
   vi) There are few natural resources in the mountains.
   viii) Other areas offered work and better prospects.

e) Write an account of the population farming and employment in the three counties, pointing out the main differences between them.

The migration of population from the mountains, which can be explained in terms of the harsh environment and the greater attractiveness of life elsewhere, reached its peak during the early nineteenth century and has since declined. It has, however, had important effects on the economic life of the many contrasting regions of Wales; effects which can still be traced in the patterns of employment in these regions. For example, the contrast between Radnorshire, with its small labour force, great dependence upon agriculture and poor development of manufacturing industry; and Glamorgan, which depends heavily upon mining and manufacturing, is obvious. In recent years changes have taken place in these patterns but they are much less clearly defined and can be identified only by a more detailed study of individual regions.

The South Wales Coalfield

The nineteenth century saw a tenfold increase in the population of the South Wales coalfield region. Much of this growth was based on the exploitation of coal and the rise of industries which depended on it.

Coal Mining

In South Wales the coal measures have been preserved in a large syncline or downfold, the axis of which runs from east to west along the edge of the Brecon Beacons and Black Mountains. The coal outcrops in the north and south, while in the centre the coal measures are concealed beneath thick layers of Pennant sandstones and grits. Mining first took place on the exposed coalfield but quickly spread to the concealed area, and, by the end of the nineteenth century, the coalfield was the most important in Britain. Many factors had contributed to this development.

FACTORS FAVOURING DEVELOPMENT	FACTORS HINDERING DEVELOPMENT
1. The coalfield is large.	1. Many seams are narrow and difficult to work.
2. It contains a variety of coal including: a) Anthracite—a hard, clean, smokeless fuel which occurs in few other coalfields. b) Coking coal which is also rare. c) Steam coal which was used to fire steam engines.	2. Folding and faulting is widespread and this adds to the difficulties.
3. The valleys cut into the coal-measures making mining easier.	3. The coalfield is remote from markets in Britain.
4. Coastal location encouraged early growth of export trade.	4. Communication from east to west, across the valleys is difficult.
5. Valleys gave easy access to the coast.	

Coal production reached its peak in 1913 when fifty million tonnes were produced. From this point, however, a decline set in which was to become rapid at times and which has not yet been halted. Many of the factors which contributed to this decline were shared by other coalfields, eg the decline of traditional markets and competition from other fuels. In addition, however, South Wales has felt the effects of other, more local problems, including the loss of export markets, remoteness from the newer industrial areas of Britain and the physical difficulties of the coalfield itself, which makes competition with pits in Yorkshire and the Midlands difficult.

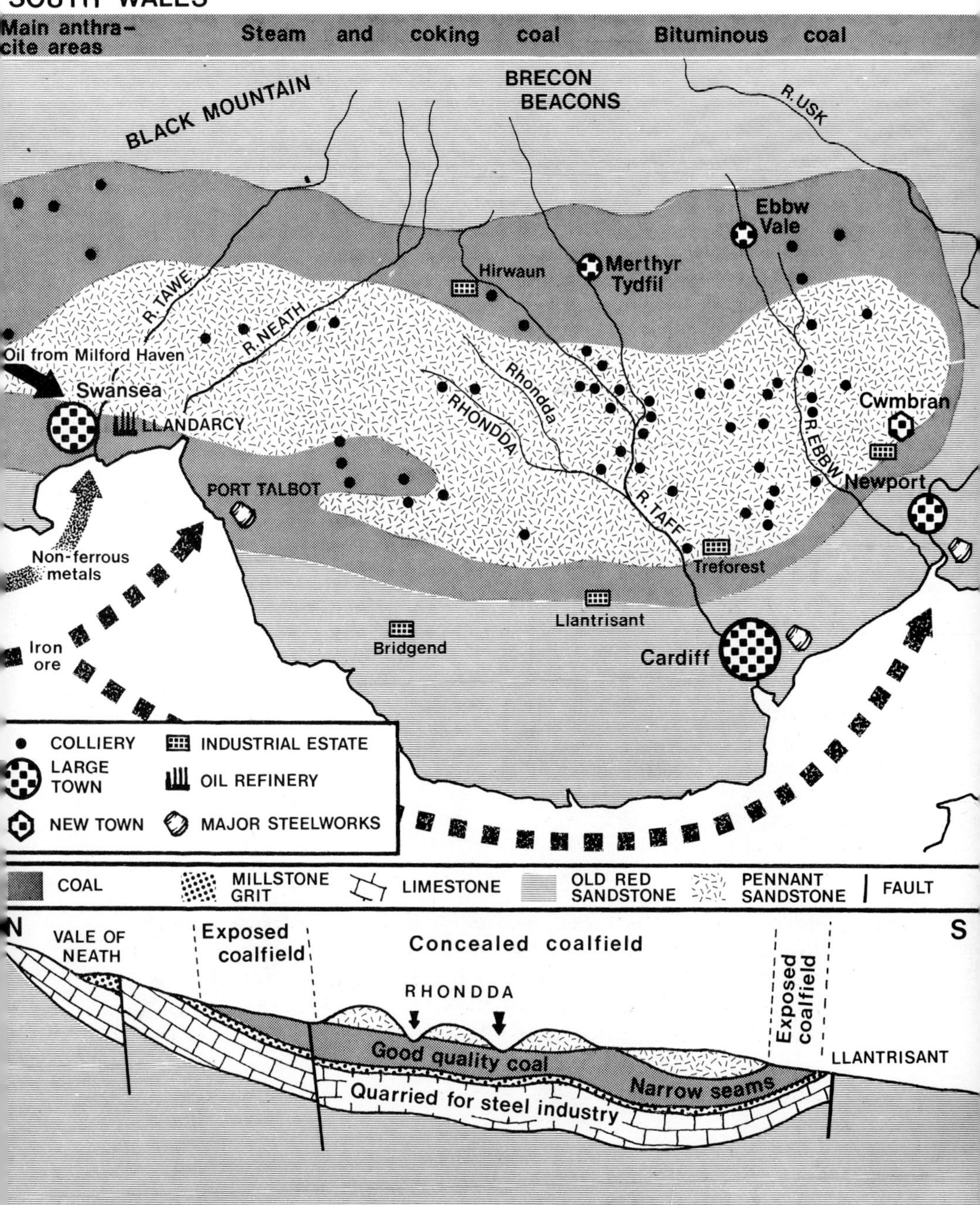
SOUTH WALES
Main anthracite areas
Steam and coking coal
Bituminous coal
BLACK MOUNTAIN
BRECON BEACONS
R. USK
Ebbw Vale
Merthyr Tydfil
Hirwaun
R. TAWE
R. NEATH
Rhondda
RHONDDA
Oil from Milford Haven
Swansea
LLANDARCY
Cwmbran
R. EBBW
Newport
PORT TALBOT
R. TAFF
Treforest
Non-ferrous metals
Llantrisant
Bridgend
Cardiff
Iron ore
COLLIERY
INDUSTRIAL ESTATE
LARGE TOWN
OIL REFINERY
NEW TOWN
MAJOR STEELWORKS
COAL
MILLSTONE GRIT
LIMESTONE
OLD RED SANDSTONE
PENNANT SANDSTONE
FAULT
N
S
VALE OF NEATH
Exposed coalfield
Concealed coalfield
Exposed coalfield
RHONDDA
Good quality coal
Narrow seams
Quarried for steel industry
LLANTRISANT

This decline has been reflected in a decline in the work force employed in mining which, in turn, has created immense social problems, particularly in the 'Valleys' where the dependence on mining was greatest. Here the situation was made worse by changes which have taken place in other basic industries.

The Steel Industry

Like coal mining, the iron industry is long established in the area and, for much of the nineteenth century, South Wales was the leading producer in Britain. This prosperity was based on the availability of basic raw materials, namely, coking coal and iron ore from the coal measures, and limestone from the rocks under the coal measures which outcrop to the north and south of them. The ideal location for the works was in the valleys, especially the valleys cut into the northern rim of the coalfield where the raw materials were most plentiful. Furnaces sprang up everywhere and

Steelworks at Ebbw Vale.

Merthyr Tydfil became the greatest iron making centre in the world. Within a short time, however, these inland sites began to suffer from serious shortages of raw materials. Local iron ores were exhausted and the industry came to depend on imported ores which had to be transported from the coast; and even coking coals were brought in by rail from other parts of the coalfield. This was expensive and, before the end of the nineteenth century, coastal sites were being used for steel works and the inland works were declining.

The twentieth century has seen the completion of this trend and today, steel making is concentrated in vast new plants on the coast while, of the inland centres, only Ebbw Vale remains, and its future is far from certain.

Refer to the information given on pages 90 and 91 and complete the following exercise:

a) Write a brief account of the processes involved in steel making.
b) Describe the advantages of the site of the Margam steel works at Port Talbot.
c) Using page 183, name the main steel making centres in South Wales. From where are they likely to obtain their raw materials?
d) What are the disadvantages of the site at Ebbw Vale? (Photograph opposite).

Other Metal Industries

Since the late seventeenth century South Wales has been an important centre for the manufacture of non-ferrous metals. Initially, copper and tin ores were shipped across the Bristol Channel from the mines in Cornwall, and the smelters were built in the Swansea area, where the coal measures outcrop on the coast. During the nineteenth century Cornish ores began to run out but the Swansea area was well situated to use other imported ores and the industry continued to grow rapidly. For most of this period, the industry simply produced metals which were sent to the English Midlands for finishing. The same period also saw the development of *plating* and *alloy* industries, both of which have grown rapidly in recent years. Plating is the process by which a cheap basic metal, such as steel, is coated with a thin layer of expensive non-ferrous metal. This gives a stronger, cheaper alternative to non-ferrous metals. For example, steel can be coated in zinc to prevent it rusting (galvanised), or in tin to limit the contamination of food, when used in canning. Both processes are complicated, involving the cleaning of the steel plate with acids and, in the case of tin plate, coating it with palm oil to provide a surface to which the molten tin will stick.

Recent trends in the growth and movement of population in

South Wales are closely related to developments in these basic industries.

POPULATION IN TWO CENTRES IN SOUTH WALES

DATE	RHONDDA	NEWPORT
1851	5,000	—
1901	115,000	69,300
1921	150,000	94,384
1971	100,000	112,286

1. For the last fifty years the area had been one of slow population growth. This reflects the decline which has taken place in basic industries, such as mining and metal working, and the failure to attract new industries.

2. Most of this growth has taken place in the coastal towns which have benefited both from the expansion of the steel industry and

Settlement in the Rhondda valley, showing the effect of relief on communication and settlement.

from the introduction of oil refining. This has been stimulated by the emergence of Milford Haven as Britain's major oil terminal. From there oil is piped to the refinery at Llandarcy which, in turn, supplies the chemical plant at Baglan Bay (see page 129).

3. During the same period, the coal field area of the interior has presented a very different picture. Here the closure of pits has resulted in serious problems. Emigration from the valleys has been high and there has been a sharp decline in population but, in spite of this, unemployment has remained above the national average.

Attempts have been made to improve the situation. Of these, the most striking are the building of a new town at Cwmbran and the establishment of a number of industrial estates, at various points throughout the region. Both have had success in attracting new industries to South Wales but they have never looked likely to reverse the decline which was taking place on the coalfield. This is not surprising in view of the fact that few of these developments have taken place in the valleys where the most serious problems exist. Instead the estates have been located nearer the lowlands, where access is easier and where materials can be assembled and goods dispatched more efficiently. As a result, migration to the coast has not been halted and, in addition, many more people now travel long distances each day from the valleys to work in coastal towns.

Study pages 183 and 188 and complete the following exercise:

a) Name three industrial estates and, in each case, state whether it is located in the valleys or on the coastal lowlands.

b) i) List the possible advantages of Treforest as a location for an industrial estate, paying particular attention to communications and the space available.

   ii) Why is this location preferable to a site in the Rhondda? (See photograph opposite.)

   iii) Describe the type of industry likely to be attracted to such an estate.

The Mountains and Coastlands

Compared with the coalfield, the problems of the mountains seem less pressing, largely because migration, spread over a period of two hundred years, has reduced the population to a very low level. Economic activity has also run down and, it is against this background, that attempts have been made to revive the region. As in so many parts of Britain these have taken two forms—the stimulation of existing industries and the introduction of new activities.

The Treforest industrial estate.

The Traditional Economy

In common with most mountain areas, this region has depended on farming and the exploitation of minerals. Both of these activities have declined over a long period of time—farming because of the harsh environment and mining because of a decline in the demand for slate, and because other minerals occur in small quantities and were uneconomical to work. In recent years this situation has tended to change.

1. Subsidies paid to hill farmers have encouraged both the use of hill pastures and the improvement of marginal land. (See page 41).

2. Improved methods of mining and recent rises in the world price of non-ferrous metals has made the exploitation of the copper and lead deposits of North Wales much more likely. In fact development has been delayed only because the main deposits occur in the Snowdonia National Park—an area of outstanding natural beauty.

Recent Developments

Early attempts to make further use of the mountain region centred on the planting of *forests* on unproductive hill sides and the use of valleys for *water supply*. Although widespread, such activities produced few jobs and did little to halt the decline of the region. Other more recent projects have suffered from the same disadvantages. For example, vast sums have been spent on developing the power sources of the mountains. *Hydro-electric* schemes were the first to be developed but these suffered from the usual problems in Britain—the smallness of the drainage basins and the lack of suitable sites for dams and power stations. As a result, even with pumped storage facilities (ie the use of surplus power generated during the night to pump water back into the reservoirs, so that it can be used during the next day), output is very small. Few sites now remain for development and, in recent years attention has switched to *nuclear power*. Attracted by the remoteness of the region which reduced safety problems, two large stations were built, but, with the entire nuclear energy programme under question, further development has been slow.

Impressive as these schemes are, it is clear that they have created few jobs and brought little prosperity to the mountain region. More important in this respect have been smaller, piecemeal developments in the fields of tourism and manufacturing industry. *Tourism* is in fact now the main industry in many parts of the mountains, particularly in the National Parks, and on large stretches of the coast line, where it provides employment for large numbers of people.

The attraction of new industries such as light engineering and textiles, has proved more difficult and the one large scheme in the mountains—the designation of Newtown as a new town—has met with little success. The failure to revive the economy of the mountains is reflected in a declining population and high rates of emigra-

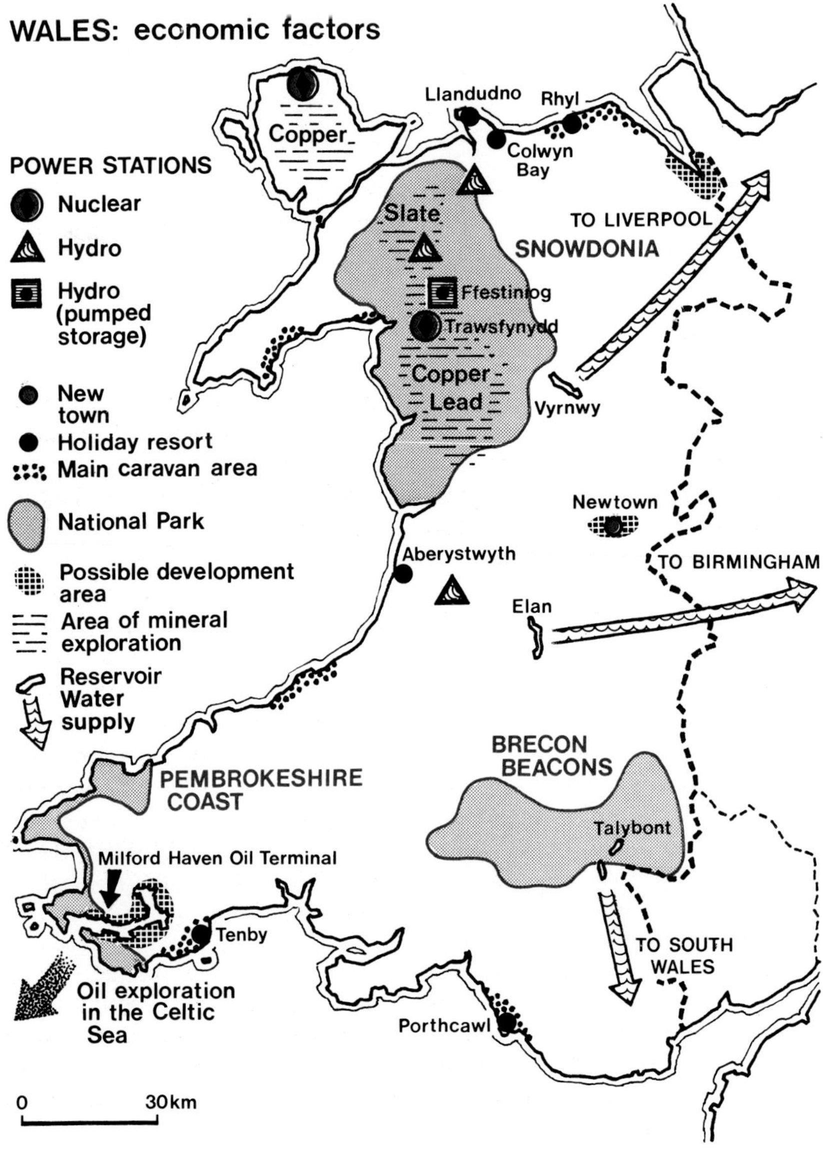

tion. Greater success has been achieved on the coastal lowlands which, because they are more accessible, have attracted a larger proportion of new industrial development than the mountains.

Using information given on map opposite, complete the following exercise:

a) List the main economic activities shown on the map and in each case state:
   i) whether it is a traditional activity or a recent development
   ii) one location for the activity.

b) State the advantages and disadvantages of:
   i) North Wales as a centre for generating nuclear power and hydro-electricity (see pages 113 and 116).
   ii) Newtown as the site for a new town (see page 227).
   iii) The North Wales coast as a centre of the tourist industry (see page 276).

Regional Study: Scotland

The population of Scotland has grown at a slow rate in comparison with other parts of Britain (see page 177). This is largely the result of emigration which has taken place on a large scale during the last two hundred years. These general trends do, however, obscure the fact that marked differences in population occur within the country as a whole. For example there is a strong relationship between the density of population and relief. A glance at page 192 shows that population is concentrated in lowland areas and that the uplands are very sparsely populated. Two major upland areas occur in Scotland. Much of the northern part of the country is occupied by the vast, complicated mountain area, generally known as the Scottish Highlands, while, in the south, the smaller and lower hill area of the Southern Uplands is found. Both have an average population density of less than one person per square kilometre. These two highland areas are separated by the wide rift valley of the Central Lowlands which is the most densely populated area of Scotland. Other concentrations of population occur on the coastal lowlands, particularly those in the east which are wider and more accessible.

Such a pattern can be explained in terms of the environment but this would be an over-simplification and, in order to understand how the pattern has emerged, it is necessary to study individual areas in more detail. This is particularly true of recent changes which are often complicated and not related to environmental considerations.

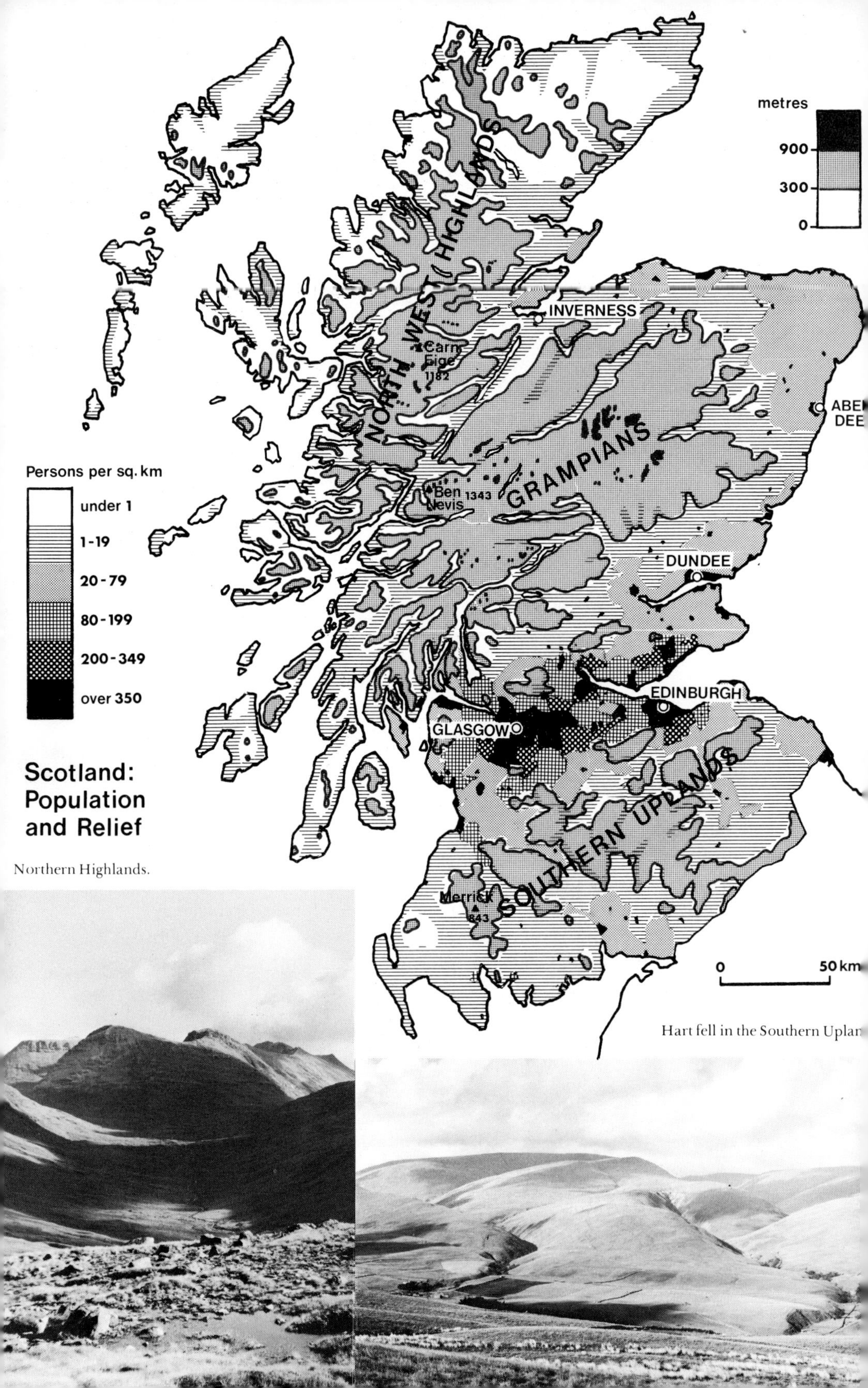

Northern Highlands.

Hart fell in the Southern Uplan

The Highlands

The Northern Highlands comprise forty-seven per cent of the land area of Scotland and more than twenty per cent of the total area of Britain. At the same time, they house less than fifteen per cent of the Scottish population and only one per cent of the British population. Nowhere else in Britain is there so large an area with such a low population density and nowhere else are the problems of depopulation and economic decline seen so clearly.

Two hundred and fifty years ago the Scottish highlands were widely settled and, in the valleys and on the coastal lowlands, population densities were quite high. Since then the population has actually declined, largely as a result of emigration. For example, since 1871 there has been a natural increase in population of more than one hundred thousand but the population itself has fallen by more than seventy-five thousand. This means that almost two hundred thousand people have left the Highlands in the last one hundred years. Most of the people moved because conditions in the Highlands were difficult, the economic activities upon which they depend were declining, and prospects seemed better in other parts of Britain or abroad.

1. *Farming*

Conditions in the Highlands are among the most severe in Britain and this has had important effects on the pattern of farming. This is particularly true of the traditional type of farming—*crofting*—which is still practised in the remote areas of northern and western Scotland.

Study page 194 and, using the information given, complete the following exercise:

a) Describe the environmental factors which make farming difficult in this area.
b) i) List the main crops grown on the farm.
   ii) What proportion of the farm is cultivated?
   iii) Where is the main cultivated area in relationship to the farm buildings?
c) i) What type of land makes up the bulk of the farm?
   ii) What is it used for?
d) i) Where are the stock reared during the winter?
   ii) What are they fed on?

This pattern of cultivating a small area of land around the farm (the *infield*) and maintaining a much larger area of rough pasture for

LAND-USE ON A CROFT

Improved land . . . 10 hectares		Rough grazing . . . 90 hectares	
Of this: Oats	18%	Dairy cattle	8
Hay	15%	Sheep	91
Silage	15%	Poultry	30
Pasture	40%		

stock rearing (the *outfield*) is typical of crofting communities and shows a clear adaptation to a difficult environment. (Compare the hill farm described on page 41.) In spite of this there has been a great decline in crofting and it has virtually disappeared from large areas of the Highlands. This started in the eighteenth century when entire crofting communities were driven out by landowners who wanted to use the land for large scale sheep and cattle rearing. In recent years, however, the decline has continued largely because of the harshness of the environment and the low incomes compared with

Carloway, a crofting community on the Isle of Lewis. Note i) the division between the infield and the unimproved rough grazing; ii) the old crofts with their low profile and thick thatched roofs.

farms in other areas; and many of the crofts which remain are run on a part-time basis.

Attempts to protect the crofting communities have met with little success and crofts today provide less than twenty-five per cent of farm income in the Highlands. The remainder comes from two main sources:

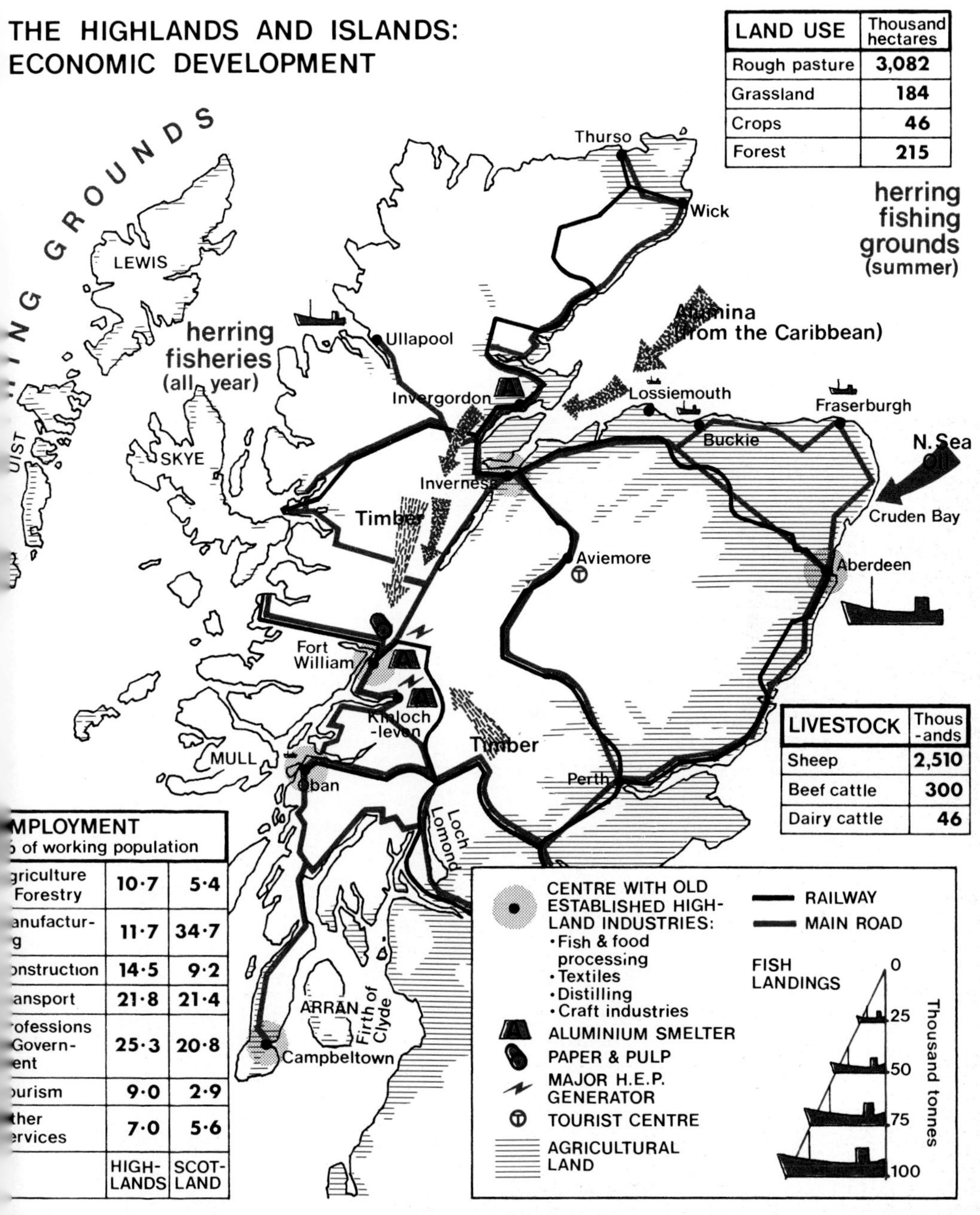

LAND USE	Thousand hectares
Rough pasture	3,082
Grassland	184
Crops	46
Forest	215

LIVESTOCK	Thous-ands
Sheep	2,510
Beef cattle	300
Dairy cattle	46

MPLOYMENT of working population		
griculture Forestry	10·7	5·4
anufactur-g	11·7	34·7
onstruction	14·5	9·2
ansport	21·8	21·4
ofessions Govern-ent	25·3	20·8
ourism	9·0	2·9
ther ervices	7·0	5·6
	HIGH-LANDS	SCOT-LAND

1. Farms in the eastern Highlands where the land is lower, climatic conditions are drier and more sheltered, and soils are more fertile. Here, particularly to the south of Glen More, oats and barley can be grown in rotation with grass and root crops. This forms the basis of a stock rearing industry with beef cattle very important.

2. Large sheep farms in the Highlands proper.

2. *Fishing*

The seas around Scotland are rich in fish. It is surprising, therefore, that the fishing industry in the Highlands has declined. The decline has been greatest in the small ports of the far north and west (see page 76) which are remote from the main markets and which have been unable to compete with the large modern trawlers of rival ports. The larger east coast ports have been more successful in resisting the challenge and fishing remains important, particularly in the Aberdeen area.

3. *Industry*

Most of the traditional industries of the Highlands started as cottage industries. Of these only two became important—textiles and distilling. Both made use of local raw materials such as wool and barley and both became organised to reach large markets.

The danger of allowing this decline in population and economic activity to continue has been apparent for a long time and many attempts have been made to reverse the trend.

Halting the Decline

Attempts to revive the traditional activities have met with limited success. For example the textile industry has been reorganised on a co-operative basis and Harris tweeds now reach world-wide markets; and distilling has become a major growth industry so that whisky now ranks as one of Scotland's leading exports. Apart from this, attention has centred on attempts to attract new industries into the region. The government has played an important part in this by providing grants for any company which is prepared to move to the Highlands.

The developments which have taken place are very similar to those seen in other mountain areas, eg.

1. Forestry—the planting of coniferous forests was an obvious use for land which was too high and exposed for agriculture; particularly in a country like Britain which depends on imported timber. Although a large area has now been planted, individual

forest areas were kept small. This allowed the benefits to be spread throughout the region and had less harmful effects on the scenery, but it also made the processing of the timber difficult, because such forests could only support small mills. It was not until 1965 that a large scale pulping plant was built at Fort William, and this is small by international standards and almost impossible to enlarge because of the cost of bringing in timber from plantations which are too small and too far away.

2. Hydro-electricity—the Scottish Highlands was an obvious area for development but, in common with the rest of Britain's mountains, there were serious problems. Most of these centred on the size of the drainage basins, which were generally far too small. As a result, schemes have been small and designed to meet local demands. Larger schemes are difficult and expensive to construct because tunnels have to be built to link catchment areas (see page 117). Even more important, future large scale developments are unlikely because the best sites have already been used and costs for other sites will be even higher.

Invergordon Aluminium Plant

3. Aluminium—the aluminium industry was attracted to the Highlands because of the prospect of cheap supplies of hydro-electricity. This is particularly important in an industry which spends more on power than it does on other raw materials. Early developments took place in the Fort William area where conditions were most favourable for the building of hydro-electric schemes. The problems of extending power supplies has limited development and it is significant that, while new smelters have been built near coal-fired power stations in North East England and a nuclear power station in Anglesey, the only recent development in the Highlands has been at Invergordon, on the Moray Firth and far away from the main hydro-electric schemes.

Complete the following exercise using page 197:

a) i) Name the main raw material used in the aluminium industry.
   ii) From where is it obtained and how is it transported?
b) Refer to the map on page 117 and explain why Fort William was chosen as the site for the first smelter.
c) Refer to the photograph of the Invergordon plant.
   i) Describe the main processes used in the manufacture of aluminium.
   ii) What are the advantages of the Invergordon site?

4. Tourism—although the tourist industry has become more important with the increased prosperity of the population and the widespread use of the motorcar, growth has been limited by remoteness, lack of roads and lack of facilities, such as hotels and entertainments.

Other developments have been on a much smaller scale, concentrating on the establishment of growth centres at which industrial development is encouraged. In particular, it is hoped to attract industries which require a large labour force—something which has not happened in the case of many of the large schemes.

These failures to reverse the decline of the last two hundred years indicate the scale of the problems presented by remoteness and the harsh environment. At present the only hope of change lies with the development of North Sea oil and it is by no means clear what effects this will have. A few drilling platforms have been built in the Highlands but orders have declined and yards have closed. Aberdeen has emerged as a major centre for servicing the rigs and the economy of the surrounding area has benefited. Further benefits will depend upon the establishment of a refining industry in the north of Scotland and even this will, in itself, create few jobs.

Similar problems can be seen on a smaller scale in the Southern

Uplands where remoteness and lack of communications and services have encouraged migration from the farms and from the textile towns of the Tweed valley.

The Central Lowlands

Many of the people who left the highlands during the eighteenth and nineteenth centuries settled in the Central Lowlands, particularly in the Glasgow region where industrial development was taking place at a rapid rate. Here the opening up of the Central Coalfield and the growth of Glasgow as a port provided a strong industrial base and, by the end of the century, the area was one of the major industrial centres of Britain, with important coal, steel, shipbuilding and engineering industries. The twentieth century has seen increasing problems in these industries and there has been a movement of population from the old established industrial areas to new centres. Reasons for this are complex and vary from industry to industry but the overall pattern is closely related to that seen in other coalfields in Britain.

1. *Coal Mining*

Mining reached its peak at the beginning of the twentieth century when the collieries of the lowlands employed one hundred and forty thousand men and produced forty-two million tonnes of coal. Today the labour force is thirty thousand and production is twelve million tonnes. The effects of this decline have been most severe in the Central Coalfield of Lanarkshire, which was once ranked among Britain's leading coalfields. Here the best seams have been worked out and production has declined to less than three million tonnes. The loss of jobs has been enormous and many miners have moved to the coalfields of Fyfe and Clackmannan which were not extensively mined during the nineteenth century and which are now being worked from large modern pits.

2. *Iron and Steel*

The iron industry grew up on the coalfield around Motherwell where coking coal and iron ore occurred in the coal measures. These iron ores were quickly exhausted and the industry came to depend on ores imported through Glasgow. This situation was far from ideal and works began to close down. Eventually steelmaking was concentrated at two large integrated plants, situated between Motherwell and Airdrie, and a new deep water ore terminal was built on the Firth of Clyde at Hunterston. In the near future it seems

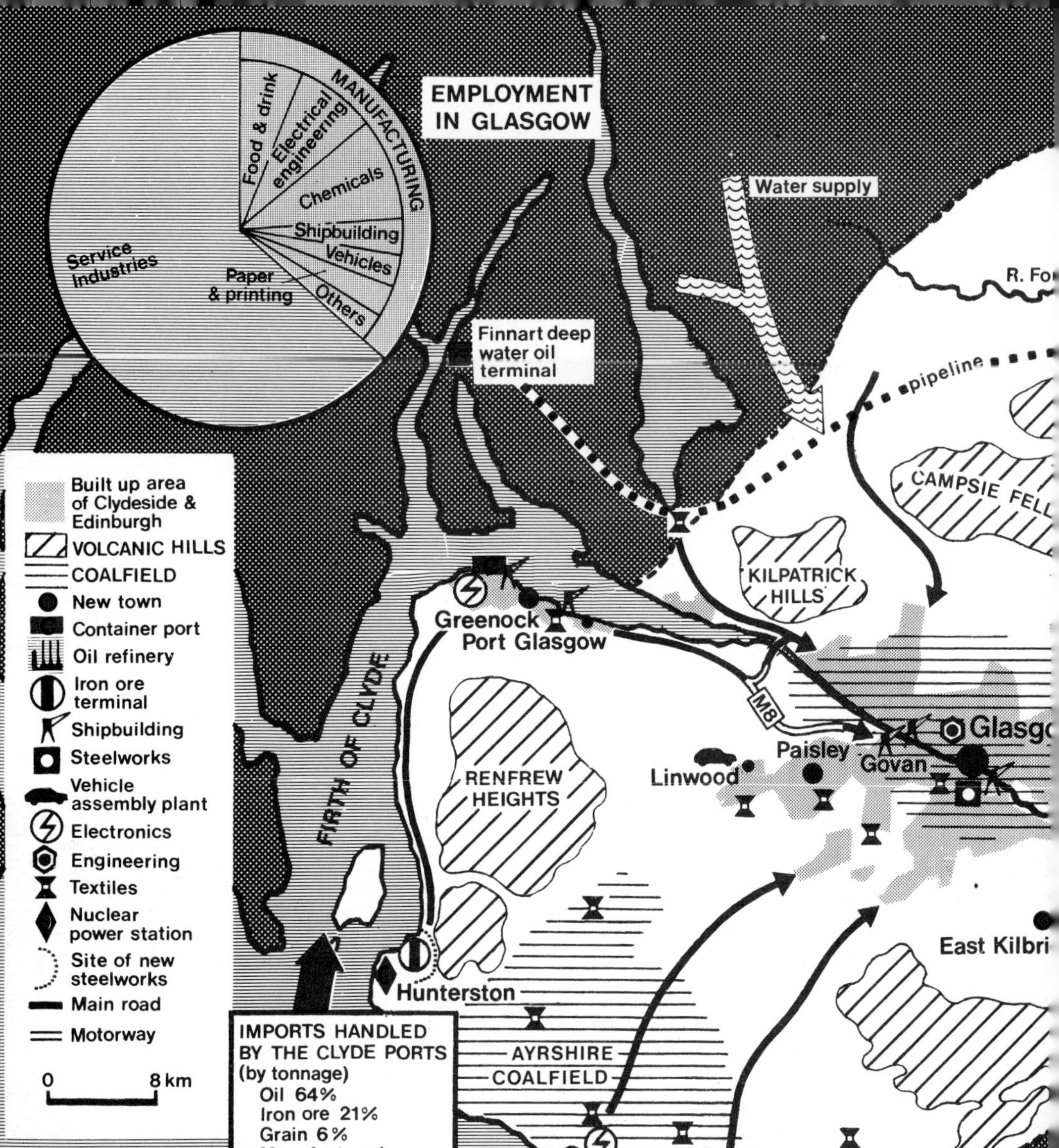

possible that a new steel works will be built near to the ore terminal and when this happens the future of the inland works will be even more doubtful.

3. *Shipbuilding*

Read pages 257 and 260 and complete the following exercise:

a) Describe the factors which encouraged the development of shipbuilding on the Clyde.

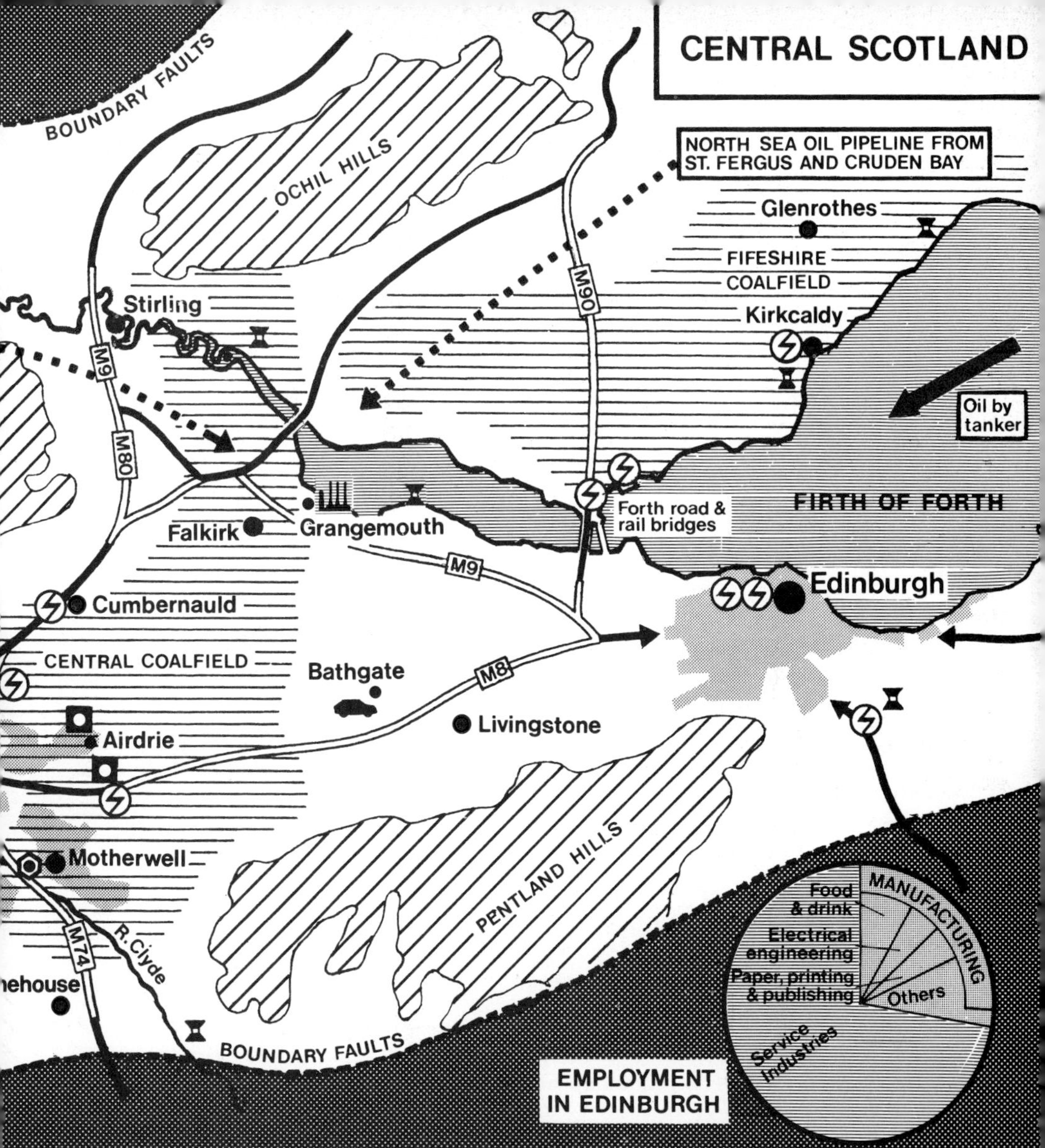

b) Name the raw materials used and explain where they are obtained from.
c) What are the disadvantages of the Clyde?
d) Describe the changes which have taken place in output since the early years of the century and explain why they have occurred.

The yards on the Clyde launched some of the most famous ships in the world but they were often small, out of date and unable to cope with the large vessels which have appeared in recent years. As

The River Clyde below Glasgow showing the main docks (a); a shipyard (b); and, in the distance, the mouth of the river where the modern port facilities are located.

foreign competition increased, yards began to close and it seemed likely that the industry would collapse. This caused the government to step in and the industry was reorganised into two large groups—one on the upper Clyde and one on the Lower Clyde. Grants were made to improve the yards but, in spite of this, the river remains too small to handle large ships and the yards are often difficult to extend.

4. *Engineering*

The shipbuilding industry is organised like the car industry, with yards assembling a large number of components to make a single ship. And, like the car industry, a large number of engineering firms have grown up to supply these components. Products range from

Glasgow region have declined, causing unemployment and serious social problems. Attempts to revive these industries have generally failed and more and more emphasis has been placed on introducing new industries. This policy has had two important effects.

In the first place the pattern of industry in the Central Lowlands has begun to change. This emerges quite clearly from the following figures.

PROPORTION OF WORKFORCE EMPLOYED

INDUSTRY	BRITAIN 1935	C. SCOTLAND 1935	BRITAIN 1975	C. SCOTLAND 1975
Mining	10·6%	10·2%	1·4%	1·6%
Steel	7·5%	10·6%	1·1%	1·6%
Engineering	6·0%	14·6%	4·7%	5·0%
Ship building	1·1%	5·2%	0·7%	2·0%
Electrical	3·4%	0·5%	3·5%	2·7%
Motor vehicles	3·9%	0·8%	3·2%	2·0%

Using the information given above, complete the following exercise:

a) i) Name three industries which have declined.
   ii) Name two industries which have expanded.

b) Work out the location quotients for the above industries in 1935 and 1975. (See page 311 for the method.)
   i) Which industries were heavily concentrated in Central Scotland in 1935?
   ii) Which industries were concentrated there in 1975?
   iii) Describe the changes which have taken place.

The most striking developments have taken place in the engineering and electrical industries. Attempts to *diversify* the engineering industry started in the 1930's but they achieved their greatest success some thirty years later when the motor industry was persuaded to come to the region. Two large plants were built at Linwood and Bathgate where it was hoped to make use of steel sheet produced in the Motherwell area to manufacture bodies and other parts. In fact, most of the components are still brought in from the Midlands and this has made development difficult. The other success has been in the electronics industry which is now important throughout the region.

The second effect has been to weaken the concentration of industry in the Glasgow area, and to spread it throughout the lowlands. Many factors have contributed to this movement.

FACTORS WHICH HAVE ENCOURAGED INDUSTRY TO MOVE AWAY FROM THE GLASGOW AREA	FACTORS WHICH HAVE ATTRACTED INDUSTRY TO THE EASTERN PART OF THE CENTRAL LOWLANDS
1. An old established industrial area with many problems.	1. Located near to the old industrial areas but with few of their problems.
2. Very heavily populated with little room for development.	2. New and expanded towns were sited in the area.
3. Dependence on coal declined and the nearby Central Coalfield was almost worked out.	3. The most productive coalfields today are in the east.
4. The port of Glasgow has also declined and new specialist port facilities have been built on the Firth of Clyde, eg the oil terminal at Finnart, the iron ore terminal at Hunterston and the container port at Greenock.	4. Scotland's largest oil refinery is at Grangemouth and this will handle oil from the Finnart terminal and from the North Sea.
5. Overspill has been housed in new towns which need industry. It has been policy, therefore, to encourage industries to move.	5. Road links between east and west have been improved and the motorway network has been extended.
	6. Edinburgh was an existing industrial centre around which growth could take place.

The results of these changes have been striking.

Using information given on pages 200 and 201, complete the following exercise:

a) i) List the main industries of the Glasgow area, arranging them in order of importance.
   ii) In each case state whether the industry is likely to be old established or a fairly recent introduction.
   iii) Which of the industries could be called 'port' industries?

b) i) Name four new towns which have taken overspill from Glasgow.
   ii) Name one industry which has been attracted to the new towns. The main requirement for this industry is a supply of labour. Why are the new towns good locations?

c) i) Why has the old port area of Glasgow declined?
   ii) Why are the new port facilities located on the Firth of Clyde?
   iii) Describe the pattern of trade of the Clyde ports (pages 8 to 14 will help).

d) Why is it possible that the steel industry will move to Hunterston?

e) Describe the pattern of industry in the Edinburgh area and explain how it differs from that of Glasgow.

f) i) What are the advantages of Grangemouth as an oil refining centre?
   ii) What are the disadvantages and how have they been overcome?

Grangemouth oil refinery, Scotland.

iii) Why is Grangemouth well situated for future development? (pages 118 to 127 will help).

Large scale changes such as these have had important effects on the population of the Central Lowlands. The pattern of very slow growth and high rates of emigration has not been reversed but, within the region, there has been a movement away from the coalfields and old port areas to the towns and cities of the east, which have attracted many of the new industries.

Settlement

The population of Britain is housed in a vast number of settlements of all shapes and sizes, ranging from the city of London, with its population of some eight million, to the isolated farmstead, occupied by a single family. So numerous are these settlements and so great their range in size and shape, it becomes difficult to identify any clear pattern of distribution across the country; a difficulty made all the greater by the fact that the present pattern of settlement owes a great deal to the patterns established in the past. Since these earlier patterns reflected a way of life which no longer exists, it is usually very difficult to interpret the fragments which have persisted down to the present day.

Allowing for this, however, it is possible to identify certain features which are widespread in the pattern of settlement in Britain.

Types of Settlement

Man appears to be naturally *gregarious.* This means that he prefers to live as a member of a group rather than on his own. As a result, his houses will tend to be grouped together to form *nucleated settlements.* It is not surprising, therefore, that over much of Britain nucleated settlements of all kinds are found. In some areas, however, particularly the highlands, the pattern of settlement is completely different, consisting largely of scattered farms. This is known as *dispersed settlement.* Many factors have contributed to the emergence of these two contrasting types of settlement. The most important are listed in the table opposite.

In fact, the division between nucleated and dispersed settlement in Britain is far from clear since the break up of the open fields and the enclosure of the land resulted in the development of isolated farms throughout the country.

Whether nucleated or dispersed, settlements exist because man has a number of basic needs which must be met. Among the most important of these are water, food, shelter, warmth and safety, and

FACTORS ENCOURAGING NUCLEATED SETTLEMENT	FACTORS ENCOURAGING DISPERSED SETTLEMENT
1. Man is naturally gregarious.	1. Plentiful water supply.
2. Defence.	2. Poor agricultural land which will not support a large population.
3. Availability of water if it is in short supply.	3. Farming methods and land holding in highland Britain favoured the development of individual farms.
4. Limited areas of dry land in marshland.	4. On the lowlands, land settled during the last five hundred years tended to be divided up into individual farms rather than kept in Open Field.
5. Rich farm land which supports a large population.	
6. For over a thousand years large areas of Britain were farmed on the Open Field system. In this the land was divided into vast fields and individual families were allocated a number of narrow strips of land, scattered about the fields. Because holdings were scattered it was easier to live in a central village and to travel out to the land each day.	

it is not surprising that these factors tended to be most important in influencing the choice of sites for early villages, eg.

1. Water supply which tended to be most important in areas of permeable rock where surface water was in short supply.

2. Arable farmland for the production of food crops, including where possible, the lighter, easily ploughed soils.

3. Pasture for animals which presented less of a problem since heavier badly drained soils could be used.

4. Shelter which was important both in terms of the *exposure* of the site and in the availability of *building materials* and *fuel*.

5. The need for defence complicated the choice and led to the occupation of hill top sites and land partially surrounded by water.

At first each family or community tried to be self-sufficient, but, as needs became more complicated, so certain goods and services had to be bought. These ranged from the services of skilled men, such as doctors and blacksmiths, to goods which could not be made either on the farm or even in the local area but which could be brought into the village. Thus settlements also became *service centres*, particularly those with good communications with surrounding areas.

Warkworth.

Cardigan.

Brandon Creek, Norfolk.

Milnathort.

The interplay of these factors can be seen quite clearly in the site and situation of certain settlements, particularly when one factor appears to be dominant.

Study the photographs opposite and complete the following exercise:

a) i) Draw four columns, one for each of the settlements named.
   ii) The following list contains points which may relate to the settlements named. Write each point down in its appropriate column. Any point can relate to more than one settlement.
   Defensive site
   Controlling a valley route
   At a cross roads
   Controlling a river crossing
   Controlling lowland routes
   Following the course of a road
   A compact village on a restricted site
   In centre of drained marshlands
   Hill top location

b) The following general terms are often used to describe settlements. Relate the terms to the settlements shown and give reasons for your choice.
   Gap town
   Route centre
   Bridging point
   Market centre.

c) Write an account of the growth of each of the settlements, paying particular attention to its site, function and the nature and extent of recent development.

Most settlements are much more complicated than those illustrated and, in many cases, later growth will have either obliterated or concealed much of the evidence relating to the establishment and early growth of the settlement. Even more important, such an approach does little to explain the distribution and spacing of settlements and their variation in size.

Size and Spacing of Settlements

Attempts to explain these aspects of the settlement pattern of Britain have concentrated on the function of towns and villages as *service centres*; each providing some kind of service to a surrounding area. (This area is hexagonal in shape.)

The way in which centres of varying size and importance emerge can be seen on page 210. Village A serves its own hexagonal service

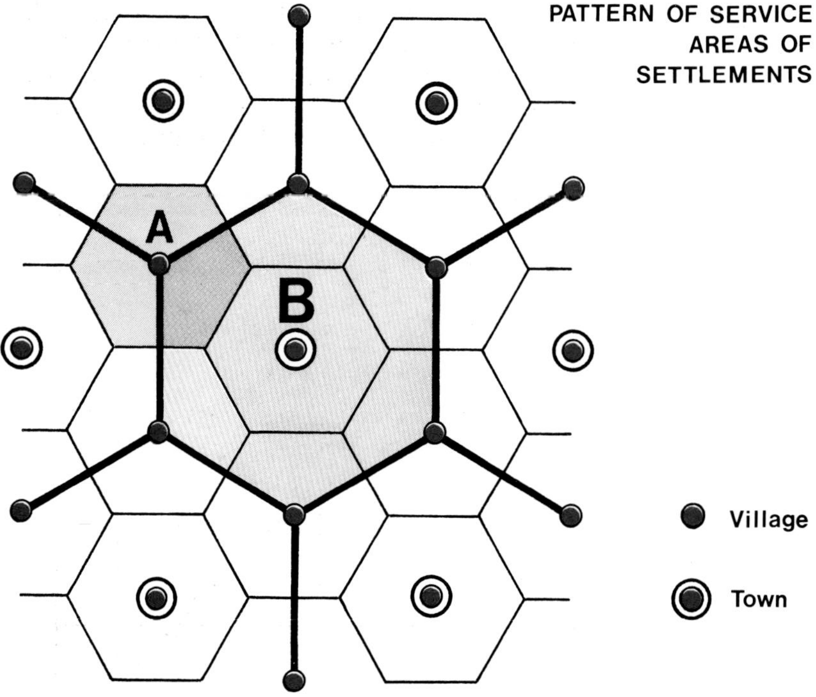

area, providing very basic amenities. The village itself comes under the influence of a larger settlement, B, which provides a much wider range of services to village A and to five other villages similar in size to A. In the same way B itself comes under the influence of an even larger centre which also has a hexagonal service area and which provides an even wider range of services.

The importance of this concept is obvious.

1. It helps to explain the great variation in size of settlements. In Britain, for example, the following pattern emerges:

POPULATION	NUMBER OF CENTRES	TYPE OF SETTLEMENT
More than 1 million	2	Millionaire city
500,000–1 million	4	Metropolis
250,000–500,000	11	
150,000–250,000	19	City
100,000–150,000	32	
50,000–100,000	117	
25,000–50,000	231	
2,000–25,000	517	Town

Lower still in the heirarchy are large numbers of villages and hamlets.

Many attempts have been made to predict the pattern which will emerge in a given area. The simplest of these is the Rank Size Rule which states that, if settlements are arranged in rank order of size, the second settlement will be half the size of the largest, the third one third of its size, the fourth one quarter, the fifth one fifth etc.

Rank size rule

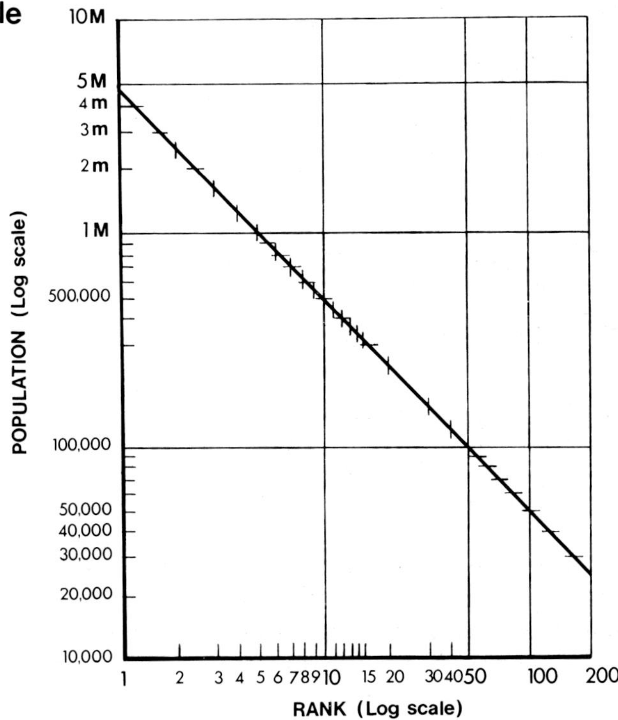

POPULATION OF BRITISH CITIES

CITY	POPULATION (000'S)	CITY	POPULATION (000'S)
London	7,281	Sheffield	512
Birmingham	1,004	Leeds	500
Glasgow	836	Edinburgh	449
Liverpool	575	Bristol	422
Manchester	531	Nottingham	295

a) i) Copy the list of British cities and their populations, arranging them in rank order of size.

ii) Using the Rank Size Graph, calculate what the population should be, a) with London included, b) excluding London.

iii) Plot these two sets of figures on a graph similar to the one shown.
iv) For which set of figures does the Rank Size Rule offer the best prediction.

b) Repeat the exercise using the towns in your county.

It is clear that in Britain enormous distortion is caused by the great size of London and that, when the exercise is carried out for cities of lower rank order, the predicted sizes are more accurate.

2. It also helps to explain the spacing of settlements. Hamlets and villages will be very close together—a study of Southern England indicated a spacing of about four kilometres, whereas small towns will be further apart (twelve kilometres) and larger towns up to forty kilometres apart. This pattern tends to reflect the difficulties of transport in Britain prior to the nineteenth century, for the service area of a settlement was limited by the distance which could be travelled in one day.

The Function of Settlements

As we have already seen, settlements have grown up for a variety of reasons and serve a variety of functions. Some of these, such as defence, are fairly easy to identify, others, such as the service function, are more difficult to understand. Most settlements provide services of various kinds and, generally speaking larger settlements will provide a wider range of services than those lower in the heirarchy. For example, a village will provide simple, basic necessities, such as a general store, a public house and a primary school; while at the other end of the scale, a city will meet specialist requirements in shopping, education and the health service. Similarly, the area served by a settlement (its sphere of influence) will vary according to the nature of the services offered and the nearness of the nearest competitor in that field. Villages, for example, offering very limited services will be located very close together and will tend to have a small sphere of influence, whereas a city will provide services for a larger regional area or, in the case of London, for a national or an international area.

SERVICES WHICH INDICATE THE STATUS OF SETTLEMENTS

HAMLET	VILLAGE	TOWN	CITY
General Stores	Butcher	Clothes shop	Department stores
Public house	Hardware shop	Secondary school	Cathedral
Post Office		Bank	University
Primary school		Supermarket	Theatre

The table gives a number of services which are found in settlements at various levels in the heirarchy.

a) i) Copy out the table, allowing ten lines for settlements.
ii) Select ten settlements, of varying size and importance, in the local area and enter them in the left hand column.
iii) Using the local telephone directory, establish which services are located in each settlement and enter a tick in the appropriate column.
iv) Establish the position of each settlement in the heirarchy.

b) i) Find out the population of each settlement.
ii) Using the classification given on page 210, once again establish the position of each settlement in the heirarchy.
iii) Point out any differences in the two classifications.

c) i) For each of the following goods and services, write down the name of the centre used when each of the following was last obtained by your family. (Leave out purchases from travelling shops or catalogues), a) sugar, b) a pair of shoes, c) a bed.
ii) Using a separate map for each item, every member of the class plot the journey from his/her house to the centre used by his/her family (plot this information as a straight line from the house to the centre of the settlement used).
iii) Study the patterns which emerge. Do they support or tend to bring into question the idea of a heirarchy in service centres?

Although such indicators can give only a rough guide as to the relative importance of settlements, it will probably become clear that in your local area, as in most of Britain, the pattern of settlement which emerges will differ considerably from that which might be expected. Reasons for this are not difficult to find.

1. The perfect pattern could only occur on a uniform surface and in Britain, there are marked differences between highland and lowland areas.

2. Resources are not spread evenly across the country and mineral deposits, such as coalfields, have tended to encourage concentrations of population and settlement in very limited areas.

3. Changes in the transport system and variations in its efficiency can seriously distort the pattern. For example the building of a motorway can increase the size of the service area of a town or city on its route.

4. As we have already seen, many other factors have contributed to the establishment and growth of settlements and, in concentrating on service functions, these tend to be forgotten.

In spite of these shortcomings, analysis of the type described can help to explain certain aspects of the settlement pattern and, in particular, the development of market towns which depend to a great extent on the services supplied to the surrounding area.

Town Study: Newcastle–under–Lyme, a market town

Newcastle-under-Lyme is a market town in North Staffordshire. It is an ancient settlement and like most market towns, it has faced periods of rapid change which have brought about great changes in the structure of the town and made it difficult to distinguish its earlier functions.

The settlement grew up on the main road from London to the North West (the present A34). Too small to warrant a mention in the Domesday Book in 1086, it must have grown rapidly during the next hundred and fifty years for, during the first half of the twelfth century, the new castle was built which was to give the town its name. The site was not a natural defensive site and, as a result, the castle was situated on an artificial mound in a man-made lake, formed by damming two streams.

High Street, Newcastle. Note the width of the street and the open market.

Security and the relatively good system of communications brought prosperity to Newcastle and it emerged, during the later Middle Ages, as one of a number of important market towns in the area. Because of the poor condition of the roads and the difficulties of road transport these centres tended to be very close together so that few villages or hamlets were more than fifteen kilometres away from a market.

It was during this period that several important features of the modern town began to develop. Among the most important of these were:

1. Open and covered markets were established.

2. The importance of these markets led to the establishment of road links between Newcastle and the neighbouring market towns. This produced the radial road pattern which is typical of so many market towns today.

3. The first industries began to grow up in the town. These included leather tanning, hat making, the manufacture of clay pipes and iron smelting—all making use of local raw materials.

4. The actual structure and lay-out of the town became clearly recognisable, with its broad high street, housing the open market, and the imposing market hall.

As a result of these developments, by the end of the eighteenth century, Newcastle was the largest market centre in North Staffordshire and the most important town in the district.

This period of growth was followed by a century of relative decline as the town was overshadowed by the rapidly growing pottery towns to the east and as road traffic declined in face of competition from canals and railways. The latter was particularly important since Newcastle was situated away from the main through routes and was served only by branch lines.

As with many other market towns, it was the development of motor transport which was to halt this decline and lead to a period of rapid growth during the twentieth century. This produced a number of characteristic developments which can be seen in market towns throughout Britain.

1. The building of new roads and the improvement of existing ones emphasised the radial road pattern of the town, which emerged as an important *route centre*. (See page 216).

2. Improved access to the town gave it an important advantage over its rivals in the area and, as a result, its service area tended to expand at the expense of smaller market centres, which tended to decline. In the modern town, therefore, the *service function* became of

major importance. This is reflected in many ways:

a) The increasing importance of *service industries* as opposed to manufacturing industries. These range from working in shops to the provision of goods and services to the farming districts to the west of the town.

b) The emergence of the town as an important district *shopping*

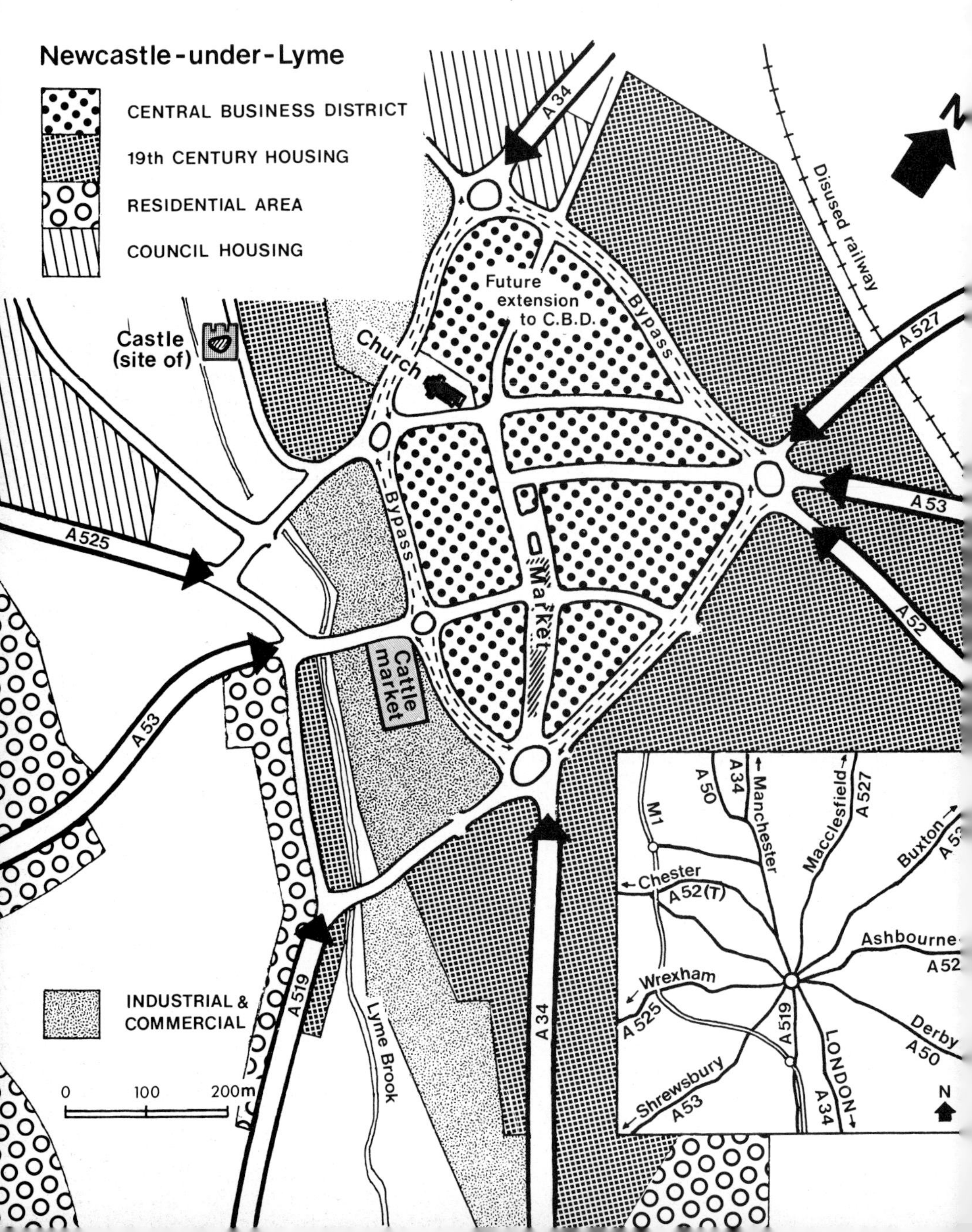

centre, meeting the demand for both convenience goods (eg groceries) and more specialised durable goods (eg hardware and furniture). Most of the shops are concentrated in the area around the old High Street and this has become the most highly rated part of the town. Sites in this area are expensive and they have tended to be taken up by large national companies which have redeveloped the area until little of the old town remains. Behind these main shop fronts rateable values decline sharply and a variety of smaller shops and businesses has grown up there. (See page 218.)

c) The traditional manufacturing industries of the town centre have disappeared and been replaced by modern industries such as

Newcastle-under-Lyme. Note i) the site of the castle (a); ii) the broad high street (b); iii) the Guildhall (c).

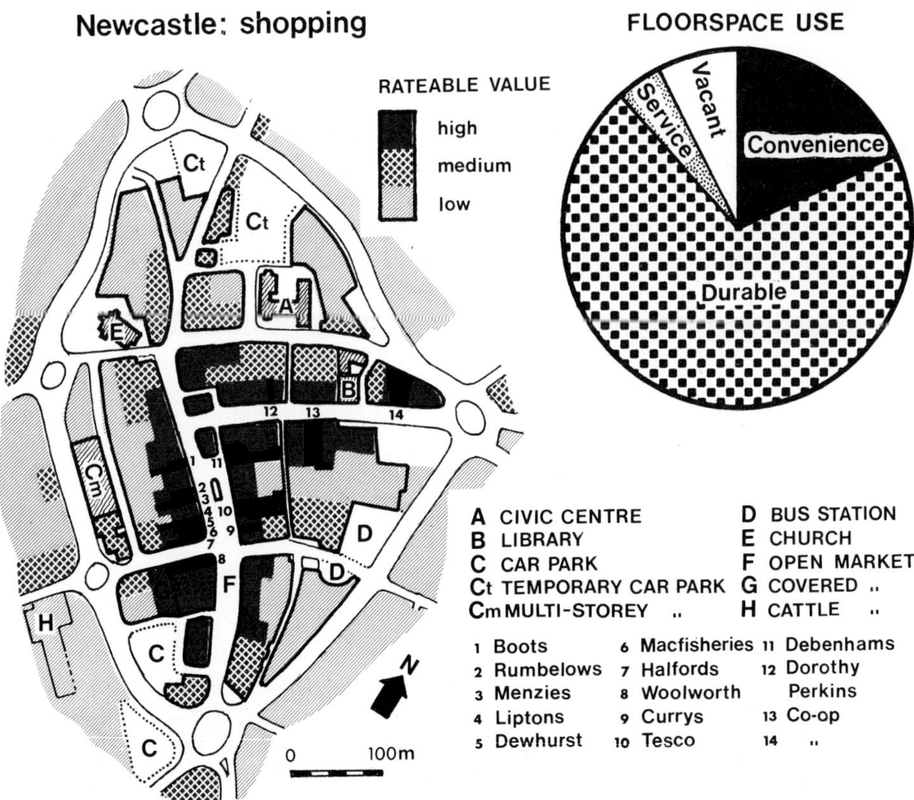

light engineering and baking which are located outside the town, along the main roads or on small *industrial estates*.

The twentieth century has also brought problems. In the first place on the radial road pattern, together with the fact that Newcastle is situated on a main north-south route, has produced serious *traffic congestion* in the town. This has been relieved by building bye-passes to the east and west of the town, which will allow the establishment of traffic free areas in the town centre. And, in the second place the building of the M6 to the west of the town, while increasing accessibility, has also encouraged the building out of town shopping centres which may threaten the prosperity of the town centre.

The City

When a settlement becomes large the problems of studying it increase enormously. It may be possible to identify the factors which contributed to its establishment and early growth but these will often seem unimportant when compared with its later development. The problems become even more complicated when the city becomes so large that it merges with surrounding towns and villages to form a continuous built-up area known as a *conurbation*. In Britain, this process has produced a group of conurbations, ranging in size from London to the comparatively small conurbation of North Staffordshire with a population of less than half a million.

CONURBATION	POPULATION (MILLS)
London	7·4
West Midlands	2·4
Merseyside	1·3
South East Lancs	2·4
West Riding	1·7
Teesside	0·4
Tyneside	0·8
Clydeside	1·8
North Staffs	0·4

In the face of settlements of such size and complexity, a possible starting point for study would be the structure of the city. For, like smaller towns and villages, cities have grown up to meet the needs of the people living in or near them, and they tend to contain similar components. Most obvious among these are:

Shops
Offices
Industries
Recreation amenities
Low cost houses
Medium cost houses
High cost houses

At first glance these components seem to be scattered haphazardly throughout the city or conurbation, which, as a result, appears to be a jumble of houses, shops, factories, offices and parks. In fact, they tend to be grouped in areas which differ considerably in character one from another. For example:

1. The main shops, offices, hotels, cinemas and theatres will tend to be found near the city centre, in an area known as the *Central Business District* (CBD).

2. Housing also tends to be grouped according to type, eg older terraced houses, council houses and more expensive private houses will form *residential areas*, each of a different character.

3. The concentration of industries is less obvious for, in towns which grew rapidly during the nineteenth century, industry and housing tended to be mixed. Even here, however, there is a tendency for industry to become separated from residential areas (because of

environmental considerations) and from the CBD because of the price of land in the city centre. This development has taken place slowly over a long period of time.

In fact, during the last hundred years, these contrasting areas have become much more clearly defined and many attempts have been made to explain their distribution within the city. The diagram below shows one such attempt for British cities.

Because a conurbation is formed by the joining together of separate settlements, a more complex pattern often appears, with several shopping centres and business districts outside the Central Business District, and with a tendency for zones to develop around these secondary centres.

The growth of the conurbations is a recent development, made possible only by the improvements in transport which have taken place since the building of the first railways in the early nineteenth

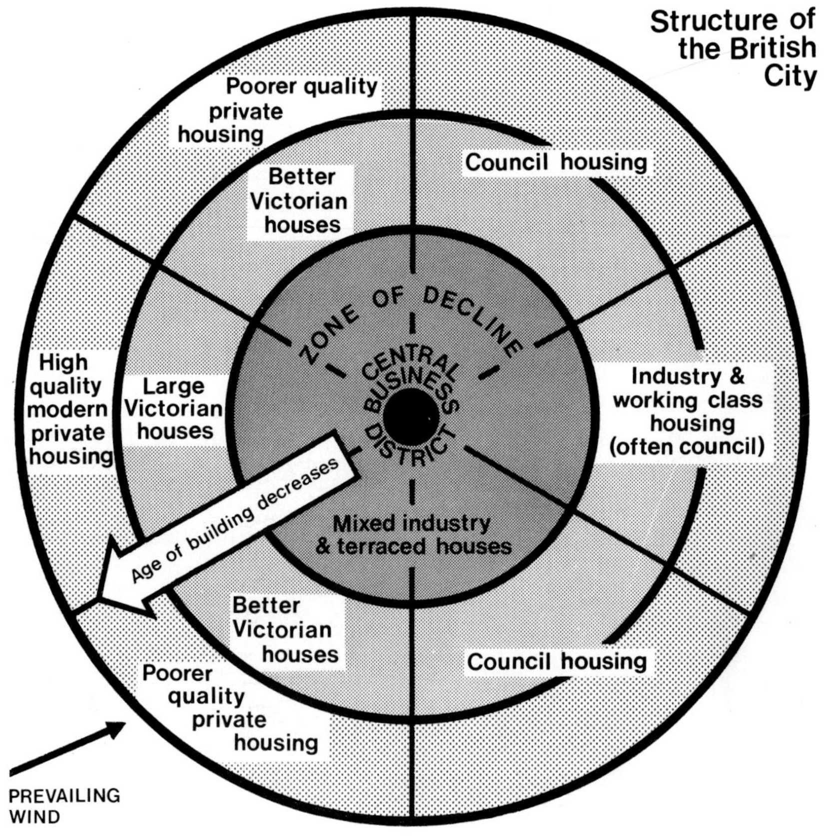

century. Prior to that time, the size of the city was limited by the problems of travel from the surrounding area to the city centre; a problem which not only limited the extent of the built-up area but also reduced the area which could be efficiently served by the city. The railways, by providing cheap and rapid transport, changed this situation and cities began to expand, absorbing into their built-up areas many smaller settlements which lay on, or near, the main lines. Improvements in road transport, with the development of the motor car, gave greater flexibility and the city spread into areas not served by the railways.

This increase in size brought with it many problems.

Problems in the City

Cities in Britain, as in most developed countries, face three major problems which are of interest to the geographer. These are:

1. Congestion.
2. Urban decay, particularly in areas near the city centre.
3. Urban sprawl or the spread of the city into the surrounding countryside.

1. Congestion

Congestion is not a new problem in cities but it has become more serious in recent years. Several factors have contributed to this development:

a) Cities, like most other settlements, have grown up to meet the needs of the people living in and around them. These people travel to the city to obtain a wide range of services.

b) Congestion is further increased by the need to supply the city itself with the goods it needs. These have to be brought into the city thus adding to the traffic and causing further congestion.

c) When cities began to grow rapidly during the nineteenth century, houses were built very close together, often near places of work. This produced a zone of high density housing around the city centre.

d) As land values near the city centre increased, so did the tendency to build upwards—a development which increased still further the density of building in the central area.

e) Much of this later building was of business and commercial premises which provided work for many people, many of whom lived in the spreading suburbs of the city. These workers (known as *commuters*) travel into the city centre each morning and leave the city centre for the suburbs each evening—a movement which adds enormously to the congestion, and strains the transport system of the city.

f) As long as cities depended upon public transport, such as trains and buses, the problem could be controlled but, in recent years there has been an increasing use of private transport, particularly the motor car, which now threatens to block the roads in and around the city centre. So serious has this problem become that in central London the average speed of movement on the roads is as low as 15 kph.

Congestion today is, therefore, likely to be caused by traffic rather than the density of housing near the city centre and attempts to control the situation have concentrated on improving the transport system—an approach which has proved to be far from easy.

The following are methods used in various cities in Britain to reduce congestion:

1. Widening existing roads (most cities).
2. Building new roads to by-pass the city centre (most cities).
3. Bus lanes (eg Manchester, London).
4. New or improved public transport facilities (eg the Tyneside Metro).
5. Banning cars from the city centre (eg Leeds).
6. Increasing parking charges.

Each has advantages and disadvantages. Using the two lists given below, write down the advantages and disadvantages of each approach.

ADVANTAGES	DISADVANTAGES
Speeds up traffic flow	Expensive because land prices are high
Makes the city centre more attractive	Widespread demolition of property necessary
Diverts through traffic	Causes resentment
Speeds up public transport	Occupies road space and thus increases congestion
Makes cars less attractive	Needs an efficient public transport system before it can work
	Causes congestion where the improvement ends
	Moves congestion further along the road

2. Urban Decay

Cities are constantly changing. Some areas within them are growing while other areas are decaying. As we have already seen, many of the major problems are found near the city centre, where congestion adds to an already difficult environment. Here, most of the buildings were constructed during the nineteenth century and are now old and ready for replacement. This is particularly true of the residential districts which generally consist of high density terraced houses, often lacking such basic amenities as inside toilets and bathrooms, and often designated as slums. Large numbers of these houses have been cleared and their inhabitants moved, usually to council estates on the city margins. This, together with clearance for road schemes, has produced a blighted area around the Central Business District; an area of low cost housing which has attracted the lower paid, including immigrant populations. In some cities large scale clearance of such areas has take place and attempts have been made to rehouse the people in the same neighbourhood. Because of the high land values so near to the city centre, however, high density housing was again built, usually in the form of flats. Such developments have created as many problems as they have solved.

In the CBD itself land values are even higher and this has tended to exclude housing developments of any kind and to encourage the building of office blocks, shops and business premises. As a result, even the most prosperous city centre tends to 'die' after working hours.

Inner city blight.

3. Urban Sprawl

The decline of the city centres has been accompanied by the development of the outskirts of the city. Many influences have contributed to this; some causing the city to spread, others, as we have seen, leading to the redevelopment of the central areas.

REASONS FOR REMAINING NEAR THE CITY CENTRE

1. Near to the amenities of the city, eg shops, cinemas, theatres.
2. Many industries and businesses grouped here, and these tend to attract others.
3. Near to places of work.
4. Inertia—ie people do not like moving.

REASONS FOR MOVING TO THE OUTSKIRTS

1. Cheaper land.
2. Land available for low density housing and industrial sites.
3. Congestion in the city centre.
4. A pleasanter environment, less traffic, noise, pollution.

Recent improvements in transport have encouraged the city to spread and this has caused many problems. The most obvious of these is the fact that any extension of the city area usually means the loss of valuable agricultural land—something that Britain can ill afford. Furthermore, development on the outskirts of the city has tended to be wasteful of land and is often unsightly. This is particularly true of some of the '*ribbon development*' which took place before the Second World War, when industry and housing spread along the main roads leading out of the city.

It was in an attempt to limit urban sprawl of this kind that *Greenbelts* were established around many of our cities. Within these areas development is strictly controlled and, before any building can take place, government permission must be obtained. Because land-use within the greenbelts is mixed, however, and because the people who live in the area must make a living, controls cannot be absolute and some development—usually in the form of planned *housing* and *industrial* estates—is allowed. Furthermore, few cities are completely encircled by greenbelts and this allows development to take place in certain well defined areas. In spite of such weaknesses, however, the system of greenbelts has slowed down the process by which one city would merge into another to form a vast built-up area. It has also enabled areas of open country to be preserved within easy reach of city centres.

London was the pioneer in the introduction of greenbelts and it is around this city that the most complete system exists. Elsewhere, the movement towards the establishment of greenbelts has been much

The outskirts of Preston, showing ribbon development along the roads with later infilling in the form of housing estates.

slower and some city authorities believe that the greenbelt should be broken to form areas of planned development separated by *green wedges*. This pattern can be seen in the West Midlands where development is planned along the lines of the motorway network.

The establishment of greenbelts has also created problems for, in spite of planning restrictions, many cities in Britain have continued to grow and pressures have built up within the city limits. One method of relieving this pressure has been to rebuild on land cleared near the city centre. Such an approach has its limitations and a more widely used method has been to move people into estates and towns beyond the greenbelt, or even into specially built *new towns*.

New Towns

Methods of rehousing surplus population (*overspill*) has varied from city to city.

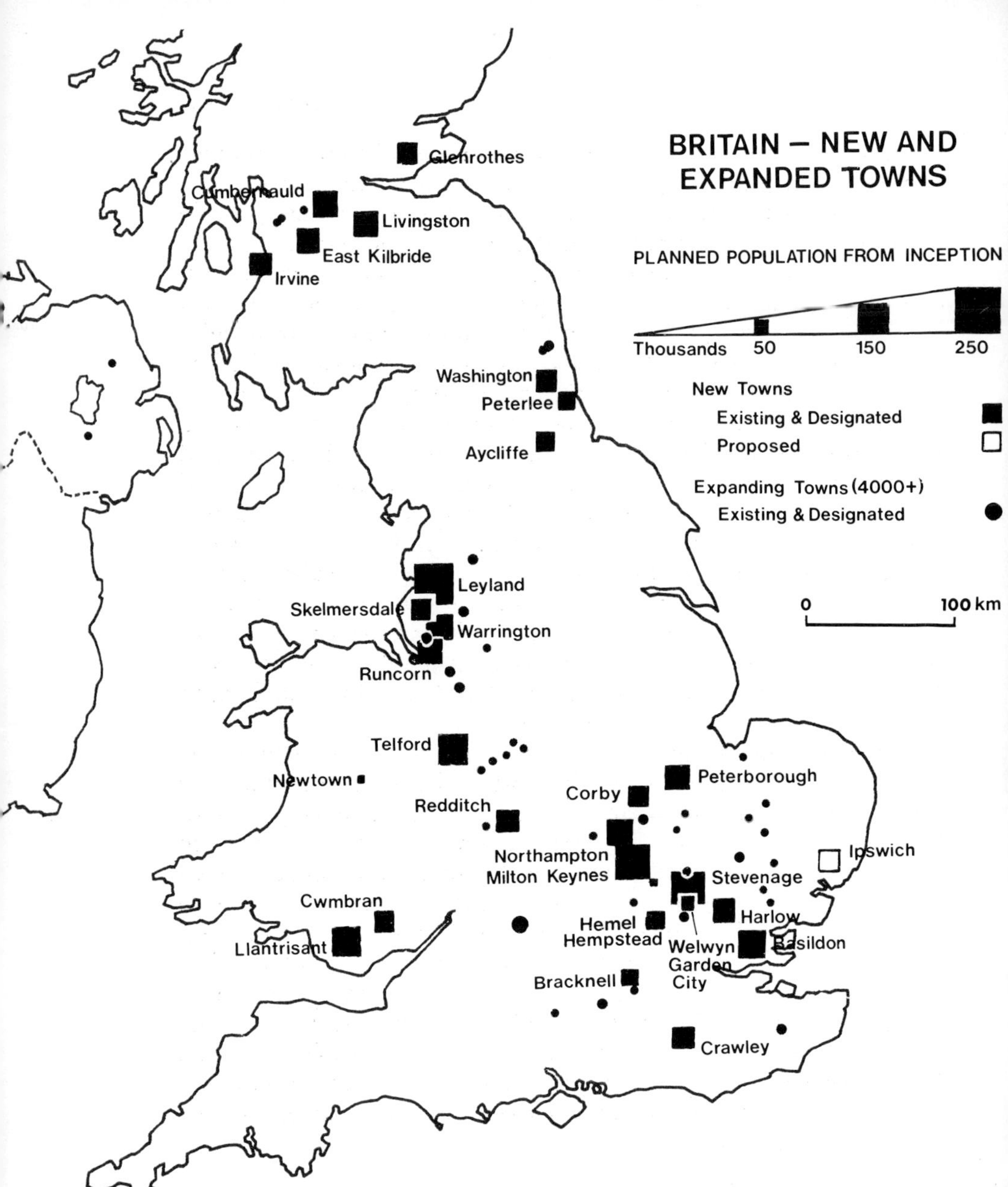

In some cases—Liverpool being a good example—large housing estates have been built outside the city area. Work was provided on nearby industrial estates but many people still travelled to Liverpool for work and links with the city have remained strong. In fact, estates such as Kirby have tended to become merely *dormitory areas*, lacking either work or leisure facilities.

Another, and often more successful approach has been to move people into neighbouring towns, some distance away from the city. These towns—known as *expanding towns*—will, it is hoped, provide both work and social facilities for the families who move.

The third approach is by far the most adventurous, involving, as it does, the creation of *new towns*, large enough and far enough away from the city to escape its influence and to exist in their own right. Such towns were designed to provide all the amenities required by the people living in them, eg work, schools, shops, hospitals etc and, since they were to be *new* towns, they could be planned so as to avoid the problems which have plagued so many of our older towns and cities. The first new town was established in 1946, soon after the end of the war, and, since then, more than twenty have been designated, mainly to take surplus population from existing cities. In some cases, however, new towns have been given additional functions. Peterlee, in Durham, for example, was built to house people from parts of the East Durham Coalfield which had been worked out. And Telford, in addition to housing overspill from the West Midlands, was sited on the worked out East Shropshire coalfield, so as to revive the economy of that area.

The development of New Towns has not been without its problems. In the first place, the cost of building the amenities required and of setting up the authorities to run them has proved to be very high. As a result, interest has tended to turn towards the alternative of expanding towns which are cheaper because they already contain many of the facilities which are needed. Even more serious has been the failure of many New Towns to attract sufficient industry to provide work for the entire population. This has led to large scale commuting to the cities from which the people have moved. And finally, the whole idea of moving people from the city centres has fallen into disrepute on account of the problems created in the areas which they have left. These problems (see page 223) are now so serious that the government proposes to divert resources into improving conditions in the inner city areas.

Refer to the map opposite.

a) i) Trace an outline map and mark on it the main conurbations.
   ii) Draw 40 km circles around each of the conurbations.
   iii) Plot the main new towns.
   iv) Describe the pattern which emerges and calculate the number of people rehoused around each conurbation.
   v) Name any new towns outside the circles and, using the index, find out why these towns grew up.

b) Refer to the photograph of Stevenage and complete the following exercise:
   i) How does Stevenage town centre differ from the central area of an old established town?
   ii) How has the problem of traffic congestion been tackled in Stevenage? Are there any disadvantages to this approach?
   iii) What evidence is there that Stevenage is a planned settlement? How does it differ from the unplanned settlement shown on page 99?

The true complexity of the British city can best be appreciated by studying a single example.

Stevenage new town centre. Note i) the segregation of the shopping centre, housing and industry; ii) the methods used to cope with traffic; iii) the road and rail links with other areas.

Settlement Study: The North Staffordshire Conurbation

North Staffordshire is unique among the British conurbations in that it is small, with a population of no more than 450,000, and its main settlement, the city of Stoke-on-Trent, was itself formed by the joining together in 1911 of six separate towns, each of which has to some extent retained its own identity. As a result, the processes which lie behind the formation of conurbations can be seen more clearly here than in most of the larger cities.

Situated on the North Staffordshire coalfield, the conurbation

EMPLOYMENT IN THE NORTH STAFFORDSHIRE CONURBATION
(as a percentage of total labour force)

INDUSTRY	STOKE-ON-TRENT	NEWCASTLE-UNDER-LYME
Primary Industries (mining & agriculture)	3	7
Pottery	17	0·5
Steel	1	1·4
Electrical Engineering	2	2·6
Engineering	8·5	11·3
Clothing	1·5	2·4
Food & Drink	1·5	2·4
Other Manufacturing	7·5	12·0
Construction & General Labouring	10	8·0
Transport	5·3	4·0
Warehouse	3·6	2·2
Clerical	10·6	10·7
Shops & Sales	8	11·7
Other Services	9	11·6
Adminstration	3·5	2·5
Profession	8	9·7

originated as a group of villages which, at an early date, became involved with the two basic industries of the area—coal mining and pottery. Changes in these basic industries have played an important part in the growth of the conurbation.

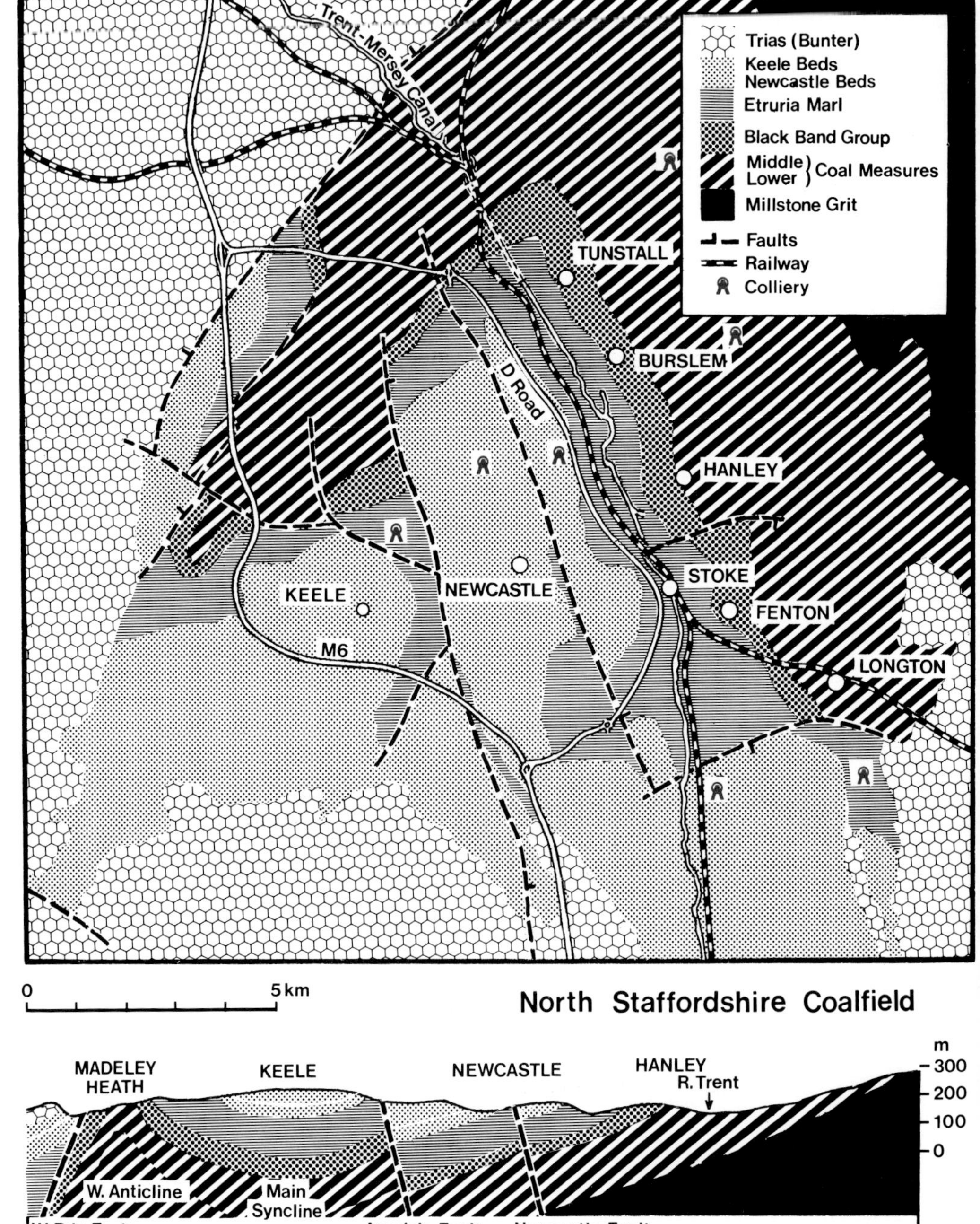

North Staffordshire Coalfield

The Industrial Base

The North Staffordshire conurbation is still generally known as 'The Potteries' and few areas in Britain today are so dependent upon a single industry. An understanding of the pottery industry is, therefore, essential to any understanding of the conurbation as a whole.

The Pottery Industry

The Early Industry BEFORE 1700	Establishment of the Modern Industry 1750 ONWARDS	The Modern Industry PRESENT DAY
1. Raw materials: Local clay Local timber Local coal 2. Location: In villages situated on the Blackband outcrop where clay and coal are found on or near the surface. 3. Production: Very poor quality ware.	1. Technical improvements led to better quality ware. 2. China clay, brought in from Cornwall, also allowed quality of ware to be improved. 3. The building of the Trent–Mersey Canal (1765–75) allowed this clay to be imported easily and ware to be exported. 4. Location: a) in the existing centres b) near the canals (and later near railways). 5. Specialisation: a) Earthenware in the northern towns of Burslem and Tunstall. b) Fine china in Stoke and Longton in the south. 6. Structure: a) many small firms b) much skilled hand labour.	1. Technical changes: a) increased mechanisation b) (oil and gas replaced coal in the firing of kilns. 2. Reduction in labour force, eg 58,000 in 1959 45,000 in 1968 3. Amalgamation of firms to finance improvements. 4. Location: a) No longer tied to raw materials but traditional sites still used. Example of industrial inertia. b) One or two firms moved to country sites, eg Wedgwood. 5. The labour force is still highly skilled.

Although there has been a reduction in its labour force, the pottery industry remains by far the most important industry in North Staffordshire and, almost alone among the traditional coalfield industries of Britain, it has remained prosperous. Its growth and prosperity have influenced the growth of other industries in the area. This is particularly true of the coalmining industry.

Coalmining

The North Staffordshire coalfield is one of the smallest in Britain in terms of size but one of the richest in terms of the thickness of the measures and the number of seams.

The coal outcrops in the east and west but is buried under more recent rocks in the centre of the coalfield. Mining first developed on the outcrops, particularly near to clay deposits, and, with the expansion of the pottery industry in the eighteenth century, many pits were sunk and production increased rapidly. These areas are now exhausted, the pits are closed and mining is concentrated in a few large pits in the centre and south of the coalfield. Naturally the labour force has declined and, with the change to oil and gas in the pottery industry, many of the local markets have been lost. In spite of this, however, production has remained steady at more than four million tonnes.

The decline of the work force of the two staple industries has been balanced by the movement of people out of the conurbation and by

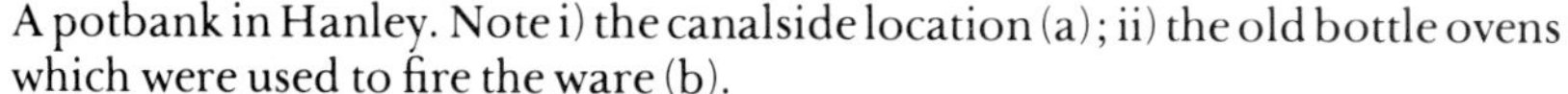

A potbank in Hanley. Note i) the canalside location (a); ii) the old bottle ovens which were used to fire the ware (b).

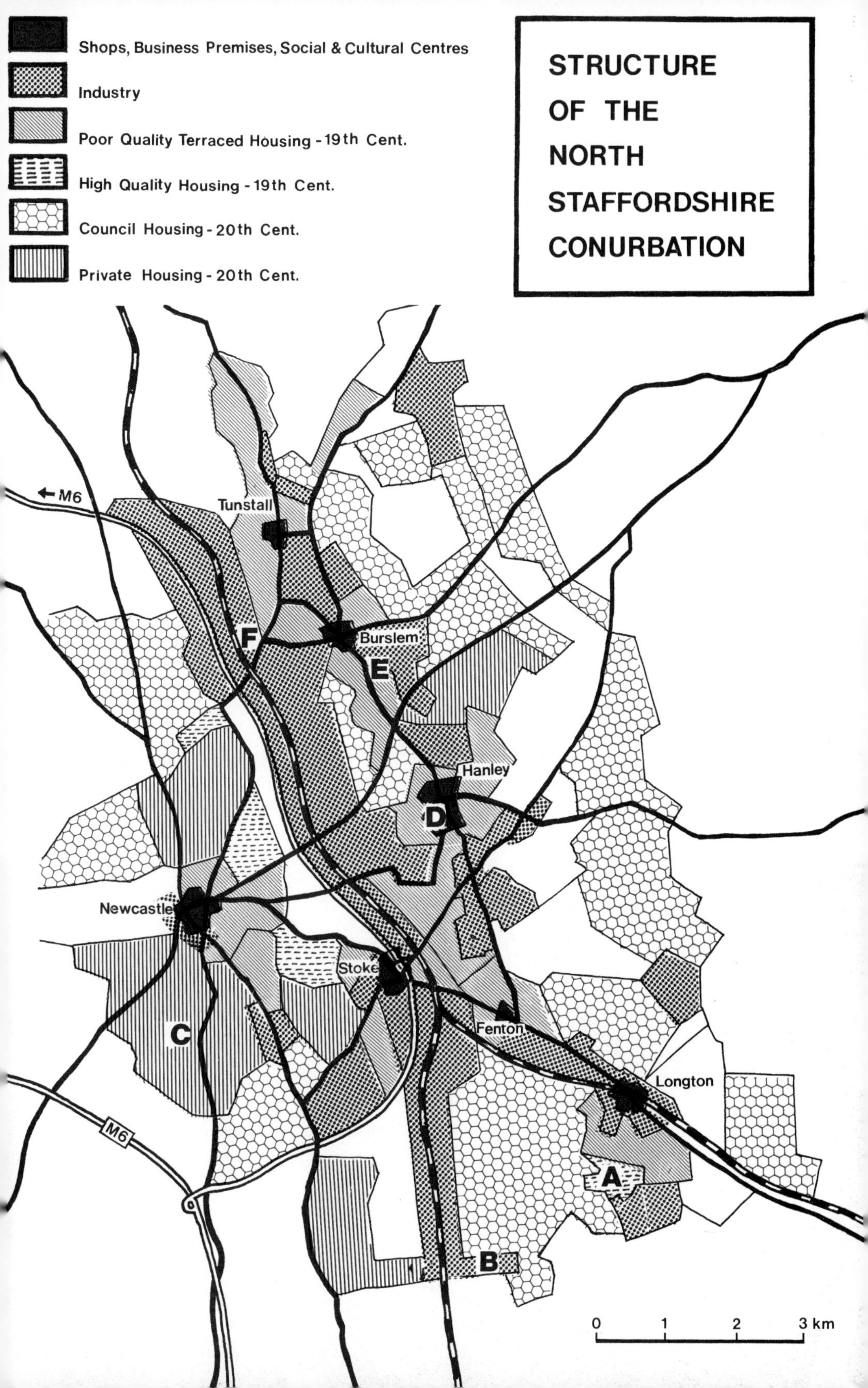
Shops, Business Premises, Social & Cultural Centres
Industry
Poor Quality Terraced Housing - 19th Cent.
High Quality Housing - 19th Cent.
Council Housing - 20th Cent.
Private Housing - 20th Cent.
STRUCTURE OF THE NORTH STAFFORDSHIRE CONURBATION
M6
Tunstall
F
Burslem
E
Hanley
D
Newcastle
Stoke
Fenton
C
Longton
A
B
M6
0
1
2
3 km

the introduction of new industries such as light engineering, tyre making, baking etc. These industries tend to be located on industrial estates on the outskirts of the city, and in particular along the main roads.

The Structure of the Conurbation

This pattern of industrial growth is reflected in the structure of the conurbation itself and this, in turn, illustrates many aspects of the growth of cities in general.

a) Refer to the model of a typical British city shown on page 220.
   i) Trace the diagram.
   ii) Place your tracing on the map of the North Staffordshire conurbation (page 233), centring it on the CBD of Hanley.
   iii) Using the following list of zones, write down those which show a good degree of fit and, in each case, say where the zone is located in the conurbation, eg in the centre, south west etc.
   Good quality twentieth century housing
   Central Business District
   Low quality nineteenth century housing
   Low cost twentieth century housing
   Old established industrial areas
   Good quality nineteenth century housing
   Modern industrial areas
   iv) Which zones do not conform to the model? Why might this variation occur?

b) Using information given in the chapter complete the following exercise:
   i) Why are the main private residential areas located in the western parts of the conurbation?
   ii) In which area of the conurbation is housing and industry likely to be mixed together?
   iii) Why have industrial and residential zones tended to become separated?
   iv) Account for the development of industries on the outskirts of the conurbation.

c) i) What evidence is there on the map that the conurbation was formed by the joining together of several separate towns?
   ii) What evidence is there that these towns have retained a separate identity?
   iii) How has this development distorted the pattern predicted by the model?

d) The photographs were taken at locations 1–6 on the map. Identify the most likely location for each photograph, state which type of area it represents and give reasons for your choice.

Areas in Stok

It is quite clear from this study that the settlements which make up the North Staffordshire conurbation have not yet been fully absorbed by it. Important shopping and business areas still exist outside the main CBD and, around each of these centres, traces of the older pattern of zones still exists. Eventually this may disappear and the structure of the conurbation will more closely resemble that predicted by the model. As it is the similarities are striking, including as they do the CBD; the zones of inner city decay; the old industrial areas near the city centre, the residential districts on the urban fringe—private housing in the west, council housing in the east where there is pollution blown by the prevailing winds, and modern industrial developments on estates established along the main roads.

Problems

The Two Nations

One of the main problems to emerge in Britain during the twentieth century has been the inbalance in prosperity between individual regions. This has been the result of the decline of the traditional industrial structure, based largely on the coalfields, and its replacement by a new structure, more closely related to accessibility and the transport network. Successive governments have attempted to slow down, or even reverse this process but they have failed and Britain has witnessed the emergence of what has been called 'The Two Nations'—one characterised by prosperity, rapid population growth and a strong industrial base; the other, which includes most of the coalfields, marked by declining industries, decaying towns and cities, and high rates of emigration.

The extent of the problem emerges clearly from a study of two contrasting regions.

Regional Study: South East England, a region of growth

The growing importance of the South East region in terms of industry and population can be seen from the following figures.

DATE	% OF LAND AREA OF BRITAIN	% OF POPULATION OF BRITAIN	% OF INDUSTRY OF BRITAIN
1901	11	27·5	18·0
1921	11	—	18·7
1931	11	29·4	23·4
1951	11	30·3	24·7
1961	11	31·0	26·9
1971	11	31·0	27·5
1975	11	30·0	29·0

Only in the last decade has the growth of population slowed—largely as a result of government policies—and even this slowing down has not been reflected in the industrial development of the region.

Reasons for this concentration in one area of Britain are not difficult to find but the overall picture is very complicated. At the centre of everything, however, is the city of London and its influence has become so widespread that the South East has, with some justification, been called the *London City Region*.

London

London is the largest city in Britain and its history stretches back to pre-Roman times.

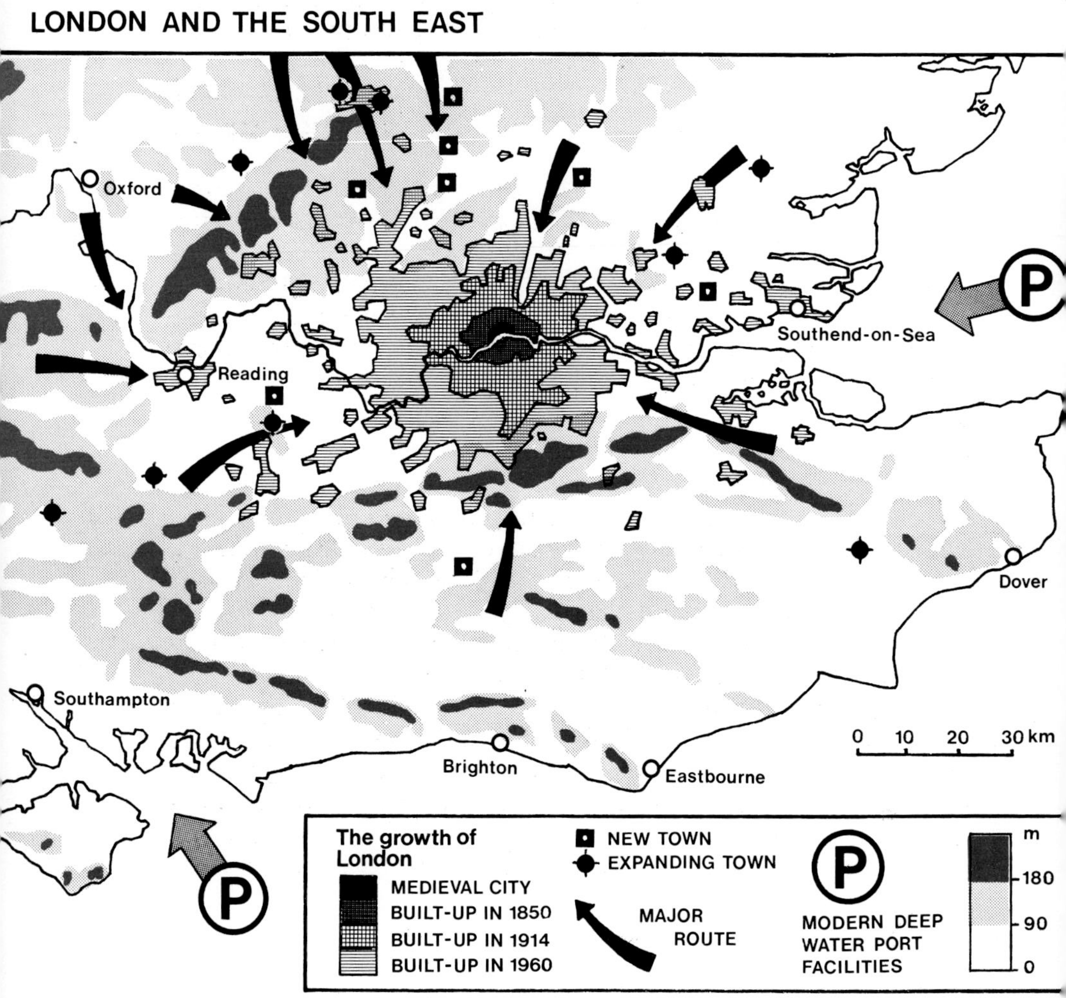

Origins

Any site which has been continuously occupied for such a long period of time must have certain natural advantages and these are reflected in the early development of the city.

1. The first settlement grew up on a dry gravel terrace overlooking the marshlands alongside the River Thames.

2. Protected by the marshes and by the streams flowing into the river, the site was easily defended.

3. London's early importance owed much to the fact that it was situated at the lowest crossing point of the Thames. As a result:

a) In Roman times London became Britain's leading port and links were established with the rest of the Roman Empire.

b) Roads converged on the site to make use of the river crossing and London became the major route centre of Roman Britain, controlling routes to all parts of the country. This radial route network has never broken down and it can be easily traced today in the roads, railways and even airways of modern Britain.

4. Because of its importance, London assumed political and administrative functions and, after the Norman conquest it emerged as the capital of England.

5. The population attracted to the centre provided both a labour force and a market for goods and, at an early date, London became an industrial centre of considerable importance.

It was the strength of these foundations which enabled London to adapt to changing world conditions and many of the early functions of the city have persisted down to the present day, and have brought prosperity to the city and the surrounding region.

The Modern City

a) *The Port of London*

London grew up on a tidal river and this caused many problems, for, although ships could enter or leave the port throughout the day, loading and unloading could take place only at high tide when small boats or lighters could carry cargoes to the shore. For the rest of the time ships had to remain anchored in the channel or were beached on the mudflats which flanked the river.

With the increase in the volume of trade handled by the port and in the size of ships, this method proved unsatisfactory and, during the nineteenth century, a complex system of docks was dug in the lowlands on either side of the river. Each of these docks could be isolated from the river when the tide began to ebb, thus allowing the handling of cargoes to continue throughout the day. For more than

London docks before the main closures took place.

a century this system met the needs of the city until in the period since 1950, technological changes have destroyed its prosperity and produced a striking change in the pattern of port activities in London.

Of the many technological advances which have taken place, the most important have been:

1. The increase in the size of vessels to over 100,000 tonnes.

2. The development of special cargo handling facilities such as container berths and roll on/roll off ferries.

Few of the old docks could be modified to cope with such developments and, as a result, they were closed down, and a new system was built lower down the river where there was room for development and the channel was deep enough to take large ships.

Using information given on pages 242 and 243, complete the following exercise:

a) i) Name two of the nineteenth century docks which have been closed.
   ii) Name one dock which remains open.
   iii) Describe the position of these docks in relation to the original port area.

b) i) Where are the modern port facilities located on the Thames?
   ii) Why have they been built there?

iii) Name the container port in this area.
iv) How does a container port differ from a normal port? What are its advantages? (See page 10).

c) i) Name two oil installations at the mouth of the Thames.
ii) Why were they built there?

d) Write a brief account of the trade of the port of London. (See page 8).

As a result of such changes London has maintained its position as Britain's leading port, handling more than one quarter of the nation's exports and one fifth of its imports. In spite of this, however, London has continued to lose ground to its main rival, the Rotterdam–Europort complex at the mouth of the Rhine which has better facilities and better labour relations.

b) *London as a Route Centre*

One factor which has contributed to the continued pre-eminence of London among British ports is the improvement which has taken place in communication systems serving the city. As can be seen from page 238, most of the main roads and railways still radiate from London and, as a result, links with the rest of the country are good. This in turn means that the hinterland of the port is larger than that of any other port in Britain and that the service area of the city is national as well as regional. The radial communications network also means that London suffers more than any other British city from the problems of congestion, both in terms of traffic and of building.

c) *London as an Industrial Centre*

The importance of London as an industrial centre depends largely upon its situation at the centre of a vast national and international network of communications. Roads, railways, searoutes and air routes converge on the city, allowing materials to be imported and the finished products of industry to be exported. It is not surprising, therefore, that the pattern of industrial development in London reflects the changing pattern of communications.

1. The oldest industrial areas are near the city centre. Here industries such as jewellery, clothing and furniture making have tended to cluster in small distinct areas, making use of specialist labour, cheap rented premises (by Central London standards) and the large local market.

2. The building of the docks near to the city centre encouraged the development of a vast range of industries which processed imported raw materials, eg food processing.

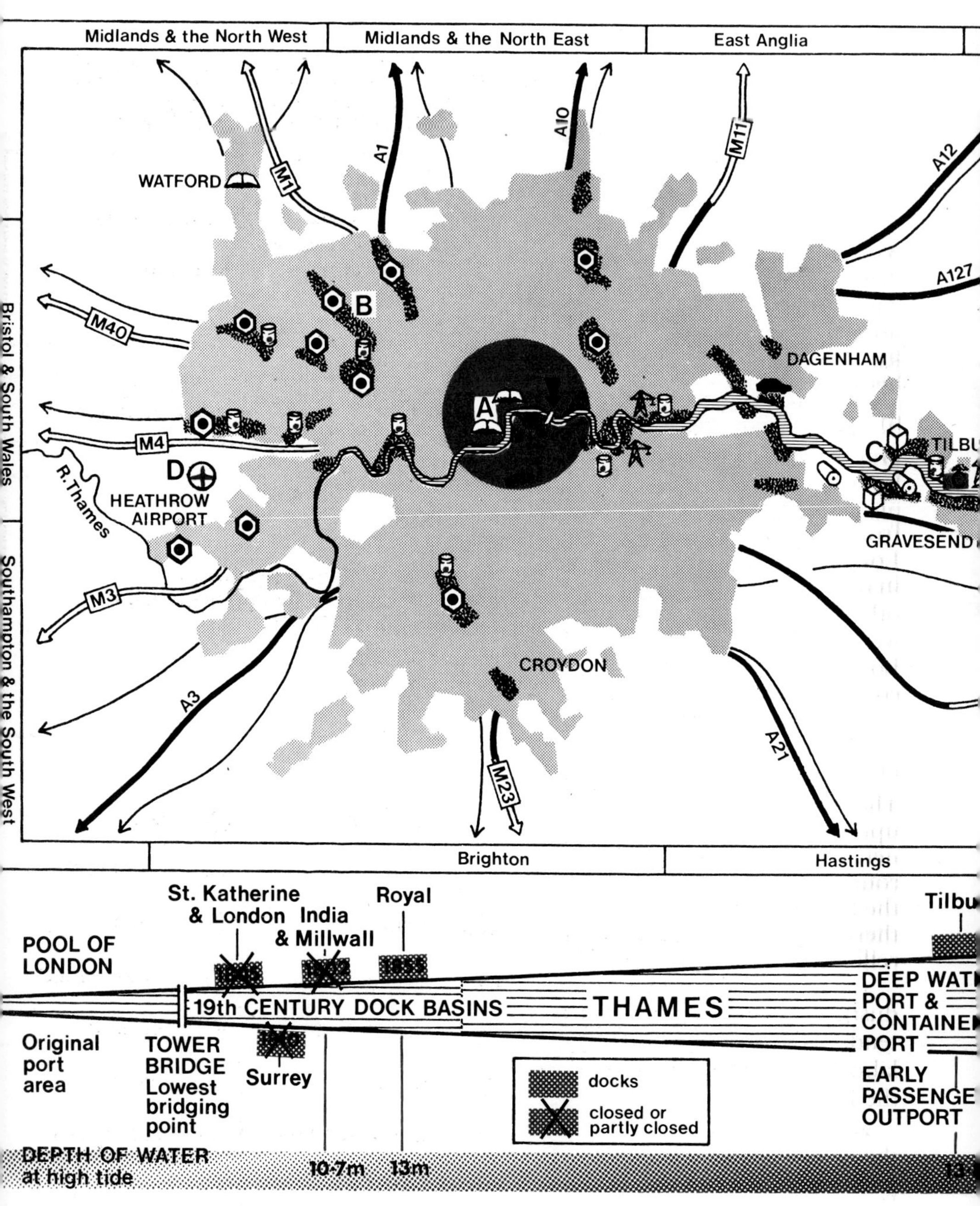

Midlands & the North West
Midlands & the North East
East Anglia
Bristol & South Wales
Southampton & the South West
Brighton
Hastings
WATFORD
M1
A1
A10
M11
A12
A127
M40
M4
M3
A3
M23
A21
B
A
C
D
DAGENHAM
TILBU
GRAVESEND
R. Thames
HEATHROW AIRPORT
CROYDON
St. Katherine & London
India & Millwall
Royal
Tilbu
POOL OF LONDON
19th CENTURY DOCK BASINS
THAMES
DEEP WAT
PORT &
CONTAINE
PORT
Original port area
TOWER BRIDGE Lowest bridging point
Surrey
docks
closed or partly closed
EARLY PASSENGE OUTPORT
DEPTH OF WATER at high tide
10·7m
13m

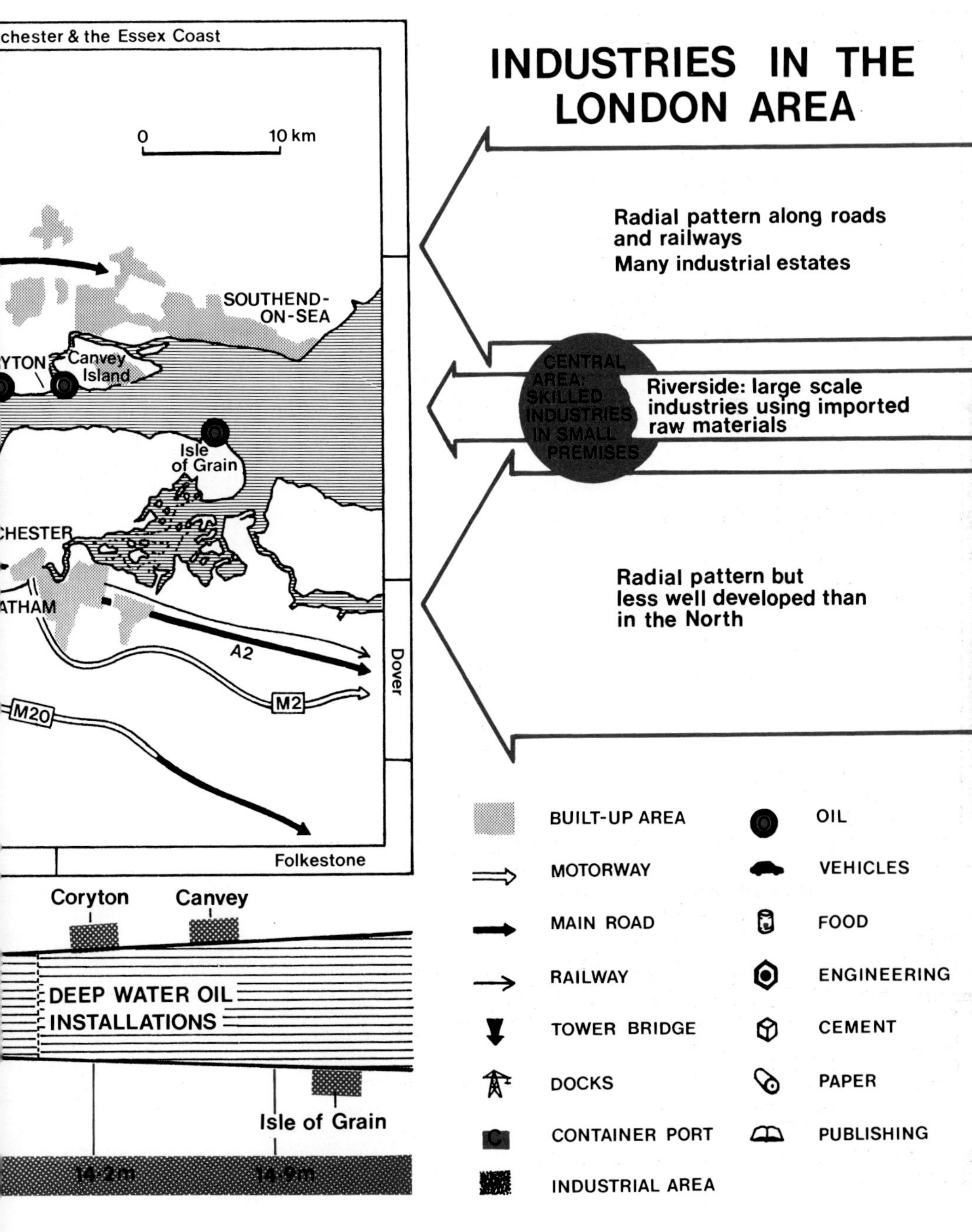
chester & the Essex Coast
0
10 km
SOUTHEND-ON-SEA
Canvey Island
Isle of Grain
A2
M2
M20
Dover
Folkestone
Coryton
Canvey
DEEP WATER OIL INSTALLATIONS
Isle of Grain
14·2m
14·9m
INDUSTRIES IN THE LONDON AREA
Radial pattern along roads and railways
Many industrial estates
CENTRAL AREA: SKILLED INDUSTRIES IN SMALL PREMISES
Riverside: large scale industries using imported raw materials
Radial pattern but less well developed than in the North
BUILT-UP AREA
MOTORWAY
MAIN ROAD
RAILWAY
TOWER BRIDGE
DOCKS
CONTAINER PORT
INDUSTRIAL AREA
OIL
VEHICLES
FOOD
ENGINEERING
CEMENT
PAPER
PUBLISHING

Both of these industrial areas have suffered from the closure of the docks and the redevelopment of the city, which has led to an enormous increase in rents.

3. The lower reaches of the river and the shores of the estuary tended to attract other types of industry.

a) Initially, the drained marshlands on the river bank were used as a site for dangerous or unpleasant industries such as munitions and chemicals. Some of these still remain.

b) Later, the extension of the dock system towards the sea led to the development of typical port industries, particularly those which require large areas of land and which, in processing the raw materials, considerably reduce their bulk, eg steel making (at Dagenham), oil refining, flour milling, paper making etc.

4. The spread of industry away from the river followed the building of the railways in the nineteenth century and the improvement of the road network during the twentieth. This development was further stimulated by the establishment, after the First World War, of large industrial estates near to the main roads and railways. These attracted a wide range of light industries, such as electrical engineering, food processing and distribution, electronics etc; all drawn by the efficiency of the communications network, the availability of labour and the vast size of the local market. Furthermore, once established, these industries tended to attract other related industries, thus adding to the importance of London as an industrial centre. This can be most clearly seen in the case of the motor industry at Dagenham. (See page 149).

Using information given on pages 242 and 243, complete the following exercise:

a) Assuming that the main factors influencing the choice of a site for a factory are:
   availability of labour
   availability of materials
   markets
   communications
   space
   but that they vary in relative importance according to the type of industry, choose one of the sites marked A–D on the map for each of the following industries and give reasons for your choice.
   Packing stationery
   Making glue
   Making watches
   Petro-chemicals

b) List the factors which have influenced the development of industries at each of the sites marked A–D.

d) *London as a Service Centre*

In addition to being a major centre of manufacturing industry, London is also the most important service centre in Britain and service industries provide employment for twice as many people as manufacturing industries. This is to some extent predictable in view of the enormous concentration of population in the city and the resultant need for services, such as shops, public transport etc. But, in addition, there are service industries which, if not peculiar to London, occur there on a scale found nowhere else in Britain. For example, more than half of the national labour force employed in banking and insurance, the civil service and scientific research, work in London. Add to these the work force catering for the tourist trade and those employed to provide services for industry, and London emerges as by far the largest service centre in Britain. And, within London, the *City of London*, occupying an area of less than three square kilometres and with a permanent population of less than twenty thousand, dominates these activities and forms a Central Business District of regional and national importance.

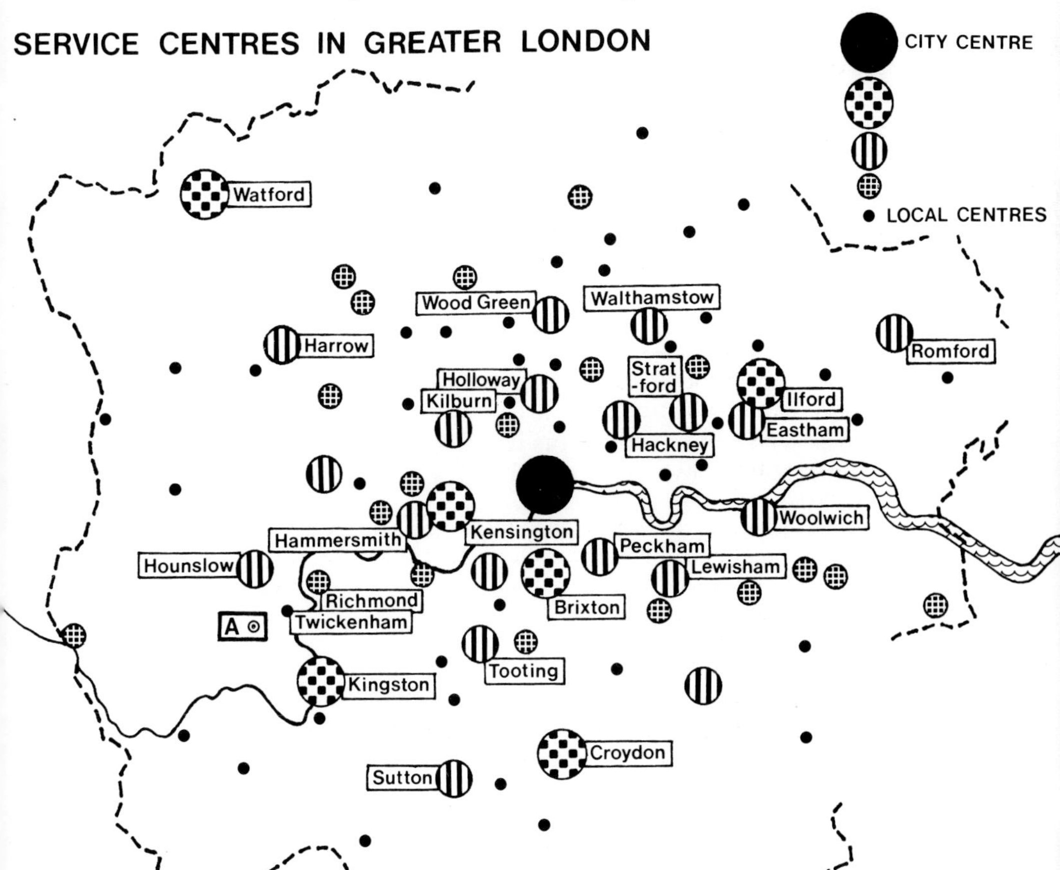

Furthermore, the enormous expansion of the city which followed the building of the railways and which has continued during the twentieth century, with the improvement of the road network, has meant that many existing towns and villages have been engulfed in the built-up area of the city. Many of these centres retained service functions and London, like most cities, does in fact consist of a large number of business and shopping districts which make up a heirarchy of service centres.

a) Refer to page 245 and complete the following exercise:
   i) Assuming that you live at location A on the map, which centres are you likely to use for the following goods and services:
      groceries
      a bank
      a department store
      furniture
      visiting the theatre
      a doctor
   ii) Which factors could distort your pattern of choice? Think of the factors which influence your choice of centre for each of the services in your local area.
   iii) All of these services can be obtained in the city centre. Why not go there for them?

b) Refer to the map on page 238:
   i) How far does the built up area of London now extend from north to south?
   ii) How far does it extend from east to west?
   iii) What effect has the development of communications had on the spread of the built-up area?

The movement of workers tends to reflect these patterns. Large numbers of commuters still travel daily into and out of the city centre, some of them from as far afield as the Midlands and the south coast. In recent years, however, commuting to the smaller business districts and to industrial areas away from the city centre has become more important. This is the result of the movement of both manufacturing and service industries out of the central area—a movement brought about by the attempts of successive governments to limit the growth of London, and by the problems created by the growth of the city itself.

e) *The Problems of London*

The problems of London are those found in every major city in Britain; the only difference being that here they occur on a much larger scale. They include:

1. Lack of space.

2. The price of land which is much higher than average for other cities.

3. A serious shortage of housing, made worse by the large scale clearance of inner areas for road widening and redevelopment.

4. The difficulty of travel within the city. This has been a problem since the nineteenth century but the increasing use of motor transport has made the situation much more serious. As a result, congestion has increased enormously and public transport has lost passengers and revenue, with a resulting decline in services.

5. Until recently there has been a shortage of labour—in spite of a high rate of immigration from other parts of Britain.

In an attempt to reduce these problems and to redistribute some of the wealth and prosperity of the London area among the other regions of Britain, governments, particularly since 1945, have placed restrictions on all new developments within the city and have offered incentives to firms to move out of the area. At first sight this policy of decentralisation appears to have been successful because, since 1950, there has been a marked decline in both population and employment in manufacturing industry in the Greater London area. This indicates a definite movement of both people and jobs out of the city, but, unfortunately much of this movement has been to the city margins or to other parts of South East England, where growth has been rapid.

The effects of the growth of London and the spread of its influence have been felt throughout the South East and they can be traced in the patterns of development in any individual area. The Darent Valley is typical.

The Darent Valley

The Darent rises to the south of the North Downs and flows northwards towards the Thames. The region through which it flows is varied both in terms of its physical features and land use. In recent years, however, changes have taken place which reflect the growing influence of London rather than the variety of the region itself.

Using information given on pages 248 and 249, complete the following exercise:

a) Complete the following passage by filling in the missing words:
The North Downs is an area of .... rock which forms .... land with a .... facing escarpment. On the upper areas of the north facing .... slope, the chalk is exposed and there is little .... drainage. This area is dissected by .... valleys which are winding

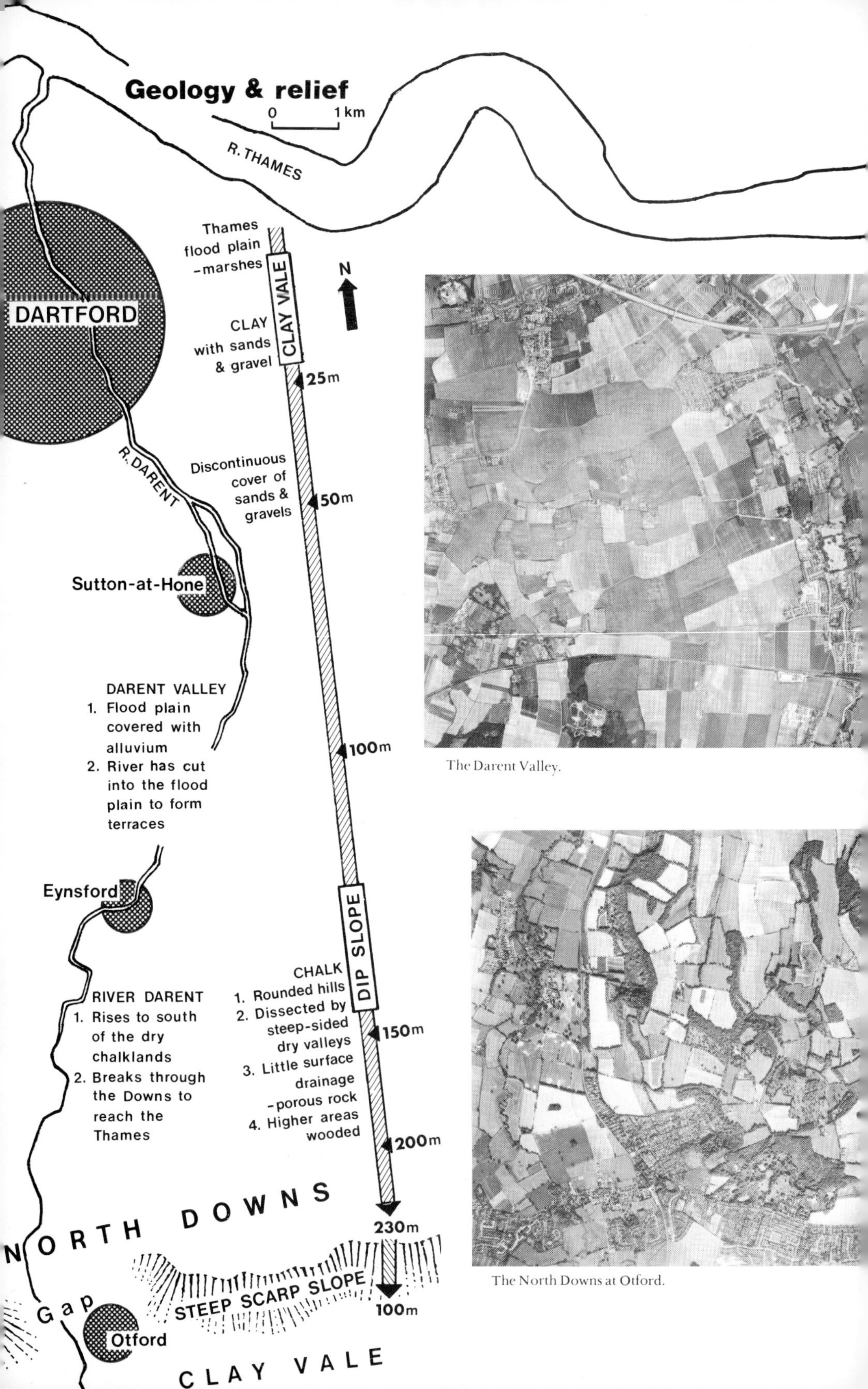

The Darent Valley.

The North Downs at Otford.

and ..... Towards the north the land becomes .... and the chalk is sometimes covered by .... and ..... Eventually the chalk disappears under the London Clays of the Thames valley and here the soils are heavy and....occur.

The Darent rises on the impermeable rocks to the south of the Downs. It has cut a gap through the escarpment at ..... and has formed a wide valley to the north. For most of its length the valley floor is covered with.... In places the river has cut into these deposits to form....

b) i) List the main types of farming in 1940, arranging them in order of importance, according to area.
   ii) Point out one way in which physical conditions influenced farming then.

c) i) List the main types of farming in 1970, arranging them in order of importance.
   ii) Describe how farming in the area has changed since 1940.
   iii) How has the nearness of the area to London influenced this development?

d) Sutton at Hone is a settlement in the Darent Valley.
   i) List the main occupations in 1940 and 1970.
   ii) Describe the main changes which have taken place.
   iii) List the places of work for 1940 and 1970.
   iv) Describe the changes which have taken place.
   v) Explain how the spread of London may have influenced these changes.

Since the Darent Valley is less than thirty kilometres away from central London, it is not surprising that the influence of the city is strong, or that the area has shared in the rapid growth which has taken place in south east England. Similar changes can be seen, however, in more distant areas; areas as far afield as the south coast, the Hampshire Basin and East Anglia. Here the most striking developments have been:

1. A rapid increase in population, by as much as three per cent a year in some places. This has been accompanied by an increase in commuting to London which has become particularly important along the south coast.

2. An equally rapid expansion of industry with growth industries, such as electrical engineering, electronics and paper and printing particularly important. Once again growth has been more rapid than in any other part of Britain, and in recent years has been more rapid than in London itself.

3. Changes in agriculture have been less striking but there has been a move towards a more intensive use of the land.

RECENT CHANGES IN PART OF KENT

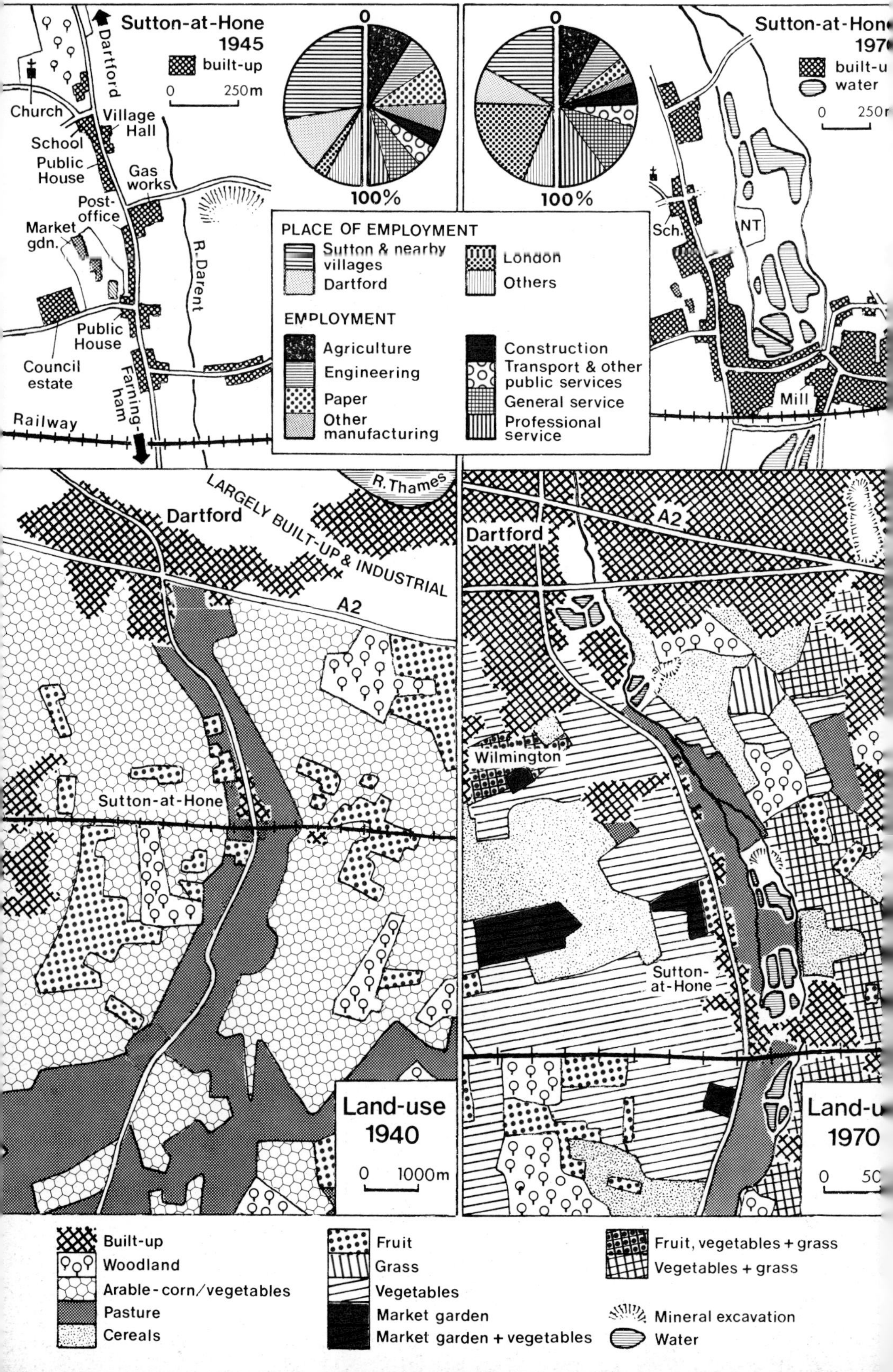

Sutton-at-Hone 1945

built-up

0 250m

Dartford

Church

Village Hall

School

Public House

Gas works

Post-office

Market gdn.

R. Darent

Public House

Council estate

Farningham

Railway

0

100%

0

100%

Sutton-at-Hone 1970

built-up

water

0 250m

Sch.

NT

Mill

PLACE OF EMPLOYMENT

Sutton & nearby villages

Dartford

London

Others

EMPLOYMENT

Agriculture

Engineering

Paper

Other manufacturing

Construction

Transport & other public services

General service

Professional service

R. Thames

Dartford

LARGELY BUILT-UP & INDUSTRIAL

A2

Sutton-at-Hone

Land-use 1940

0 1000m

Dartford

A2

Wilmington

Sutton-at-Hone

Land-use 1970

0 50

Built-up

Woodland

Arable-corn/vegetables

Pasture

Cereals

Fruit

Grass

Vegetables

Market garden

Market garden + vegetables

Fruit, vegetables + grass

Vegetables + grass

Mineral excavation

Water

The Hampshire Basin

The Hampshire Basin clearly illustrates these developments. Here the influence of London was felt at an early date and, from the opening of the railway in 1836, *Southampton* began to develop as an outport for London. Situated on the sheltered Southampton Water, the port grew up at the confluence of the Test and Itchen rivers where the deep water channel approached the east bank. The waterfront is long, giving ample room for docks, and, even more important, the Isle of Wight obstructs the tidal currents offshore, causing double high tides and a low tidal range, which enables the docks to be used for long periods each day. During the nineteenth and early twentieth centuries, Southampton's importance rested mainly on its development as a liner port. It handled passengers, many of whom were travelling to London and found it quicker to leave the ship there and to continue their journey by rail. The general trade of the port was also highly specialised, with perishable goods such as fruit and vegetables—often destined for the London market—particularly important. This pattern still exists but it has been broken down by

The waterfront at Southampton with the mouth of the river Itchen in the background.

the development of air transport and by the emergence of Fawley as a major oil terminal. As a result, oil imports now dominate the trade of the port and the development of the refinery at Fawley has led to the growth of new industries such as chemicals and synthetic rubber.

Outside Southampton the pattern has been more typical of South East England as a whole, with industries such as light engineering, electronics and printing being attracted from the London area to locations in relatively small country towns. This development has been matched by the movement of office workers, including branches of the civil service to the same area.

The same is true of agriculture, as can be seen from the table below. The main changes have been:

CHANGES IN FARMING IN HAMPSHIRE: 1925–1975

LAND-USE	AREA (1925 AREA = 100)
Rough grazing	84
Permanent pasture	53
Arable land	125
Wheat	159
Barley	600
Oats	16
Potatoes	115
Vegetables	200
Fruit	33
Clover and sown grass	153
STOCK	**NUMBER (1925 = 100)**
Cattle	197
Sheep	45
Pigs	330

1. An increase in the area of arable land at the expense of rough grazing and permanent pasture. This was largely the result of the ploughing of the chalk uplands during the Second World War.

2. Closely related to this has been the decline in the number of sheep which were reared on these upland pastures.

3. The increase in the number of cattle—often dairy cattle—reflects a more intensive use of pasture on the clay lowlands and the more widespread use of sown pasture.

4. The increase in pigs and poultry which has taken place since the introduction of factory farming methods.

5. Among the crops the enormous increase in the area of barley and the decline in the area of oats reflects trends seen in British farming generally.

6. This, together with the increased importance of wheat, vegetables and potatoes, is an indication of the development of more intensive farming methods in the area.

Some of the produce, particularly vegetables and milk is destined for the London market but the influence of the city is not easy to separate from the general trends in British farming.

Regional Study: North East England, a region in decline

For centuries the North East depended for its prosperity on the Northumberland and Durham coalfield and the industries associated with it. During the twentieth century this coalfield—in common with most British coalfields—has faced serious economic problems, many of which have yet to be overcome. Basic industries, such as coal mining, steel making and ship building have contracted, and new industries have not emerged to replace them. As a result, the industrial structure of the area is weak, unemployment is high and population growth very slow—a far cry from the situation in the London area. The extent of the problem can be seen from a brief examination of the basic industries of the region.

Coal Mining

One hundred years ago Northumberland and Durham was the leading coalmining district in Britain, with a production of more than forty million tonnes a year. Today only fifty pits remain and output is less than half this figure.

Many factors have contributed to this early prosperity and equally early decline.

FACTORS ENCOURAGING GROWTH

1. A large coalfield with a large area of exposed coal measures.
2. Seams outcropped on the sides of the river valleys which were cut into the coalfield.
3. Position near the sea encouraged

FACTORS CAUSING DECLINE

1. Because of the early rise of mining, most of the accessible seams have been worked out.
2. Loss of overseas markets. This is reflected in the decline of the coal trade from the Tyne ports which shipped 20 million tonnes

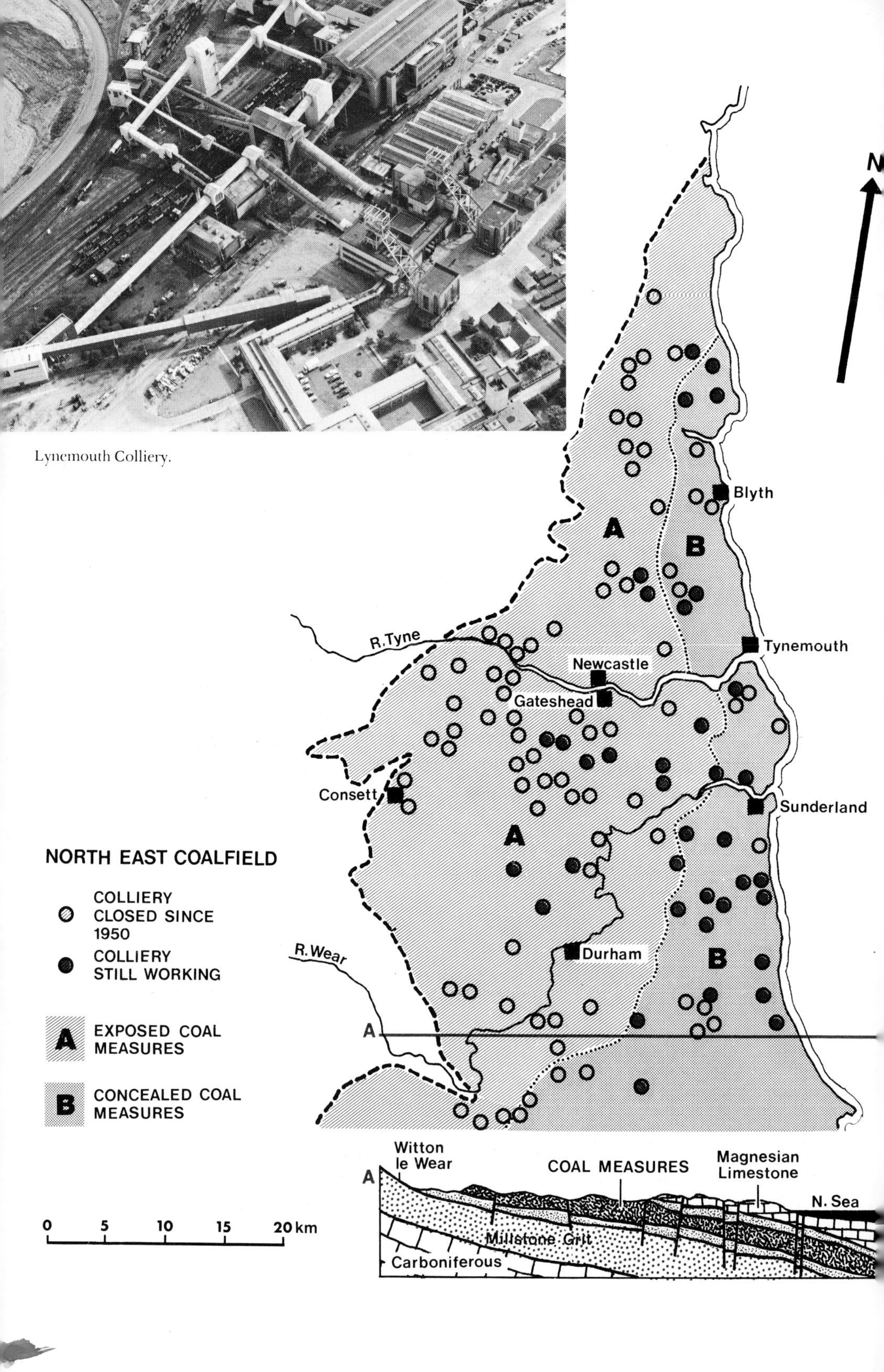

Lynemouth Colliery.

the early rise of the coal trade. For example, in the thirteenth century ships from the North East were supplying London with coal.

4. Much of the coal was of very high quality, particularly the coking coal.

of coal in 1920 and which now handle none.

3. Competition from other fuels, particularly oil and gas.
4. Lack of a growing local market brought about by the weakness of the industries in the North East.

The effects of the contraction of the coal mining industry can be seen throughout the coalfield but they are particularly apparent in certain areas.

Using information given on page 254, complete the following exercise:

a) i) Which part of the coalfield has suffered most from the contraction of the coal mining industry?
   ii) Why has this area suffered more? (The factors given above will help in answering this).

b) i) Where are most of the working pits now located?
   ii) Explain why this pattern has developed.
   iii) Give one geological factor which hampers further development of the coalfield.

As with most British coalfields, the decline in the labour force employed in the pits has been more rapid than the decline in coal production. In the North East, therefore, people have been encouraged to leave the old mining areas, allowing some villages to become derelict before they are eventually cleared. Attempts have been made to provide work in newer pits on the concealed coalfield or in new industries within the region, but this has often proved impossible and many mining families have moved to other coalfield areas where prospects are better.

Iron and Steel

The iron industry also grew up at an early date in the North East because all of the basic raw materials were available locally—iron stones in the coal measures, limestone and charcoal. During the eighteenth century the industry began to make use of the high quality coking coals available in the area and, as a result, the coalfield became the obvious location for the ironworks. This pattern persisted until the 1850's when the discovery of rich iron ore deposits in the limestones of the Cleveland Hills, to the south of the Tees, led to the rapid expansion of ironmaking in the Middlesbrough area. Even then many coalfield works remained in production, making use of these ores, just as Middlesbrough made use of the coking coals of the coalfield.

Hamsterley, a mining village on the west Durham coalfield.

The collapse of the coalfield industry came with the gradual exhaustion of the Cleveland ores, towards the end of the nineteenth century. It then became necessary to import large quantities of ore and this favoured sites near the coast, such as those on Teesside. Today, in fact, only one steelworks remains on the coalfield—at Consett—and its future is far from certain.

Using information given on pages 254 and 261, complete the following exercise:

a) i) Why did Consett develop as a steel making centre?
   ii) From where does it obtain its raw materials today?
   iii) Consett is a perfect example of *industrial inertia*. What does this mean (see page 81) and why is Consett such a good example?

b) Referring to page 92 explain why Teeside is a good site for a modern steelworks.

Although Consett has been kept open, largely because the town is completely dependent upon the steelworks and closure would cause great hardship in the area, all modern development has been concentrated on Teesside which has become one of the major steelmaking centres of Britain.

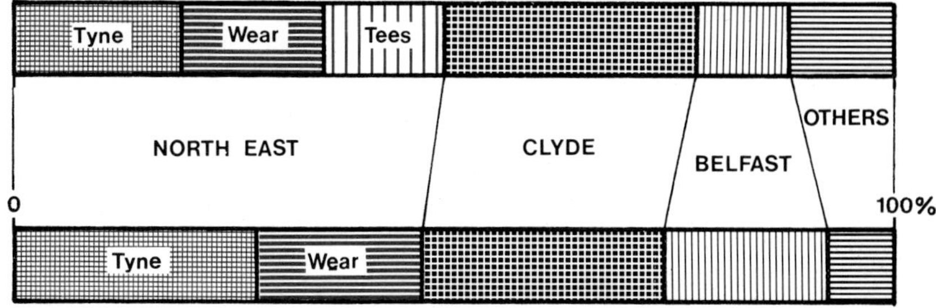

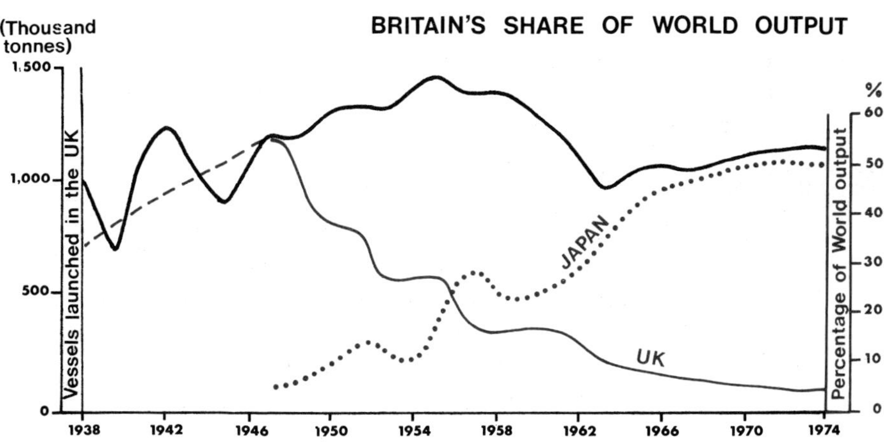

Shipbuilding

Among the major traditional industries of the North East, shipbuilding alone was a finishing industry, ie an industry which produced finished goods rather than materials for other industries. As such it required a large, highly skilled labour force and gave rise to a large number of subsidiary industries which supplied parts to the shipyards. Any decline in the shipbuilding industry, therefore, has important effects on the region as a whole and, since 1945 the industry has been in a steady decline.

Using information given above, complete the following exercise:

i) In 1946 British shipyards built 5/20/50/70% of the world's ships.

ii) Of these, almost 5/20/50/70% were built in the North East.

iii) Today British yards build 5/20/50/70% of the world's ships and of these almost 5/20/50/70% are built in the North East.
iv) The period of most rapid decline in the industry was during the 1940's/1950's/1960's/1970's.
v) The decline in the position of the British industry was the result of declining production/static production at a time when world production was increasing/slow growth at a time of rapid world growth.
vi) Which country replaced Britain as the main shipbuilding country?
vii) The North East's share of British output has declined/remained the same/increased since 1945.
viii) The tonnage of shipping launched in the North East has increased/remained the same/decreased since 1945.

The emergence of North Eastern England as one of the leading shipbuilding areas in the world, during the nineteenth century, was the result of a combination of favourable circumstances. For example:

1. There was an existing industry in the area building wooden ships for the fishing fleets and the coal trade.

2. When iron ships appeared the raw materials were available from the iron works of the coalfield and the Middlesbrough area.

3. The estuaries of the rivers which flowed through the coalfield provided water, wide enough and deep enough to cater for the size of ship built at that time.

As a result, the second half of the nineteenth century saw the building of many small shipyards, particularly along the Tyne, Wear and Tees, and the establishment of the many industries required to serve them.

During the twentieth century, however, conditions changed rapidly and British shipyards failed to keep pace with the change. For a long period there was little money to invest in the yards and much of the equipment in them became out of date. This left the companies in a poor condition to meet competition from foreign shipbuilders and, as a result, the British share of the world market began to decline. Furthermore, British yards were slow to change over to mass production methods, continuing to build individual ships long after the market for them had ceased to grow. And, finally, the rivers of the North East were too small to cope with the large vessels which began to appear after the Second World War.

These changes weakened the position of the North East and, in spite of the availability of raw materials in the area, its share of the world market declined at an ever increasing rate after 1950. By 1960

Shipyards at Walker on the river Tyne.

there was a danger that the industry would completely collapse, in face of competition from Japan and West European countries. It was at this time that the government decided to intervene to save the industry, both in the North East and in Britain as a whole. The following changes were made:

1. The remaining small yards were combined to form large groups—one on Tyneside and two on the Wear. This also happened on the Clyde.

2. Parts were standardised to cut costs and yards began to specialise in a limited range of vessels.

3. Grants were given to improve out of date yards and to build new berths capable of handling super-tankers.

Nothing could increase the size of the rivers, however, and launching large ships remains a problem.

Study the photograph above and complete the following exercise:

a) What evidence is there from the photograph that this is a nineteenth century industrial area?

b) i) What are the disadvantages of the Tyne as a site for a modern shipyard?
   ii) Why is the shipyard in the centre of the photograph well situated?
c) How would the super-tanker, moored outside the yard, have been launched into the river?

Modern techniques of launching ships sideways or assembling entire sections of the ship when they are afloat make it possible to build large vessels on the rivers of the North East. But it seems likely that many of the yards on the upper reaches of the rivers will be closed and replaced by yards built nearer to the river mouth.

Halting the Decline

The decline in these basic industries has resulted in high levels of unemployment in the North East and this, in turn, has led the government to designate the area a Development Region. As in most of the declining industrial areas in Britain, this has resulted in two major types of development:

1. Attempts have been made to improve the position of the traditional industries of the area, usually by making available large amounts of capital. As we have already seen, this has led to the development of Teesside as a major steel working area and to the reorganisation of the shipbuilding industry. But such measures can only slow down the decline, they cannot reverse it and more drastic solutions have been necessary.

2. New industries have been attracted to the area; usually to the new towns and industrial estates which have been developed since 1945. Much of this development has taken place around the Tyne and Wear, in an attempt to reverse the decline of these areas.

Refer to the map opposite and complete the following exercise:
a) i) Name four New Towns in the region.
   ii) In each case give its present and projected populations (see page 226) and state where this population is likely to have come from.
b) Describe and explain the distribution of New Towns and industrial estates of the North East.

This policy of attracting new industries—usually light industries such as electrical engineering, the manufacture of clothing etc, faces many problems, particularly:

1. The remoteness of the area from the main industrial markets in Britain. This adds to the transport costs of any industry in the area.

2. A declining coalfield area does not provide the best environment for new industries.

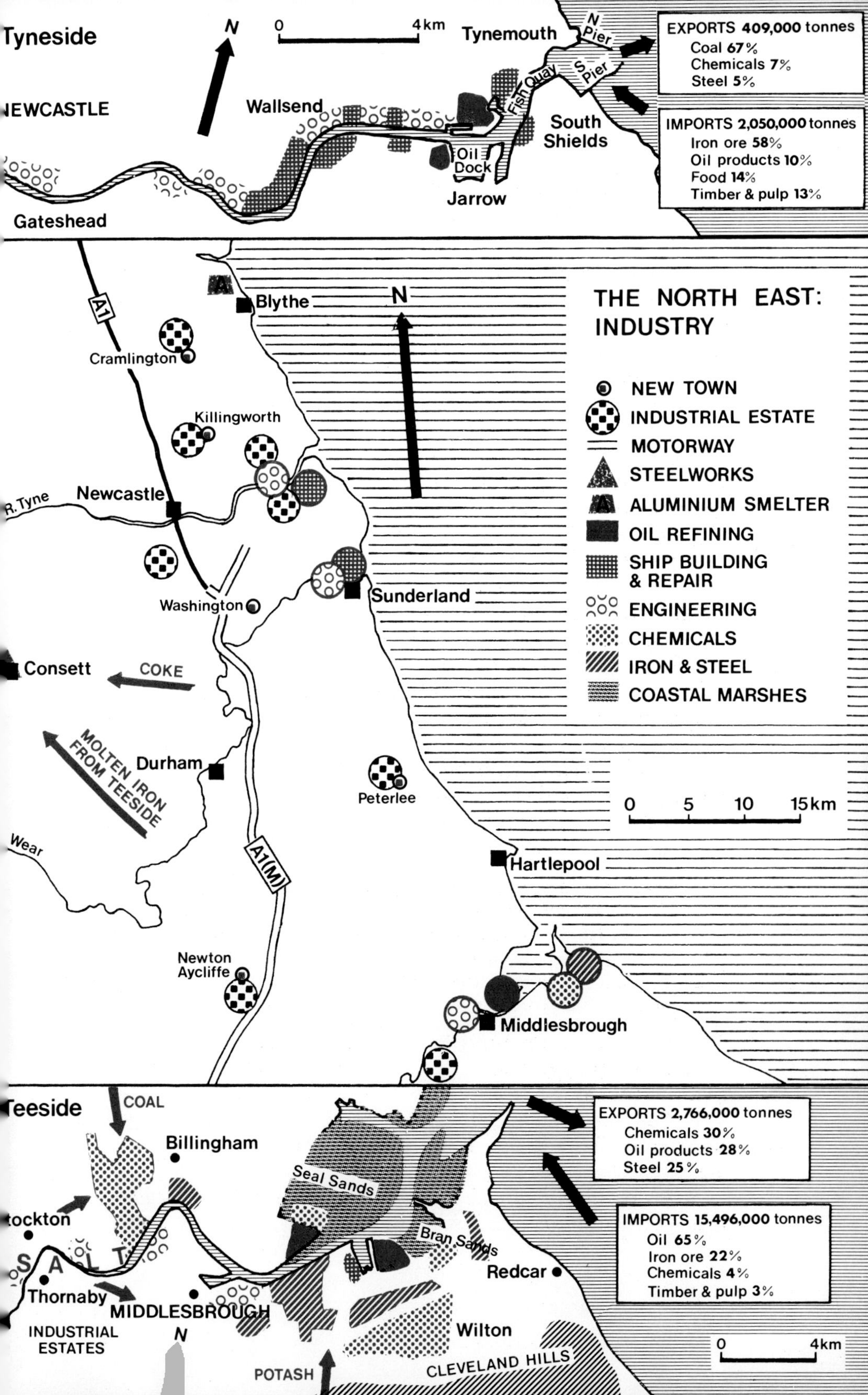

Tyneside
N
0
4km
Tynemouth
N Pier
S Pier
NEWCASTLE
Wallsend
Fish Quay
South Shields
Oil Dock
Jarrow
Gateshead
EXPORTS 409,000 tonnes
Coal 67%
Chemicals 7%
Steel 5%
IMPORTS 2,050,000 tonnes
Iron ore 58%
Oil products 10%
Food 14%
Timber & pulp 13%
A1
Blythe
N
Cramlington
Killingworth
R. Tyne
Newcastle
Washington
Sunderland
Consett
COKE
MOLTEN IRON FROM TEESIDE
Durham
Wear
A1(M)
Peterlee
Hartlepool
Newton Aycliffe
Middlesbrough
THE NORTH EAST: INDUSTRY
NEW TOWN
INDUSTRIAL ESTATE
MOTORWAY
STEELWORKS
ALUMINIUM SMELTER
OIL REFINING
SHIP BUILDING & REPAIR
ENGINEERING
CHEMICALS
IRON & STEEL
COASTAL MARSHES
0
5
10
15km
Teeside
COAL
Billingham
Seal Sands
Stockton
SALT
Thornaby
MIDDLESBROUGH
INDUSTRIAL ESTATES
N
Bran Sands
Redcar
Wilton
POTASH
CLEVELAND HILLS
EXPORTS 2,766,000 tonnes
Chemicals 30%
Oil products 28%
Steel 25%
IMPORTS 15,496,000 tonnes
Oil 65%
Iron ore 22%
Chemicals 4%
Timber & pulp 3%
0
4km

3. The decline of the area has been so rapid that new industries have never been attracted on a scale large enough to improve the situation.

4. Many of the new industries require very small labour forces and, therefore, do little to improve the unemployment situation.

Teesside

In one area, however, the policy has been successful, for *Teesside* has emerged as one of the industrial growth points of the north of England. Here the decline in the traditional industries has been less marked than in the rest of the North East.

Shipbuilding has virtually disappeared but the other early industry—*iron and steel* has actually expanded, benefiting from the continued availability of good quality local coking coal and from a coastal location, which allows iron ore to be imported cheaply, particularly from Sweden. There are also large areas of lowland on the shores of the estuary which provided ideal sites for the building of modern steelworks. As a result Teesside produces about five million tonnes of steel a year.

The petro-chemical industry at the mouth of the Tees, showing the area of drained marshland available for development.

Furthermore, the *chemical industry* which grew up during the nineteenth century, making use of local salt deposits, has also expanded rapidly. This started with the building of a coal based petro-chemical plant at Billingham and it has continued at an ever increasing rate since the end of the Second World War. A vast new complex has been built at Wilton and an oil terminal has been created at Teesport to handle the imported oil which has replaced coal in the petro-chemical industry. Once again the lowlands and marshes on the shores of the estuary have provided ideal sites for such developments. Further development seems inevitable as North Sea oil production increases and Teesside is already linked by pipeline to some of the major fields.

Unfortunately the benefits of such industries to the region are less than may first appear. Investment in them has been enormous but the labour force required is often small. As a result unemployment remains high and people continue to move out of the North East to the more prosperous areas of London and the Midlands. In short, the problem of the 'Two Nations' remains.

Water Supply

Britain has, with some justification, been regarded as an area of abundant rainfall, and it seems inconceivable that the country could ever run short of water. Simple statistical evidence appears to support this view for, while precipitation over Britain provides, on average, 73 million cubic metres of water per day, less than ten per cent is actually consumed (ie used and not returned to rivers or lakes). Unfortunately such a view is far too simple and the picture is complicated by many factors, particularly:

1. As can be seen on page 264, not all of the precipitation which falls on Britain is available for use. A considerable proportion of the water which percolates underground and collects in permeable rocks can be retrieved and ground water does in fact supply about one sixth of the public water supply (the piped water supply as opposed to the general supply which includes the vast quantities of water drawn by industry from the rivers for cooling purposes and for washing materials). The water lost by evaporation and transpiration from plants cannot be retrieved and this presents serious problems in some parts of Britain.

2. Rainfall is not evenly distributed across the country and, as a result the average figure of seventy-three million cubic metres per day is virtually meaningless. A glance at page 30 shows the extent of

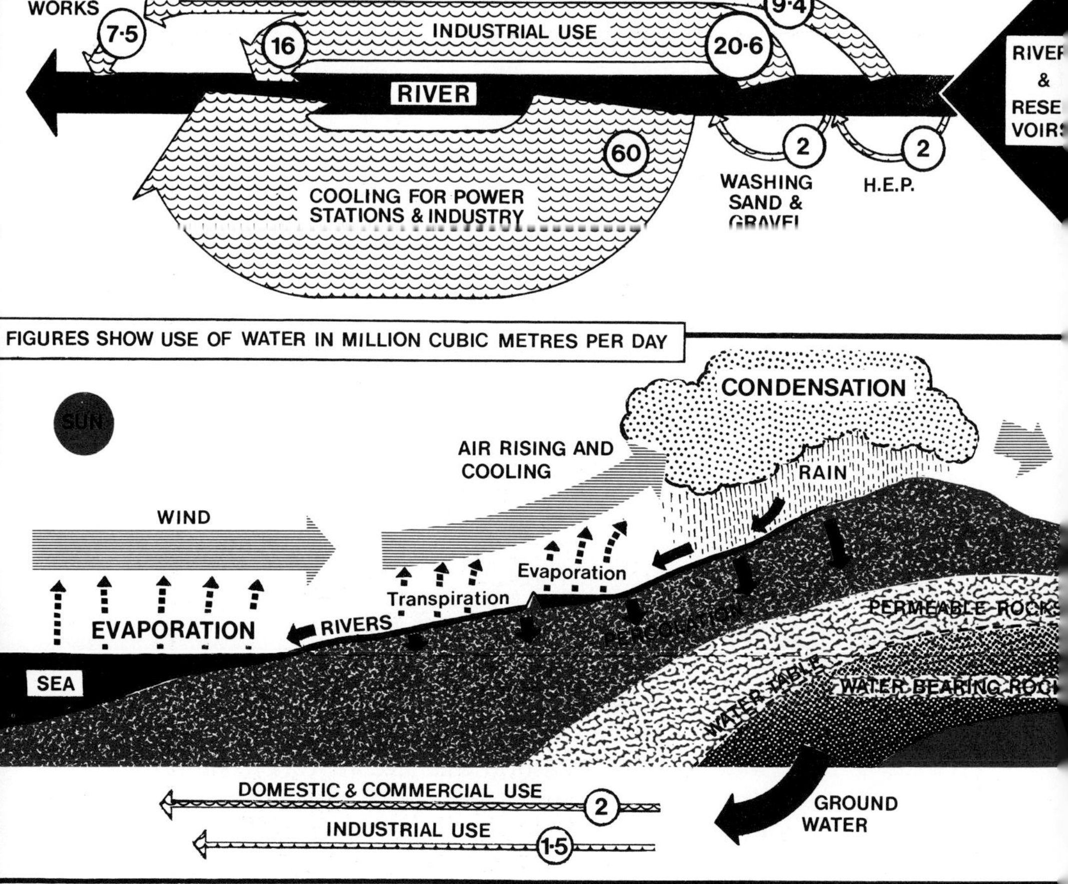

the problem. In the north and west, particularly over high land, there is heavy rainfall and a surplus available for the water supply system. In the south and east rainfall is much lower and it is here that problems have developed.

3. These problems are increased by the fact that evaporation and transpiration are also highest in this southern and eastern region and, as a result, already scarce water resources are reduced still further. (See page 265).

4. Finally, and perhaps most important, industry and population are not evenly distributed across the country. Both have tended to concentrate in the south and east of Britain; the area which has the lowest rainfall and the highest rates of evaporation.

It is these factors which have made necessary the complex pattern

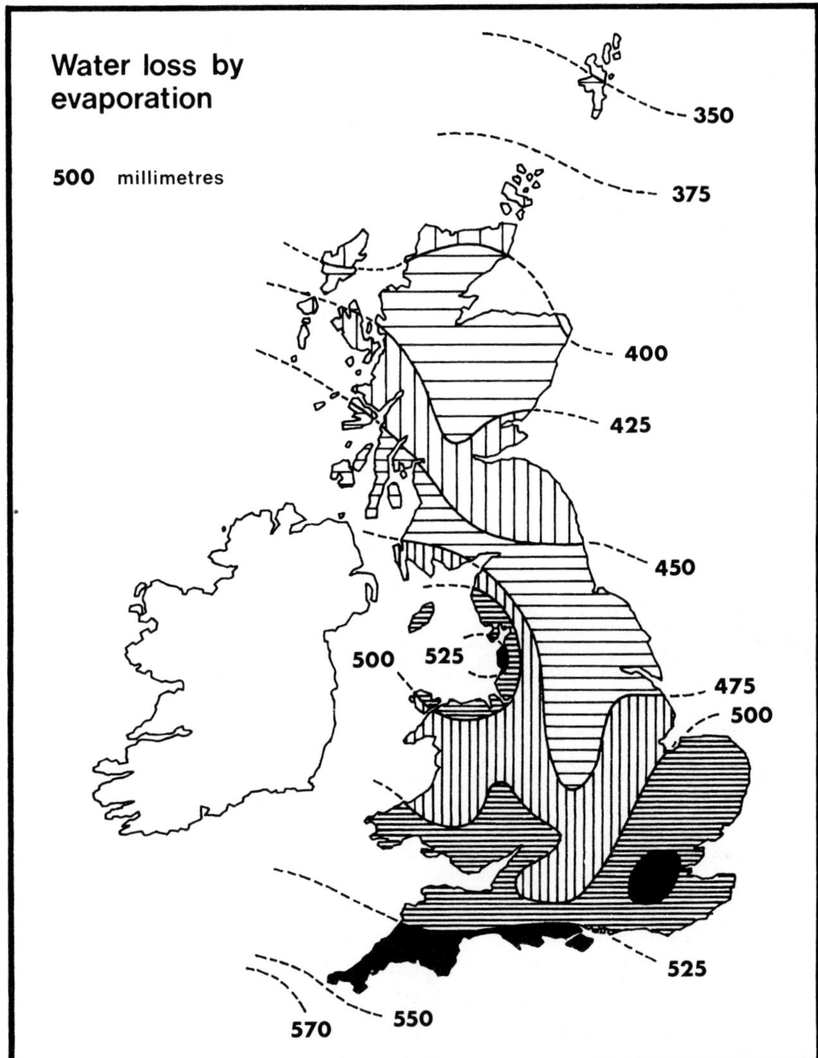

of water supply which we see in Britain today.

a) Refer to the map on page 267 and for each of the following cities Glasgow, Manchester, Liverpool, Newcastle-upon-Tyne, Birmingham, Belfast find:
   i) The area or areas from which it draws its water.
   ii) The name of one reservoir supplying the city.
   iii) Give any possible alternative sources available to the city.

b) List the physical characteristics which each of the main areas of water supply has in common. (The maps above and on page 30 will help).

Britain's water supply problems are not yet serious, affecting the general public only during periods of exceptional weather conditions, such as the drought of 1976. When one examines future trends, however, the picture is very different for, by the end of the century, it is possible that demand for water will have more than doubled. Meeting this demand presents many problems and several strategies have been put forward. These centre on the development of traditional sources and the introduction of new sources.

The Development of Traditional Sources

Britain depends almost entirely upon rivers, reservoirs and ground water for the maintenance of its water supplies. Of these, surface supplies are of overwhelming importance (see below) and this importance becomes even more marked when supplies to industry are taken into consideration. It is not surprising, therefore, that most of the strategies proposed to meet the future demand look to an expansion of the traditional surface and groundwater supplies.

SOURCE	% OF SUPPLY
Reservoirs and rivers	84·0
Groundwater	15·9
Desalinisation	Less than 0·1

a) *Rivers and Reservoirs*

The most obvious and widespread sources of water in Britain are the rivers and, if these could be fully used, there would be no water shortage, either now or in the foreseeable future. Furthermore, since most of Britain's major cities stand on rivers, the need to transport water would be reduced and costs would be low. Unfortunately these rivers are usually polluted by industrial and domestic waste, which makes them unsuitable as sources of drinking water. Cleaning and recycling water is possible but it is expensive and is rarely attempted until other cheaper sources have been exhausted. It is not surprising, therefore, that most cities have tried to tap the rivers in their upper courses before pollution begins.

This approach is not without its problems, however, for few rivers in their upper course can provide a large enough, or regular enough, supply of water to satisfy the demands of a large water authority. One method of overcoming these problems is to build dams across the rivers, so as to store clean water and to regulate its flow. This immediately increases the cost of the water and limits the

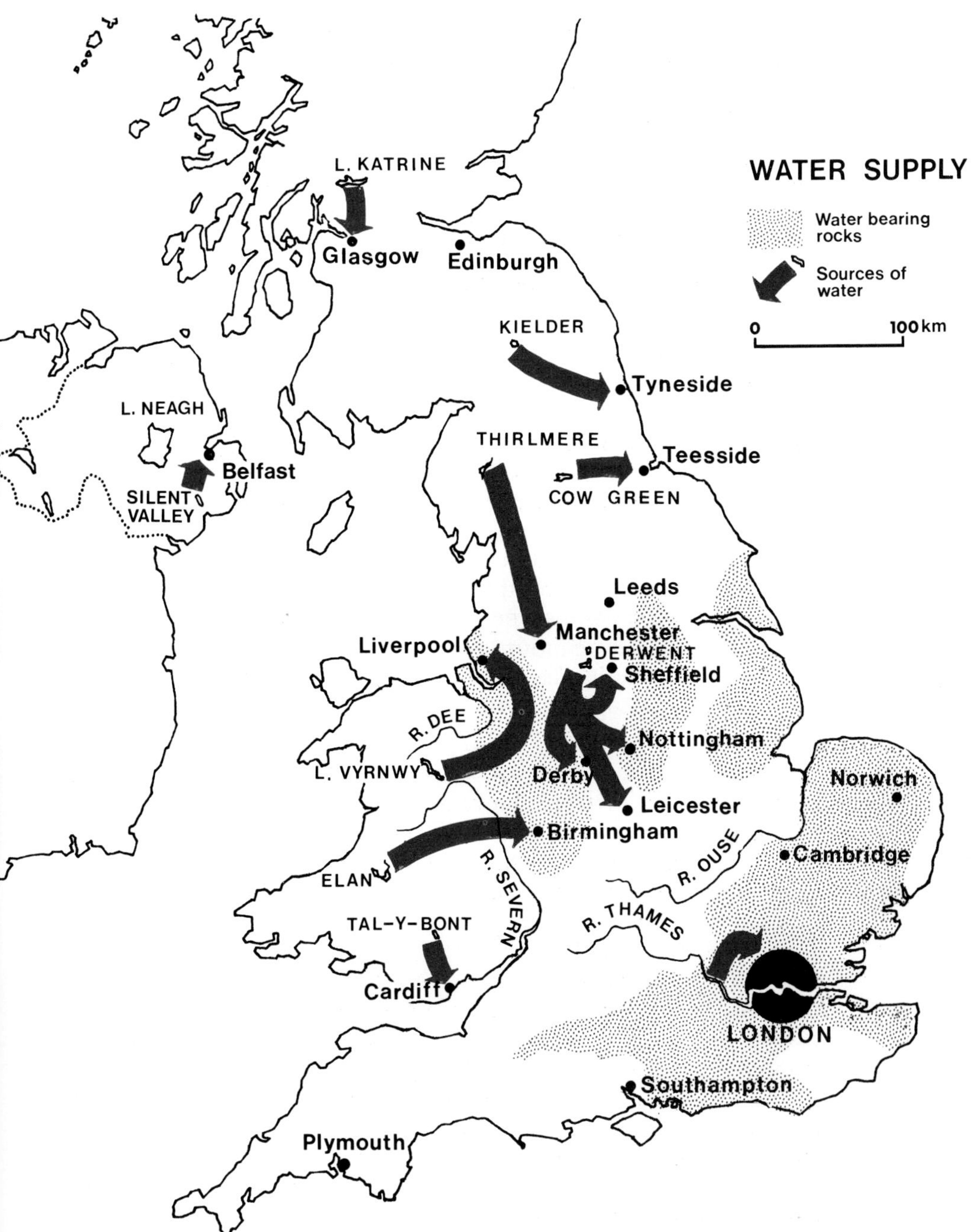
WATER SUPPLY
Water bearing rocks
Sources of water
0
100km
L. KATRINE
Glasgow
Edinburgh
KIELDER
Tyneside
L. NEAGH
Belfast
SILENT VALLEY
THIRLMERE
Teesside
COW GREEN
Leeds
Manchester
Liverpool
DERWENT
Sheffield
R. DEE
Nottingham
L. VYRNWY
Derby
Leicester
Norwich
Birmingham
ELAN
R. SEVERN
R. OUSE
Cambridge
TAL-Y-BONT
R. THAMES
Cardiff
LONDON
Southampton
Plymouth

number of places from which it can be obtained, largely because dams are expensive to build and cannot be built just anywhere. Ideally the following conditions should be met:

1. Geological conditions should be such that the underlying rocks are suitable both for building the dam and storing water. Permeable rocks, for example, may allow water to be lost underground, while wide, open valleys can be difficult to dam.

2. Rainfall should be heavy and regular, and run off into the streams serving the reservoir should be high.

3. Evaporation should be low since water loss can be heavy from the surface of a large lake. For example, in Eastern England evaporation exceeds rainfall for a large part of the year and water loss from the surface of a reservoir can be as high as 0·6 metre each year.

4. Valleys should be narrow and steep sided because they are easier to dam and large amounts of water can be stored in lakes with small surface areas. As a result, evaporation will be reduced and the loss of agricultural land will be small.

5. The reservoir should be near to the centre of population which it serves, since this reduces the costs of moving the water.

6. If this is impossible, then highland valleys are to be preferred since the water can be moved to the cities without pumping.

Given such limiting factors, it is obvious that the use of a river system for water supply can be very complicated.

CONDITIONS IN TWO RIVER BASINS

	DERWENT	NENE
Average precipitation	1,350 mm	600 mm
Run off	950 mm	150 mm
Water loss	400 mm	450 mm
Evapo-transpiration	525 mm	625 mm

a) Refer to page 271 and complete the following exercise:
   i) The dotted line indicates a watershed. What is a watershed?
   ii) Name two tributaries of the Trent which are polluted virtually from source and, in each case, name the town or city responsible for the pollution.
   iii) Name three tributaries which are relatively unpolluted and which can be used for water supply.
   iv) What effect has pollution had on the use of the river for water supply?
   v) What is the name given to water bearing rocks?

Empingham reservoir.

Reservoir in the Elan Valley, Wales.

vi) Name one city which depends to some extent on groundwater.
vii) Which city depends upon water sources outside the area? Where are these sources?
viii) Write a brief account of water supply in the Nottingham area.

b) Study the photographs on page 269 and, for each reservoir, list the advantages of its site and the disadvantages.

Because the creation of new reservoirs involves the flooding of large areas of land, objections to them have grown. In mountain areas these objections have centred on environmental considerations, such as the destruction of plant and animal life, and the changing of areas of great natural beauty; while, in lowland areas, they have concentrated on the loss of valuable agricultural land. Empingham is a case in point, where some of the richest farm land in England has been taken to create a vast, shallow lake. Problems such as these can only increase as new reservoirs are planned, for the best sites have already been developed and those which remain, almost invariably, have serious disadvantages. In spite of this, the low cost of water from reservoirs ensures that further development will take place.

b) *Groundwater*

Fifteen percent of the water supply of Britain comes from beneath the ground, where the water has collected in porous or permeable rocks. For centuries man depended upon such supplies, and settlements grew up where the water emerged from the ground in the form of springs. Such natural sources are, however, limited and, for the last one hundred and fifty years, man has exploited supplies of groundwater by drilling wells and by pumping the water to the surface. As a result of this development, groundwater became available in many parts of Britain, particularly in the south and east where, as we have seen, other sources are limited. The water produced is generally so pure that little treatment is needed and, as a result, costs are very low. It is not surprising, therefore, that during the nineteenth century development was rapid and the water level in the rocks underlying many of our cities was lowered at an alarming rate. In the London area, for example, the water table in the chalk was lowered some seventy metres before pumping was restricted.

This is in fact the major problem with groundwater. It is a cheap source of very pure water, but it is slow to build up and, if reserves are used too quickly, the wells can run dry. Pumping must not, therefore, exceed the rate at which the water builds up in the rock; a

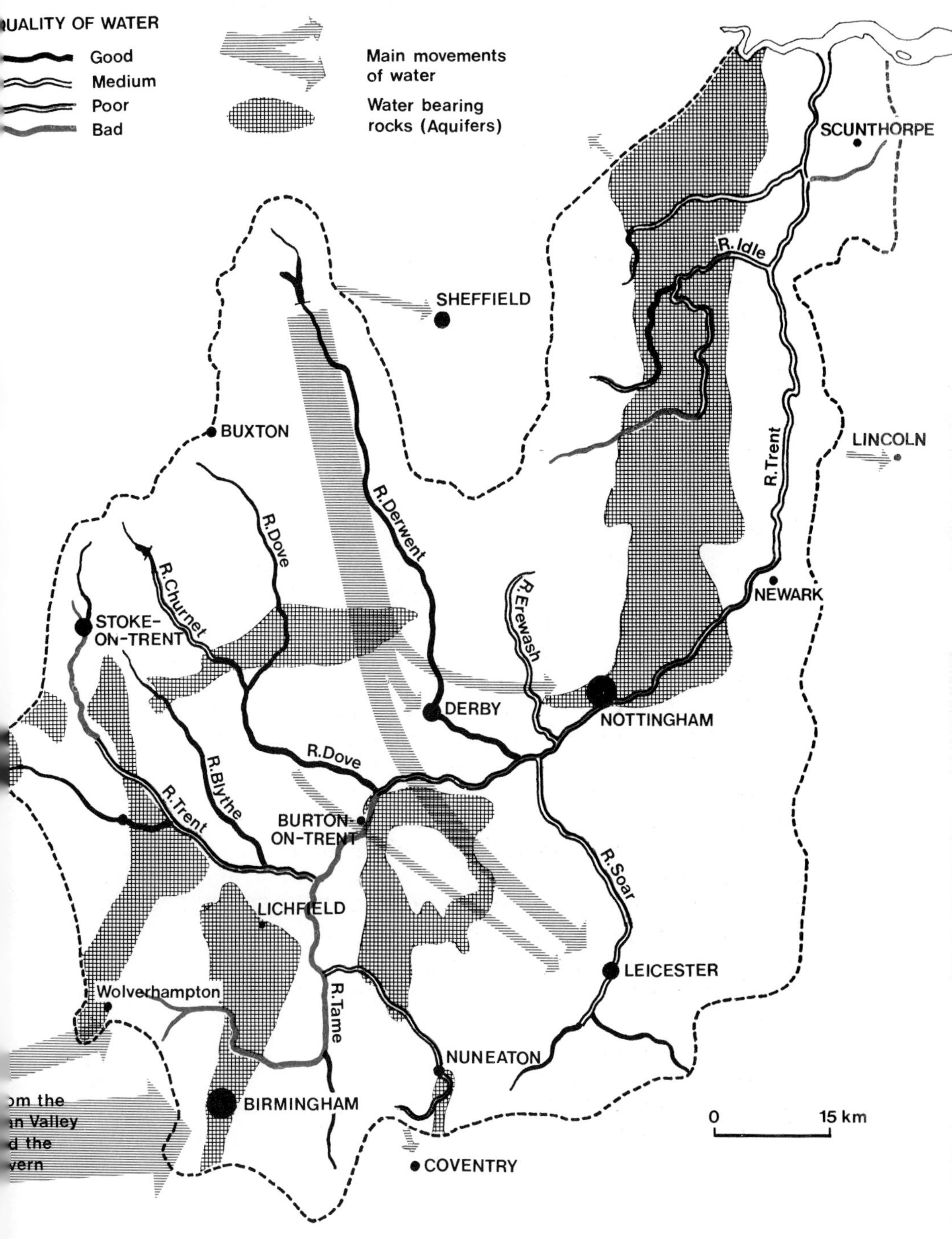
'RENT VALLEY: WATER SUPPLY AND POLLUTION
UALITY OF WATER
Good
Medium
Poor
Bad
Main movements of water
Water bearing rocks (Aquifers)
SCUNTHORPE
R. Idle
SHEFFIELD
BUXTON
LINCOLN
R. Trent
R. Dove
R. Derwent
R. Churnet
STOKE-ON-TRENT
R. Erewash
NEWARK
DERBY
NOTTINGHAM
R. Dove
R. Blythe
R. Trent
BURTON-ON-TRENT
R. Soar
LICHFIELD
LEICESTER
Wolverhampton
R. Tame
NUNEATON
om the
n Valley
d the
vern
BIRMINGHAM
COVENTRY
0
15 km

fact which would appear to impose severe limits on any future development of groundwater supplies. In an attempt to overcome this problem, some water authorities have started to pump river water into boreholes so as to store it underground. Such a development has much to recommend it:

1. A reservoir of water would be formed without flooding valuable land. This is particularly important in areas like London where pressure on land is severe.

2. Percolation through the rocks would purify the water and make further treatment less expensive.

3. Evaporation would be nil, thus removing an important source of water loss.

The Development of 'New' Sources of Water

Pressure on existing water supplies, particularly in south east England and in the conurbations, has resulted in a search for new sources of water. Two alternatives have attracted serious attention:

a) *Coastal Barrages*

The building of barrages (low dams) across coastal inlets is an attractive idea. Sea water would be excluded and this would allow river water to build up behind the dam until eventually a fresh water reservoir would be formed. Unfortunately there are also problems. It is not surprising in view of this, that the proposals to build barrages across Morecambe Bay, the Dee Estuary, Solway Firth and the Wash have not progressed beyond the study stage.

ADVANTAGES

1. Large amounts of fresh water could be stored without flooding land.
2. The lakes would provide useful leisure facilities.
3. Roads can be built on the barrages, thus improving communications across the inlet.
4. Barrages could be built near areas of greatest shortage, eg the Wash in SE England.

DISADVANTAGES

1. Water would be taken at the river mouth where many British rivers are heavily polluted.
2. Many of the least polluted inlets are in areas of natural beauty.
3. Schemes are expensive and there are technical difficulties in closing dams across stretches of open sea.
4. Water is stored at sea level and pumping is necessary to move it to the cities.
5. The environment will be changed, eg silting may take place and wild life may be destroyed.

b) Desalinisation

The one apparently limitless source of water available to Britain is, of course, the sea. Unfortunately, before sea water can be included in the public water supply system, salt has to be removed from it. This process is known as *desalinisation* and it can be achieved either by distillation or by freezing. Both methods effectively remove the salt but both require enormous amounts of power, and this makes the water produced much more expensive than water from any other source. At one time the development of desalinisation seemed inevitable since it would provide water without requiring vast reservoirs. But, with rising costs and a threatened world shortage of energy, this development now seems less likely, particularly in areas such as Britain where rainfall is plentiful.

Each of these alternatives, whether traditional or new, has serious disadvantages and in recent years the idea of water *conservation* has become popular. In Britain water is still regarded as a freely available commodity and, as such, it is all too often wasted. In such a situation, conservation is not only possible, it is essential, and a variety of measures have been proposed to save water. These include, at one end of the scale, the proposal that industry should, wherever, possible, make use of less pure and, therefore, more plentiful supplies of river water, and, at the other end, the introduction of a smaller toilet cistern which could halve the amount of water used each day. Measures such as these, regardless of scale, would appear to be essential if the future demand for water is to be met without destroying large areas of Britain.

Case Study: Water Supply in the London Area

It is in South East England, and in the London area, in particular, that problems of water supply are at their greatest. Here the combination of comparatively low rainfall and a rapidly increasing population has forced the authorities to explore every source of supply available and to use a good many of them.

Refer to pages 274 and 275 and complete the following exercise:

a) List, in order of importance, the main sources of London's water.

b) i) Why is water drawn only from the upper reaches of the rivers?

ii) Describe the change which takes place in the Thames at Teddington. How does this affect the use of the river for water supply?

c) Describe the stages by which river water reaches the consumer as drinking water.

d) i) From which rock strata is the bulk of the groundwater obtained?
ii) Why is it found there?
iii) The London area is part of an artesian basin. What does this term mean?
iv) Name three important pumping areas. Why are the wells located here?

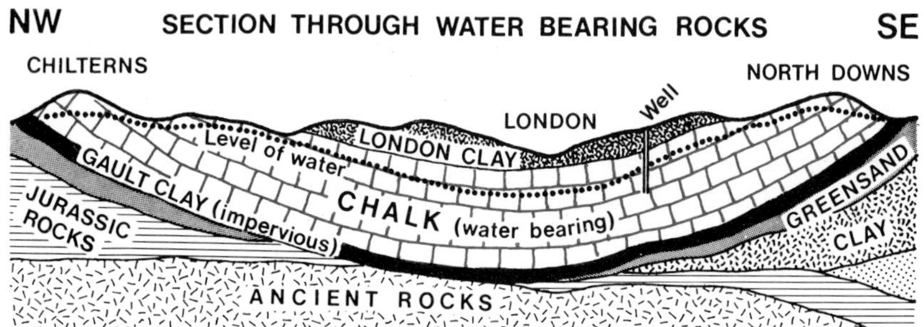

Meeting future demand in the London area is a serious problem and many alternatives have been studied.

1. It seems unlikely that ground water supplies can be substantially increased. As we have already seen, the water table in the chalk has been lowered by excessive pumping and, even if river water is pumped underground, ground water supplies will not be able to meet the expected rise in demand.

2. River water is the obvious alternative but even here there are problems. Most of these stem from the fact that demand for water is increasing in the Thames valley above London and this reduces the amounts available for the city. Furthermore, large amounts of water are returned to the river from sewerage works and this has to be cleaned and recycled before it can be used. Such an operation is expensive, particularly in the London area where it is said that the water consumed by the people has already been drunk several times before.

3. Water is brought into the area from neighbouring water authorities and this is likely to increase. In fact it seems possible that a barrage will have to be built across the Wash if the water supply problems of London and the South East are to be overcome.

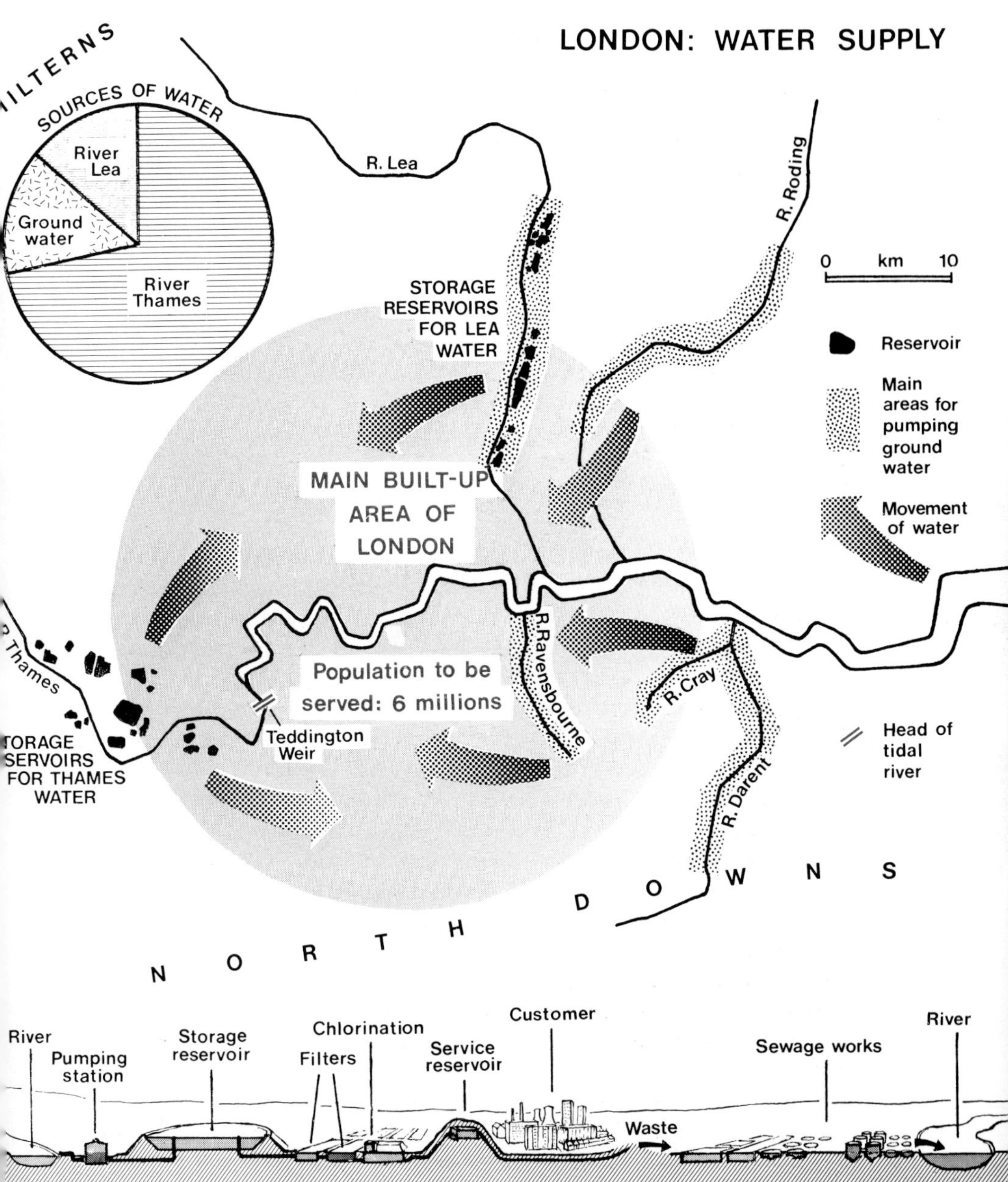

LONDON: WATER SUPPLY
SOURCES OF WATER
River Lea
Ground water
River Thames
R. Lea
R. Roding
0 km 10
STORAGE RESERVOIRS FOR LEA WATER
Reservoir
Main areas for pumping ground water
Movement of water
MAIN BUILT-UP AREA OF LONDON
Population to be served: 6 millions
R.Ravensbourne
R.Cray
R.Darent
Thames
Teddington Weir
FOR THAMES WATER
Head of tidal river
N O R T H D O W N S
River
Pumping station
Storage reservoir
Filters
Chlorination
Service reservoir
Customer
Waste
Sewage works
River
WATER SUPPLY SYSTEM

Leisure

The last hundred years has seen an ever increasing demand for leisure facilities in Britain. This has been the product of the growing affluence of the population which has gone hand in hand with a reduction in the length of the working week (from almost sixty hours in 1850 to the present average of about forty hours) and the introduction of holidays with pay. Initially needs were met by the provision of amenities in the local area and by the development of large holiday resorts. Since the Second World War, however, improvements in the standard of living and in the mobility of the population has led to new developments such as the introduction of cheap package holidays abroad, the emergence of new holiday resorts and the creation of the National Parks.

Holiday Resorts

Holiday resorts as we know them, date from the second half of the nineteenth century. Prior to 1850, the few resort towns which existed were spa towns, where a small number of wealthy people could 'take' the waters in comfort. At first most of the spas were situated inland and the 'waters' were obtained from mineral springs. The most famous of these towns was Bath. Here the settlement dates back to Roman times when the hot springs were known and exploited, but the town reached a peak of prosperity during the eighteenth century when it was the leading resort town for the English aristocracy.

Towards the end of the eighteenth century, however, fashions changed and seaside towns began to replace the inland spas as the main resorts. At first the drinking of sea water, for medicinal purposes, was practised but gradually this was replaced by sea bathing. Scarborough was the first of these new resorts but it was quickly overshadowed by the rise of Brighton.

Most of these early resorts were very small, partly because they catered for only a minority of the population and partly because the difficulties of transport meant that each could serve only a very limited area. It was, however, from these towns that our modern resorts developed when the building of the railways in the mid-nineteenth century produced a massive influx of trippers; first on one day excursions and later, as holidays were granted in various industries, for longer periods. The resorts which responded to these new demands shared two advantages:

1. They were situated near to large centres of population.
2. They were easily accessible by rail.

This pattern persisted for almost one hundred years, during which time the 'railway' resorts such as Bournemouth, Brighton, Blackpool and Rhyl reigned supreme. The challenge to their supremacy, when it came, stemmed from changes in British society, for, during the twentieth century people became more affluent, and, at the same time, motor transport was growing and the railways were declining. As a result, hitherto remote areas of Britain became accessible and new resorts developed there. These resorts did not possess the advantages of the old established holiday centres. They were often far from the major cities and were accessible only by road. But they were usually situated in areas of great natural beauty and were often in the parts of Britain where the best weather might be expected.

Blackpool: a traditional resort

In 1851 Blackpool was a village with a few hundred inhabitants. By 1871 it had grown into a town with a population of more than ten thousand and today it houses more than one hundred and fifty thousand inhabitants. Much of this growth can be attributed to the rise of Blackpool as a major holiday resort and this in turn has been influenced by a number of factors.

1. *General Situation*

In this respect Blackpool was an ideal location for a holiday resort. By 1850 Lancashire had emerged as Britain's leading industrial region, containing almost one quarter of the nation's population. It is not surprising, therefore, that holiday resorts grew up on the stretch of coast nearest to this concentration of population.

2. *Environmental Factors*

At first sight, environmental factors would appear to have contributed little to the growth of Blackpool, for the attractions of this stretch of the Lancashire coastline are limited and certainly equalled or excelled by other coastlines which have never achieved its popularity. Such a conclusion would, however, be misleading for environmental factors have played an important part in the dominance of Blackpool over the other Lancashire resorts. For example, deposition varies considerably along the coast. In the south it has taken place on a large scale and wide sandy beaches have been formed, backed by sand dunes. Unfortunately, the sea here is so shallow that, at low tide, the water line can retreat more than one kilometre, creating serious problems for resorts, such as Southport. To the north, in Morecambe Bay, mud is deposited rather than sand and beach formation is limited. Only in the central area, around Blackpool, therefore, are sandy beaches combined with water deep enough to limit the retreat of the sea at low tide.

BLACKPOOL CLIMATIC STATISTICS

	J	F	M	A	M	J	J	A	S	O	N	D	TOTAL
Mean temperature (°C)	3·3	3·9	5·7	8·0	10·9	13·9	15·5	15·5	13·9	10·9	7·0	4·8	—
Rainfall (millimetres)	76	53	48	55	60	60	72	90	89	87	84	83	857
Sunshine (av. hours per day)	1·5	2·6	3·9	5·7	6·9	7·3	6·1	5·7	4·3	3·1	1·9	1·4	—
Number of days with rain	16	11	14	14	15	14	12	13	14	16	19	15	173

3. *Exploitation*

Even more important has been the way in which Blackpool has exploited the advantages of its situation. Almost from the first the resort catered for middle and working-class people and, as holidays with pay became established, so Blackpool began to overshadow most of its rivals. In keeping with this approach, Blackpool's few natural attractions were supplemented by a whole range of man-made attractions which helped to make it the largest and most popular resort in Europe.

Referring to the information given above and opposite, complete the following exercise:

a) Describe the natural advantages enjoyed by Blackpool as a holiday resort.

b) i) Study the photograph and state in which direction the camera is pointing.
   ii) What evidence is there on the map and photograph that Blackpool is a holiday resort?
   iii) Describe how the town has been adapted to cater for tourists.

c) i) Give the age of building in the centre of the town.
   ii) When did the railway reach Blackpool?
   iii) During which period did the town grow most rapidly?
   iv) When were most of the town's amenities built?
   v) How does this pattern reflect the history of the town, as described above?

ackpool

OWTH

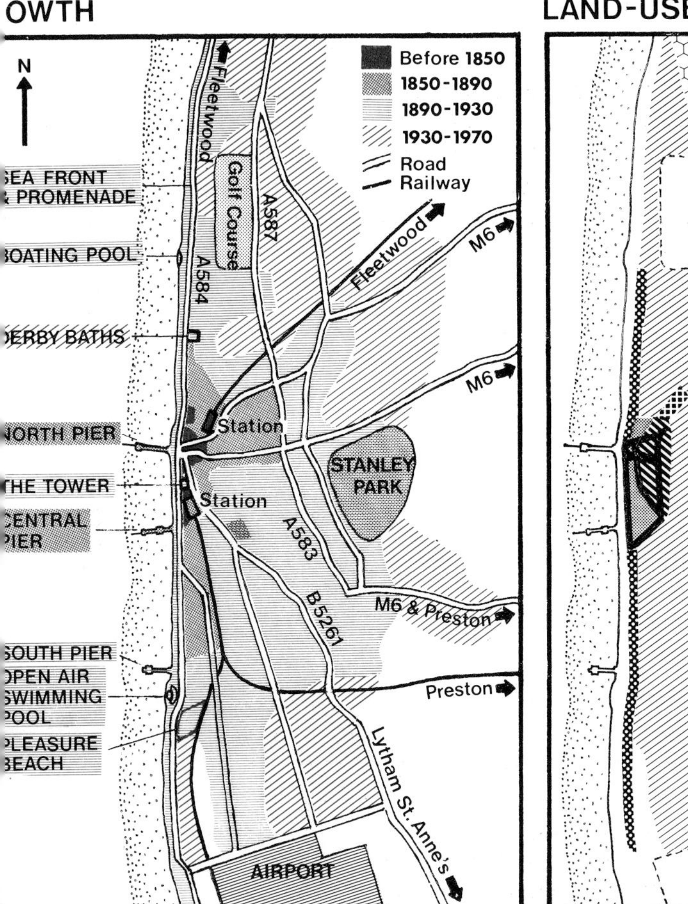

LAND-USE (SIMPLIFIED)

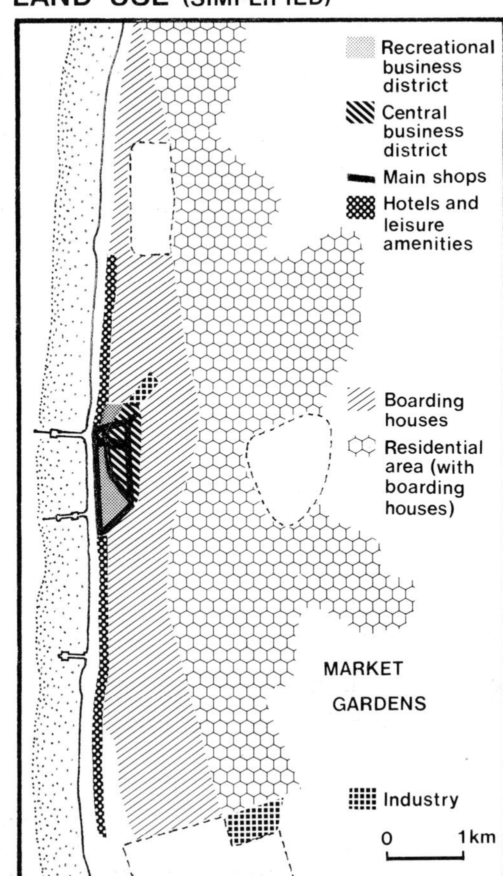

Blackpool has exploited to the full its situation on a stretch of coast near to a large centre of population and expansion has continued down to the present day, in spite of changing holiday habits and changing patterns of communication. The latter is particularly important since the resort developed during the railway age and has maintained its position into the motor age. This has been made possible largely because Blackpool is well situated in relation to the motorway network and the building of the M6 and the M62 has improved links with large areas of the country.

The impact of the motorways is clearly apparent in the origins of visitors to Blackpool and the methods of transport used by them to reach the resort. Gone are the days when Blackpool was the destination of trainloads of holidaymakers, escaping from the Lancashire cotton towns or the cities of the Midlands to enjoy their annual 'wakes' holiday. Instead tourists are drawn from further afield and the majority of them arrive by road. Furthermore, the 'season' has been extended into the late autumn by the introduction of the illuminations and into the spring by the offer of cheap accommodation.

TOURISM IN BLACKPOOL

TYPE OF ACCOMMODATION		TYPE OF VISITOR	
Registered Hotel	28%	Holiday Makers	2,360,000
Unregistered Hotel	48%	Day Trippers	3,840,000
Self-catering	19%	Total	6,200,000

The Structure of a Holiday Resort

The unusual demands of the tourist industry produce an urban structure which is peculiar to resort towns. At its simplest this is seen in the amenities provided but it can also be traced in the structure of the town and its population.

EMPLOYMENT IN BLACKPOOL

INDUSTRY	% OF WORKFORCE
Food, drink, tobacco	6·9
Manufacturing	16·9
Service: All	71·4
Distribution	21·0
Wholesale	2·1
Catering	14·3
Miscellaneous	26·9

BLACKPOOL: AGE STRUCTURE OF POPULATION

NUMBER MALE	AGE	NUMBER FEMALE
9,300	0–9	9,000
9,000	10–19	8,800
7,600	20–29	7,700
7,100	30–39	7,300
8,900	40–49	9,700
9,600	50–59	11,400
10,000	60–69	13,700
5,400	70–79	9,900
1,300	80–89	3,400
700	over 90	1,300

Study the information given in the two tables and complete the following exercise:

a) i) Draw a graph to show the age structure of the population of Blackpool. (See page 174).
ii) How does this pattern differ from that for Britain as a whole?
iii) Why do these differences occur?

b) i) What is the most important type of industry in Blackpool?
ii) Why is this type of industry so important in the town?
iii) How does the industrial structure of Blackpool compare with that of a market town and of an industrial city? (See page 229).

c) How does the structure of the city compare with that of other British cities. (See page 220).

The high proportion of old people is characteristic of resort towns and it is caused by a tendency for people to retire to the seaside. In fact, Blackpool is less seriously affected by this trend than many south coast resorts which have so many retired people that serious social problems have arisen, and the hospital and health services have been placed under great strain.

Similarly, the industrial structure of most holiday resorts is dominated by service industries. This is to be expected in view of the need to provide accommodation, food, shops, entertainment and other services for a vast number of visitors.

Finally, the actual shape and structure of the town is affected by its function as a holiday resort. This can be seen most clearly in:

1. The presence of an entertainment and recreation district.
2. The elongation of the central recreation and business districts along the sea front.

3. The absence of any major industrial development near the business districts.

4. The large areas of open space which are required for recreation in a resort town.

5. The mingling of residential and business functions in the area around the town centre. This is largely the result of the growth of the hotel and boarding house industry.

St Ives: a 'New' resort

Many of the resorts which have developed in recent years display characteristics very different from those seen at Blackpool. They tend to be smaller in size, and the comparatively late development of tourism means that the features of a resort town have been grafted onto an existing structure.

a) The following table contains a list of factors which have played a part in the development of holiday resorts. Those which have been particularly important in the case of Blackpool are marked.

FACTORS	BLACKPOOL	ST IVES
Scenery		
Climate		
Good beaches	*	
Near to major cities	*	
Good rail links with cities	*	
Good road links	*	
Entertainment		

TOURISM IN WEST CORNWALL

ORIGIN OF TOURISTS

REGION OF ORIGIN	% OF TOTAL VISITORS
Scotland	2
North	2
Yorkshire	7
North West	9
East Midlands	6
West Midlands	13
East Anglia	2
South East	39
Wales	3
South West	15
Foreign	2

METHOD OF TRAVEL

METHOD	% OF ALL VISITORS
Car	80
Rail	10
Coach	10

ACCOMMODATION USED

ACCOMMODATION	% OF ALL VISITORS
Hotel/Boarding House	27
Camp, Caravan, Flat	49
Friends/Holiday Camps	24

i) Copy out the table and using information given on pages 26, 29, 30, 284 indicate the factors which have contributed to the development of St Ives as a holiday resort.

b) What was the main function of St Ives before the tourist industry developed?

c) What evidence is there on the photograph that St Ives is an important resort?

d) Describe the main differences between Blackpool and St Ives as holiday resorts paying particular attention to the origin of the tourists, the type of holidays and the structure of the towns.

e) Using information given in this section, describe the main features of a resort town, concentrating on its functions, population structure and urban structure.

The very success of the new resorts has brought with it many problems. Because they often originated as small fishing villages, they have tended to become overcrowded and to outstrip their resources. Furthermore, because they are in remote areas and are, therefore, highly dependent upon motor transport, they have suffered seriously from traffic congestion, an inadequate road

St. Ives, Cornwall.

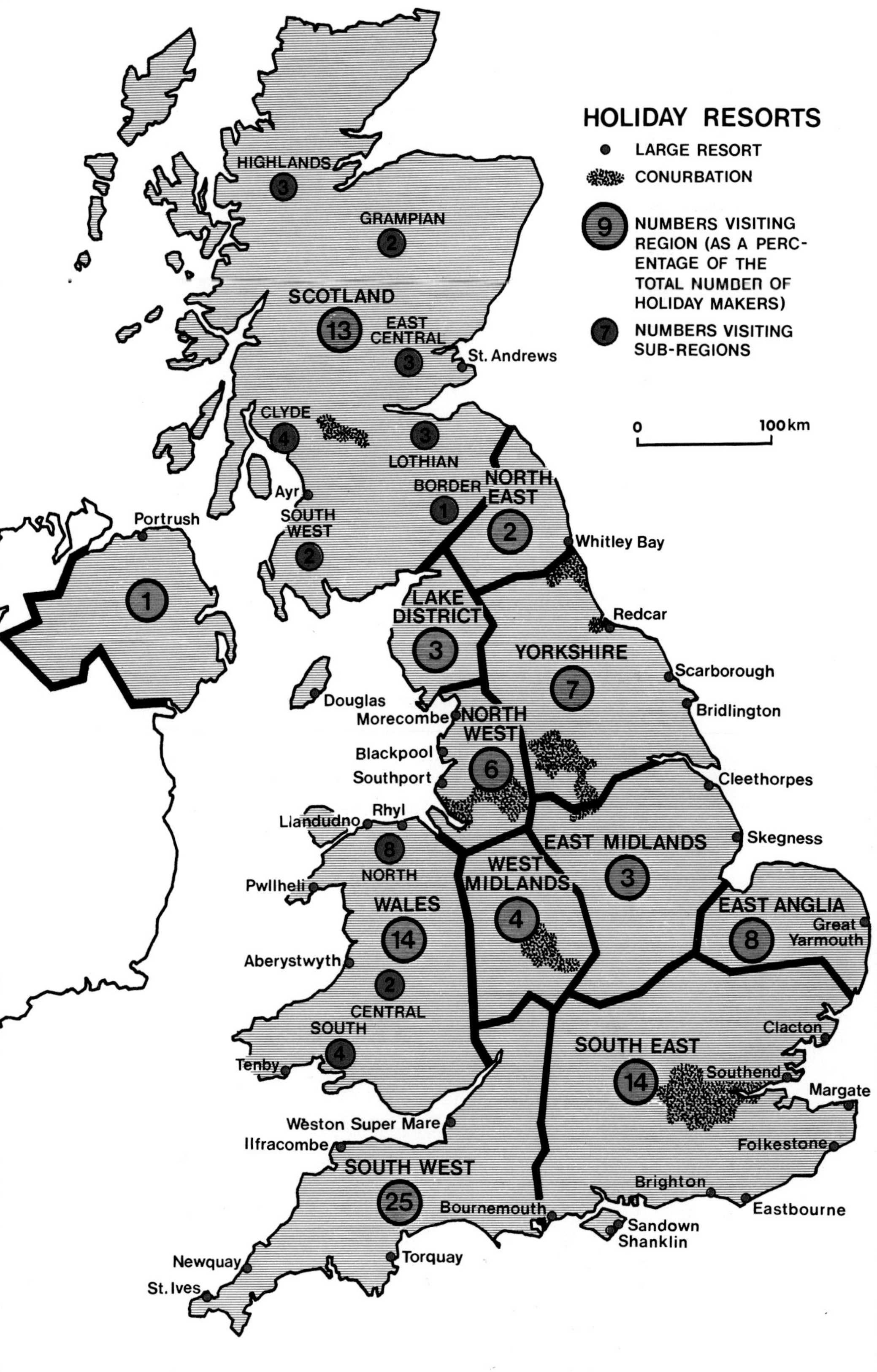

HOLIDAY RESORTS
LARGE RESORT
CONURBATION
9 NUMBERS VISITING REGION (AS A PERCENTAGE OF THE TOTAL NUMBER OF HOLIDAY MAKERS)
7 NUMBERS VISITING SUB-REGIONS
0 100km
HIGHLANDS 3
GRAMPIAN 2
SCOTLAND 13
EAST CENTRAL 3
St. Andrews
CLYDE 4
3 LOTHIAN
BORDER 1
Ayr
SOUTH WEST 2
Portrush
1
NORTH EAST 2
Whitley Bay
LAKE DISTRICT 3
Redcar
YORKSHIRE 7
Scarborough
Bridlington
Douglas
Morecambe
NORTH WEST 6
Blackpool
Southport
Cleethorpes
Llandudno
Rhyl
8 NORTH
Skegness
EAST MIDLANDS 3
WEST MIDLANDS 4
Pwllheli
WALES 14
EAST ANGLIA 8
Great Yarmouth
Aberystwyth
2 CENTRAL
SOUTH 4
Tenby
Clacton
SOUTH EAST 14
Southend
Margate
Weston Super Mare
Ilfracombe
Folkestone
SOUTH WEST 25
Bournemouth
Brighton
Eastbourne
Sandown
Shanklin
Newquay
Torquay
St. Ives

network and lack of parking. Such problems are difficult to overcome since the resorts are often in areas of great natural beauty where development is restricted.

The Distribution of Holiday Resorts

The distribution of resorts reflects the changing patterns of holiday making in Britain. Traditional resorts cluster along stretches of coastline near to the main urban centres, while the newer resorts have grown up in less accessible areas, particularly in the highlands. This development can be seen from the map opposite which shows the emergence of the South West, one of the most remote areas of England, as the major tourist region.

These changes in the pattern of tourism in Britain are, in part, a reflection of changes in society as a whole. Many of the 'new' holiday areas have attracted large numbers of motorists by offering camping, caravan and other self-catering holidays. In this way, they have grown at the expense of the traditional resorts which offered hotel and boarding house holidays. Furthermore, the increased affluence of the population was reflected in the number of British holidaymakers taking holidays abroad, particularly in the Mediterranean area where sunshine can be virtually guaranteed during the summer, and where costs have been kept down by the introduction of the block booking of hotels and aeroplanes by tour operators. Competition of this kind has produced changes in even the most conservative of British resorts and attempts have been made to extend the season, so as to make more economic use of accommodation and amenities, and to develop cheaper self-catering holidays. Changes such as these may well pay off in the future for rising fuel costs have almost brought to an end the cheap continental package holiday and it seems likely that Britain's traditional resorts may benefit from its demise.

National Parks

In 1949 an act of Parliament was passed setting up the first *National Parks*. It had two main aims:

1. To preserve the character of certain areas of countryside; usually areas of great natural beauty.
2. To give people access to these areas.

In passing this bill, Parliament was attempting to meet a growing demand, particularly among city dwellers, for access to the countryside—access often denied them because so much land in Britain was privately owned. This demand was the product of the increasing amount of leisure time available to working people and the improvements in private transport which were taking place with the growing popularity of the motor car.

The Peak District of Derbyshire was the first area to be designated a national park and since then more land has been acquired until there are now ten national parks, covering an area of 13,482 sq km. Most of these parks are located in upland areas which can offer striking scenery and, where possible, in upland areas which are within easy reach of conurbations. Logical as this development may

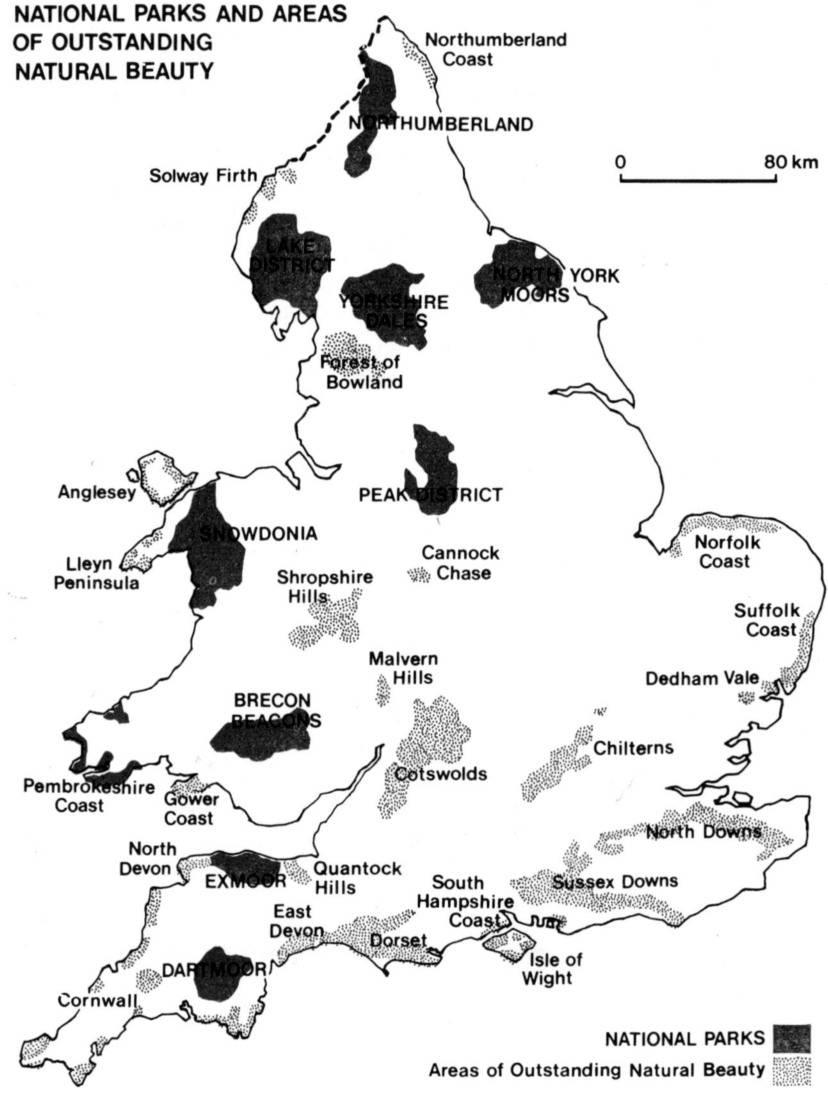

be, it has tended to produce an uneven distribution of national parks in England and Wales (the scheme has not yet been extended to Scotland) and certain densely populated areas have been left without adequate access to the countryside. In an attempt to overcome this problem some areas of country have been designated Areas of Outstanding Natural Beauty, with a view to controlling development within them and to providing increased access.

In 1975 nearly 80 million people visited the national parks and demand is expected to continue to rise during the next decade. Such success has also brought problems, for demand has tended to outstrip existing resources; a situation which is difficult to rectify since development within the parks is strictly controlled. This conflict of interests between providing amenities such as car parks and toilets and preserving the character of the park is one of the main problems faced by the planning authorities. The same conflict of interests can be seen in the economic development of the parks, for people live and work within their boundaries and they, like the rest of the population of Britain, look forward to improved job prospects and improved standards of living. Whether this can be provided without destroying the park is sometimes open to doubt. And, finally, some of the parks are finding it physically impossible to cope with the number of visitors. So serious has this problem become that, around certain 'beauty spots' soil erosion has taken place and plastic grass has been put down to halt it!

These problems can best be understood by examining them in relationship to one of the national parks.

Regional Study: The Peak District National Park

Designated in 1949, the Peak District was the first of Britain's national parks. There were many reasons for this choice but among the most important were the nature of countryside and its situation near to the main centres of population.

a) **The Land**

The national park is made up of two distinct landscape regions; The Millstone Grit region which, in the north around the Kinder Plateau, forms the Dark Peak, and the Carboniferous Limestone region which forms the White Peak in the south. The contrast between these two areas is striking.

Using information given on pages 288 and 289, complete the following exercise:

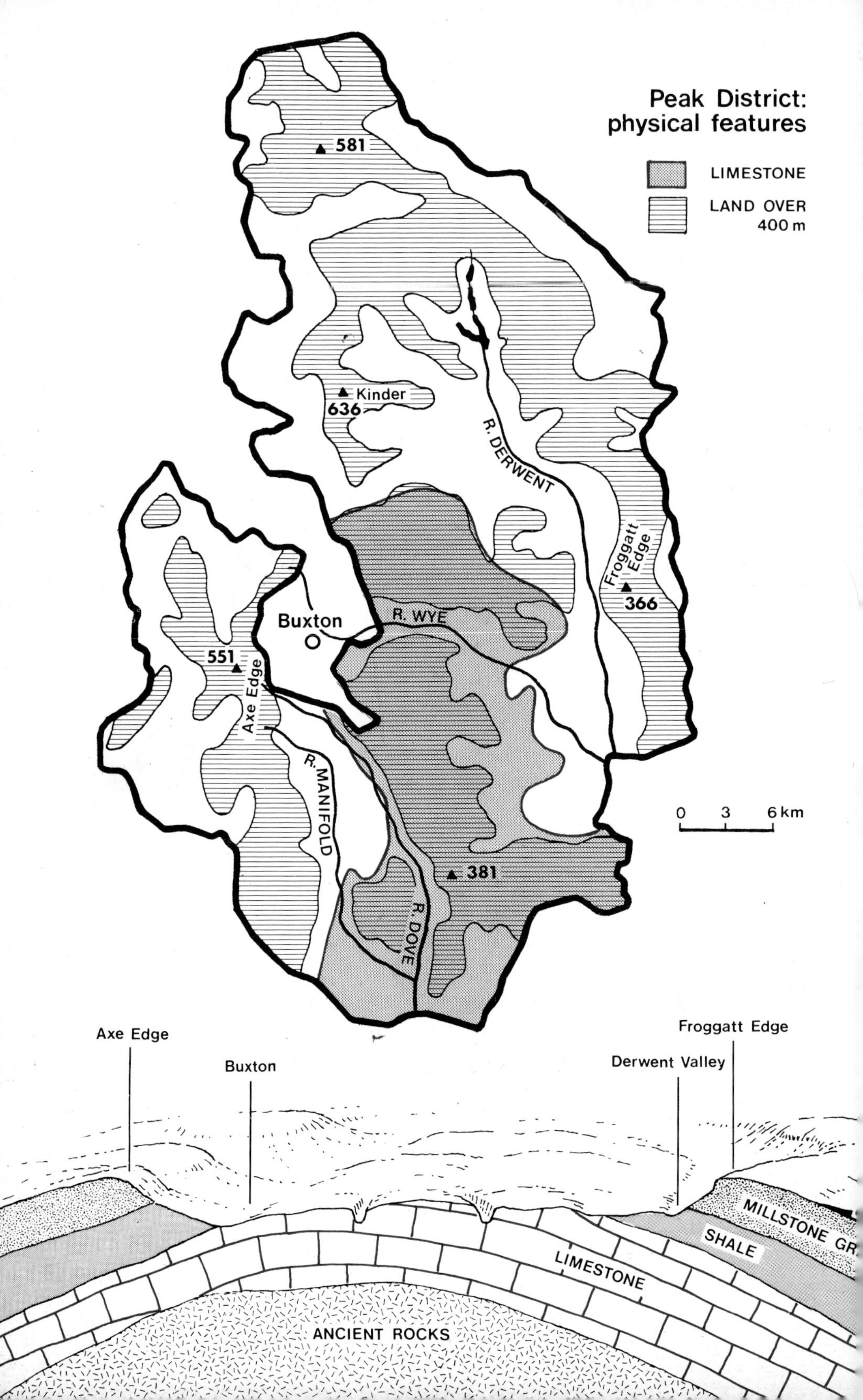

Peak District: physical features
LIMESTONE
LAND OVER 400 m
581
Kinder
636
R. DERWENT
Froggatt Edge
366
Buxton
R. WYE
551
Axe Edge
R. MANIFOLD
R. DOVE
381
0 3 6 km
Axe Edge
Buxton
Derwent Valley
Froggatt Edge
MILLSTONE GR
SHALE
LIMESTONE
ANCIENT ROCKS

Peterdale in the Limestone Peak showing the gentle relief of the plateau surface and the narrow steep sided valleys which characterise the area.

The Roaches on the Gritsone Peak, showing the 'edges' and the massive rock outcrops which are typical of the area.

a) Complete the following description of the Peak District by filling in the missing words.
The Peak District comprises the southern part of the .... hills. The rocks are of .... age and they have been folded to form a vast .... Erosion has removed the .... and .... which once covered the region, exposing the underlying .... in the centre.

b) Using the two photographs:
   i) Describe the landscape on the gritstone moors.
   ii) Describe the landscape of the limestone plateau.
   iii) List the main differences between the two landscapes.

TOURISM IN THE PEAK NATIONAL PARK

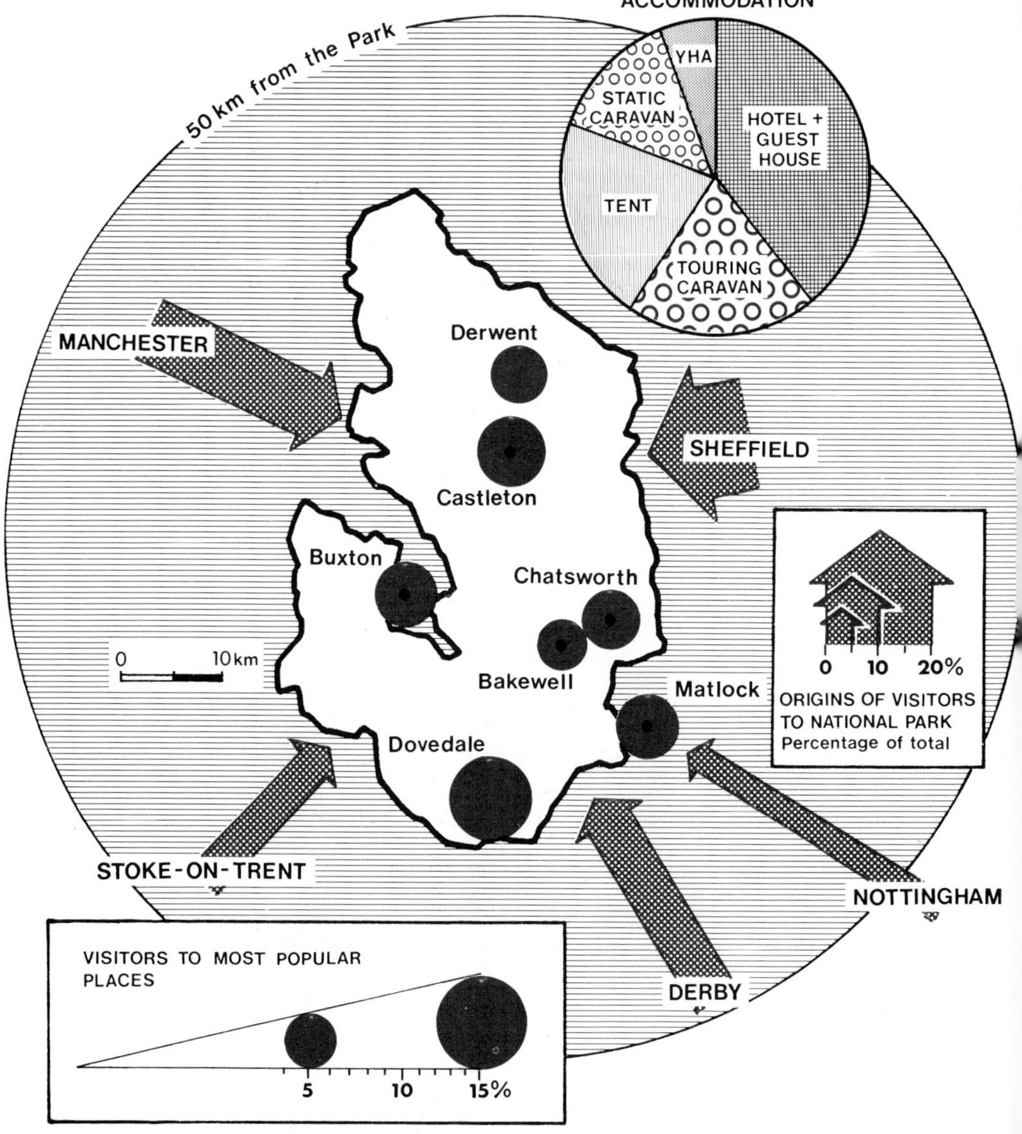

It is the limestone district which has proved most attractive to tourists. Here the plateau is much lower than on the Millstone Grit moorlands, but its surface is dissected by narrow steep sided valleys which are sometimes extremely beautiful. Furthermore, many of the features usually found in limestone areas are well developed here, including cave systems, underground rivers and gorges.

b) **Use of the Park**

National parks are intended to provide a service for people, and, in this respect, the Peak District is ideally situated, for the major conurbations of Lancashire, Yorkshire and the Midlands are all within an easy journey of its boundaries. In fact, more than eight million people live within fifty kilometres of the park. This, more than anything else, explains the success of the venture for, even today, more than three quarters of all visitors come from within this area. In the summer of 1963 four million people visited the Peak District National Park. Ten years later this figure had risen to twelve million. Of these, ninety-four per cent were day trippers and, on a typical Sunday afternoon in the summer, more than one hundred and sixty thousand visitors arrive in the park in fifty thousand cars. If these visitors were spread throughout the area, pressure would be considerable but, as can be seen from the map opposite, more than half of them were attracted to a handful of places. Here pressure becomes enormous and many of the problems of the national parks become clearly apparent. Around Dovedale, for example, access roads are often overcrowded, cars have to find parking spaces in the surrounding fields and basic facilities such as toilets and cafés are lacking. At first sight, the answers are obvious—new facilities should be provided. But this immediately raises questions which underlie one of the basic dilemmas facing the national parks namely:

1. Can such developments take place without destroying the natural beauty of the area?
2. Will further development simply attract more tourists to an already overcrowded beauty spot?

Viewed in these terms there are no easy answers, and the national park authorities are left facing the almost insoluble problem of providing access to the countryside for motorists who are hardly prepared to leave their cars, let alone walk anywhere.

c) **The Effects on the Economy of the National Park**

Tourism has brought great benefits to the region, ranking with farming and quarrying as a source of income. It has also brought serious problems, particularly with regard to the balancing of the needs of the national park and the needs of the local economy.

EMPLOYMENT IN THE PEAK

EMPLOYMENT	% OF WORKING POPULATION
Agriculture	3
Mining	8
Textiles	21
Other manufacturing	20
Tourism	9
Other services	37

This is the second basic dilemma of the national park system. The people living within the boundaries of the park (126,000 in the case of the Peak District) expect industrial development to continue so as to maintain or improve their standards of living. But is such development possible without destroying the park itself? Three economic activities provide most cause for concern.

1. *Agriculture*

Although only a small proportion of the labour force is employed on the land, the development of agriculture is of vital importance to the well being of the park. This is because the pattern of farming has played a large part in creating the landscape of the Peak District, and any major changes could have serious consequences. At present the balance is good, both in terms of maintaining the landscape and of providing access to the countryside.

Refer to the map opposite and complete the following exercise:

a) i) Describe the pattern of agricultural land use in the park.
   ii) Describe how it changes between the gritstone moorlands of Staffordshire and the limestone plateau of Derbyshire.
   iii) Why is this pattern of land use ideal for a national park?

b) Name the main types of farming in the park and explain how they are influenced by the physical make-up of the land.

The environment is difficult, farms are small and lacking in resources and, as a result, change has been slow. But, where it is taking place, the effects can be striking. For example, on the Staffordshire moorlands, farmers are leaving the land and, without grazing, the vegetation is changing rapidly.

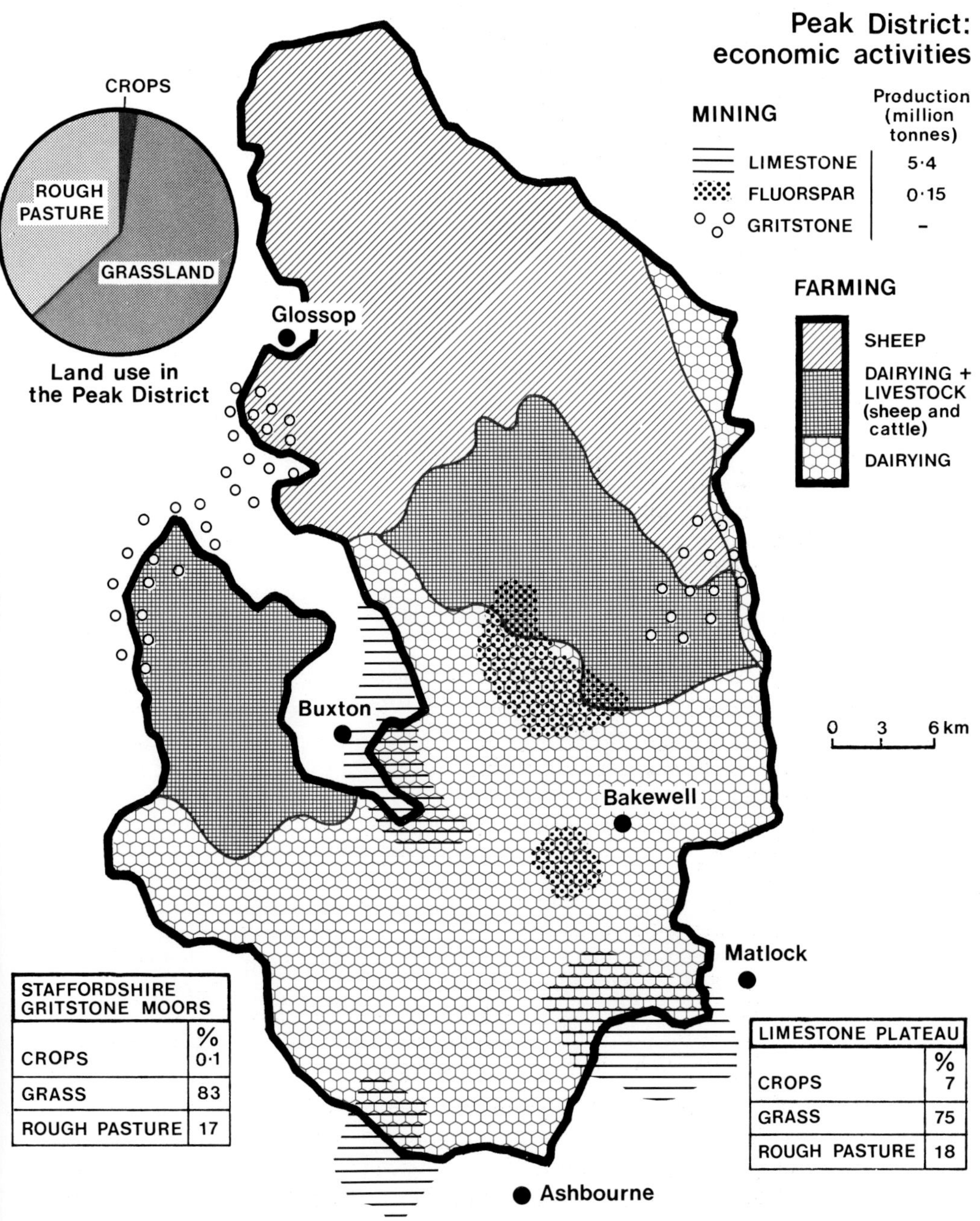

STAFFORDSHIRE GRITSTONE MOORS	%
CROPS	0·1
GRASS	83
ROUGH PASTURE	17

LIMESTONE PLATEAU	%
CROPS	7
GRASS	75
ROUGH PASTURE	18

2. *Mining and Quarrying*

Of all the economic activities in the area of the park, none has presented more problems or created greater controversy than the mining and quarrying industry. For centuries the Peak District has been an important source of limestone and today it supplies material to the steel, cement and chemical industries, as well as aggregate to the construction industry. Most of the major quarries lie outside the area of the national park but the reserves of high grade limestone stretch across the boundary and, with the ever increasing demand for the rock, it seems likely that these will be exploited. Exploitation of the other main mineral resource of the Peak District—*fluorspar*—is inevitable. Demand for this material which is used in the aluminium industry and as a flux in the steel industry, is increasing rapidly and there are few alternative sources of supply. As a result, production within the area of the national park is likely to increase and this will add to the problems which already exist. These include:

a) Dust pollution around the quarries and cement works.

b) The disposal of waste material which, in the case of fluorspar, is in the form of liquid sludge.

c) The destruction of the natural landscape.

d) Heavy traffic and noise.

Given such problems it is not surprising that many people feel that quarrying should be excluded from a national park.

3. *Water Supply*

Similar points of view have been expressed about the building of reservoirs in the park, although objections are not so strong because many people feel that lakes—even man-made lakes—improve the scenery. As in most upland areas, the water resources of the southern Pennines have been exploited for more than a century and large reservoir schemes, including the complex at the head of the Derwent, were opened long before the national park was established. Since that date development has been restricted but, if demand continues to rise, further building may take place eg the proposed development at Carsington between Ashborne and Matlock.

The need to control development is obvious but the effects of doing so are far from clear. For example the Peak District is continuing to lose population. Is this simply because it is an upland area (most upland areas are losing population) or have the restrictions imposed by the National Park authorities contributed to the decline?

Northern Ireland

Of all the problem areas of the United Kingdom, Northern Ireland is unique for, in addition to economic problems similar to those seen in other regions, there are religious and political divisions which threaten to destroy the province. As a result, in the six counties which make up Northern Ireland:

NORTHERN IRELAND: Relief and Counties

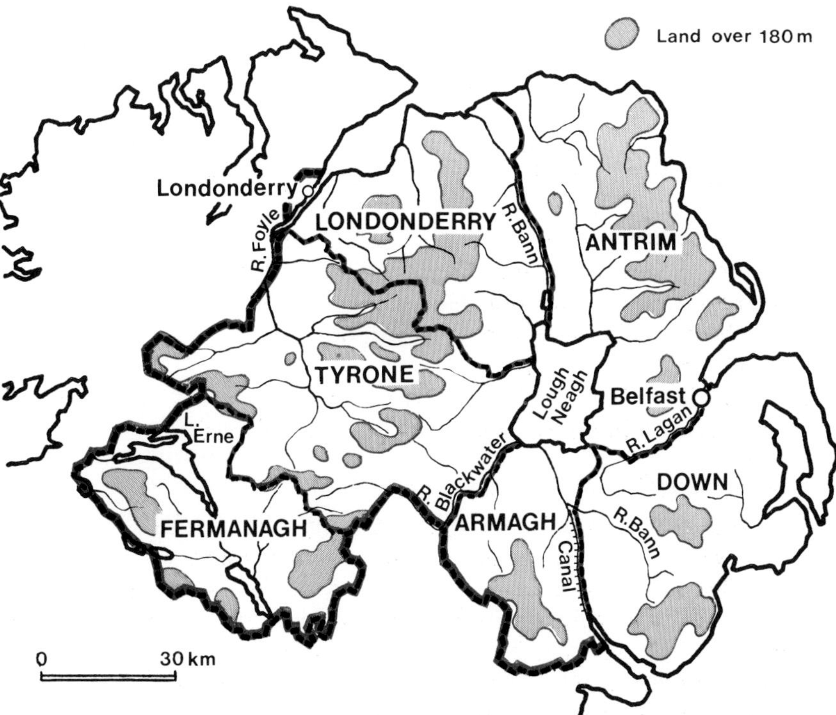

1. Unemployment is up to three times greater than the average for the United Kingdom.

2. Wage levels are only 87% of the average for Britain.

3. Emigration is higher than from any other part of Britain and since 1950 more than one hundred and fifty thousand people—often the young and most able—have left the region. In spite of this, the population has continued to increase, largely because the birth rate is the highest in Britain and the death rate is low.

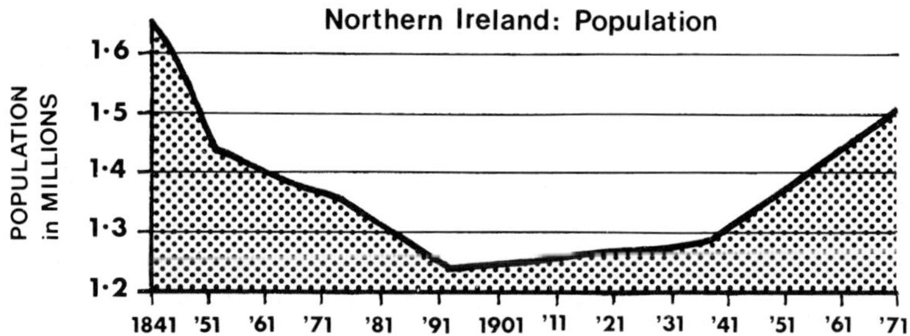

4. Industrial growth has been very slow.

Many factors have contributed to the emergence of this gloomy picture but the two most important are:

a) The economy depends too heavily upon a very limited range of industries.

b) The religious and political differences between catholics and protestants, those favouring a united Ireland and those wishing to remain in the United Kingdom, have produced violence and civil disturbance on a scale seen nowhere else in Britain.

All attempts to improve the situation in Northern Ireland have concentrated on changing these factors.

The Economy

In 1950 the economy of Northern Ireland depended almost completely upon three activities—farming, shipbuilding and the manufacture of textiles. Since then each of these industries has faced serious problems which have produced striking changes in them.

1. Agriculture

The situation in agriculture is typical. In 1950 nearly one quarter of the total labour force was employed in farming. Today that figure is twelve per cent—a loss of more than seventy thousand jobs. Such a large reduction in the labour force has meant major changes in the pattern of farming in Northern Ireland. Among the most important of these are:

a) Farms in Northern Ireland are generally very small. Many are less than ten hectares in size and, of these, a large number cannot keep even the farmer in full time work. Attempts have been made to improve this situation and farms have become larger by a process of amalgamation or by renting land from neighbours, who have given up farming but are not prepared to sell their land. (This is known as *conacre* and it is very important in Northern Ireland, where most of the land is owned by the farmers who work it.) As a result of these developments, a large number of farmers have lost their land and have left agriculture.

b) Given a system of small holdings and a plentiful supply of labour (usually the farmer's family), it is not surprising that farming in Northern Ireland was much less highly mechanised than farming in other parts of Britain. This has begun to change in recent years as the size of farm has increased, and there has been a sharp reduction in the labour force employed in agriculture.

c) Physical conditions, such as the heavy, well distributed rainfall, the mild winters and cool summers, have always favoured the establishment of pasture in Northern Ireland; and, for most of the twentieth century, the province has concentrated on dairy farming, producing butter and cheese for export to the British market. This dependence on pastoral farming remains but dairying is less important than it was, and other forms of stock rearing have replaced it in many areas, eg beef cattle, pigs and poultry. Such activities require less labour than dairy farming and this has added to the loss of agricultural jobs in the region.

Although these general characteristics can be traced in the pattern of farming throughout Northern Ireland, it is important to remember that factors such as relief, rainfall, soil type and accessibility have produced marked variations within the province.

Using information given on page 298, complete the following exercise:

a) i) List the main agricultural products of Northern Ireland, arranging them in order of importance.
   ii) For each item, state whether it is the product of dairy farming, stock rearing, or other farming activities.
   iii) Which branch of farming is most important today?

b) i) List the crops grown in order of importance.
   ii) Why do crops not figure among the agricultural products of Northern Ireland shown on the other graph?
   iii) What are the crops used for?

c) Write a brief account of farming in Northern Ireland, describing the changes which take place within the province and explaining why they occur.

It is clear from the map that farming is most highly developed in the eastern counties of Antrim and Down where the environment is less difficult and where farms are larger. At the same time, it is in the more remote, mountainous west that the dependence on farming is greatest and it is here that the decline in employment in agriculture has caused most problems.

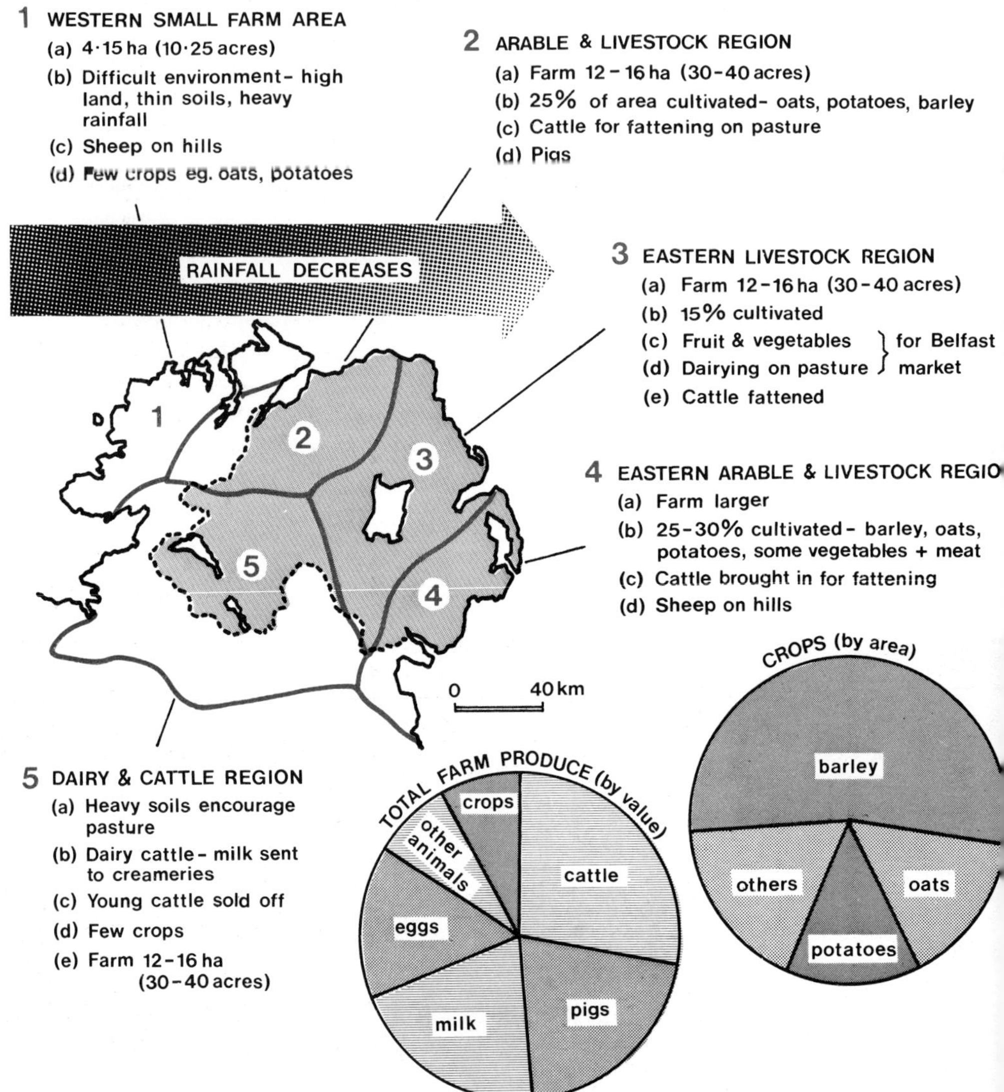

2. Textiles

The textile industry of Northern Ireland came to depend almost entirely on *linen* which was made from *flax* grown in the province. The industry started as a cottage industry, scattered throughout the region wherever flax and soft water were available. It was not until the nineteenth century that the first large mills were built and then the Belfast region began to emerge as the main centre of production. Even then the scale of operation was smaller than in the other branches of the textile industry and this helps to explain the fact that linen production continued in many other parts of Northern Ireland.

By 1950 this traditional industry was still of major importance, employing 60,000 workers, which was more than half of the total labour force employed in manufacturing industry at that time. Furthermore, many of these workers were women—and employment for women was scarce in Northern Ireland—and many lived in areas which had no other industry. Within ten years one third of the factories had closed and the labour force was less than 30,000—a collapse on a scale seen only in the Lancashire cotton industry. Many factors contributed to this decline, some of them shared by the cotton industry:

Linen mills on the river Lagan, near Lisburn.

1. The industry was out of date in terms of machines and organisation. The latter was most important for it had remained a small scale craft industry at a time when mass production was the rule.

2. Irish linen was a luxury product and its market was declining.

3. Competition from artificial fibres was growing.

4. Improvements in cotton cloths meant that they could compete in the same market.

5. This meant that there was strong competition from overseas countries.

Since 1960 the decline has continued, albeit at a slower rate, and the linen industry survives as a craft industry producing specialised luxury goods, on a very small scale. The effects of the collapse of the industry have been felt throughout the region but they have been most severe around Belfast, which was the centre of the industry, and in the west where there was no alternative employment.

3. Shipbuilding and Engineering

The engineering industry of Northern Ireland has been dominated by shipbuilding and it is a dominance which is difficult to understand. Shipbuilding originated on a very small scale and made use of local raw materials, such as timber. During the nineteenth century it grew very rapidly, and it is this growth which is difficult to explain for, with the introduction of iron ships, the industry was forced to import, from Britain, most of its raw materials, including coal, iron and steel. This increased costs and made competition with other British yards difficult. It was during this period that the industry became centred on the shores of the Belfast Lough where there were deep–water anchorages and where the large area of flat land surrounding the lough provided sites for docks and yards. The twentieth century has seen the continuation of this process and, by 1950, there was one large shipbuilding concern in Northern Ireland—Harland and Wolff—which employed 20,000 men, produced ten per cent of the total British output and was capable of building the largest vessels afloat.

Since then the industry has faced problems similar to those experienced in North East England and on the Clyde, eg:

1. Out of date facilities.

2. Poor labour relations which meant that delivery dates were missed.

3. Competition from foreign countries, particularly Japan.

4. A concentration on individually built ships, especially liners

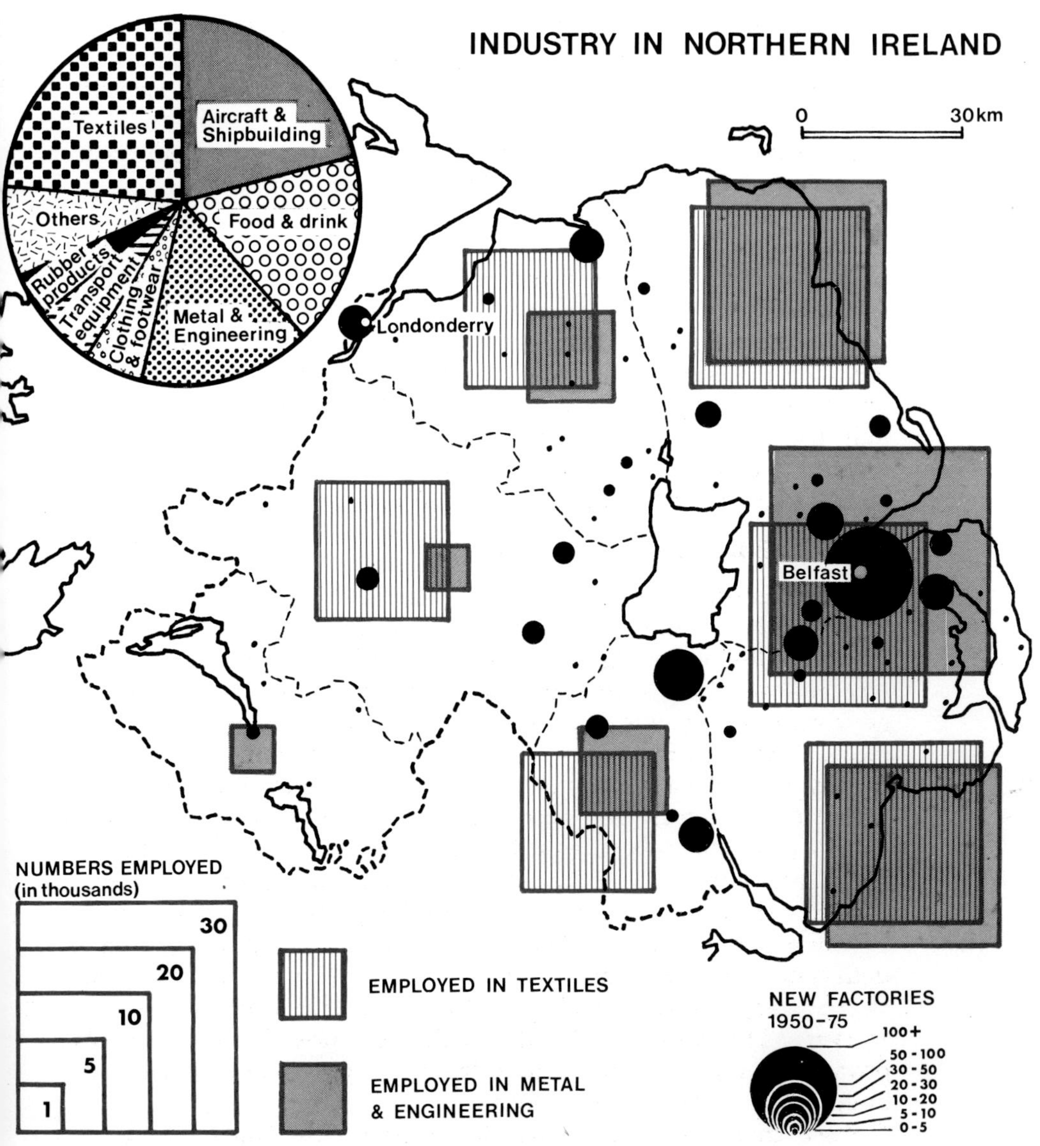

and war ships, at a time when mass production methods were being introduced.

As a result, output has declined and many jobs have been lost—more than ten thousand in the case of Harland and Wolff alone. Attempts have been made to reverse this trend and the government has put money into the firm and has financed the building of a new dry dock and facilities for the construction of ships as large as any afloat. So far this has met with little success and the decline continues.

Futhermore, since modern shipbuilding is an assembly industry, its decline has seriously affected the engineering industry which supplies the yards with components. This situation in the engineering industry has been made worse by the problems of the aircraft industry which is also located in the Belfast area. Too small to compete with its rivals in Britain and abroad, this industry has also required considerable government help simply to survive, and today it depends largely on government contracts for military aircraft.

In spite of government help and attempts at reorganisation the three basic industries of Northern Ireland have declined alarmingly and since 1950 more than one hundred thousand jobs have been lost in them. This has meant that the need to attract new industries to the region has been urgent.

Annsborough industrial estate at Craigavon, on the shores of Lough Neagh.

Recent Industrial Development

The policy of attracting new industries to the province has met with considerable success and, until quite recently, new jobs were being created at the rate of more than six thousand a year. Many factors influenced the choice of Northern Ireland as a site for new industry:

Advantages

1. Low labour costs.
2. A pool of unemployed labour—this was important during periods of full employment.
3. Generous government grants.
4. Industrial estates and factories provided.
5. Restrictions on development in other areas of Britain.

Disadvantages

1. Few local raw materials.
2. Remoteness from the main centres of population and industry in Britain. This is particularly true in terms of transport for any link involves a sea crossing, which means that there is a break of bulk at each of the ports involved.
3. Threat of civil disturbance.

Three types of industry have been attracted:

1. Those which can make use of local produce, eg meat packing and food processing. Such industries often already existed in the province and have merely expanded in recent years.

2. Those which make use of existing skills, eg new branches of the textile industry, clothing, engineering and tobacco.

3. The so called 'foot loose' industries, ie those industries which are not strongly limited by location factors, such as the availability of raw materials. Typical of such industries are electronics and light engineering both of which have been attracted to locations on industrial estates.

The effects of these developments can now be seen in the structure of industry in Northern Ireland today.

Using information given on page 301 and the graph on page 304, complete the following exercise:

a) i) List the main industries in Northern Ireland, arranging them in order of importance.
   ii) In each case state whether the industry is likely to be a traditional industry or a new one.
   iii) Which group is most important?

b) i) Name three industries which have declined since 1955.
   ii) Name three which have grown.

iii) In each case state whether the industry is likely to be traditional or new.

c) i) Describe the distribution of industry in Northern Ireland.
ii) Describe the location of new industries.

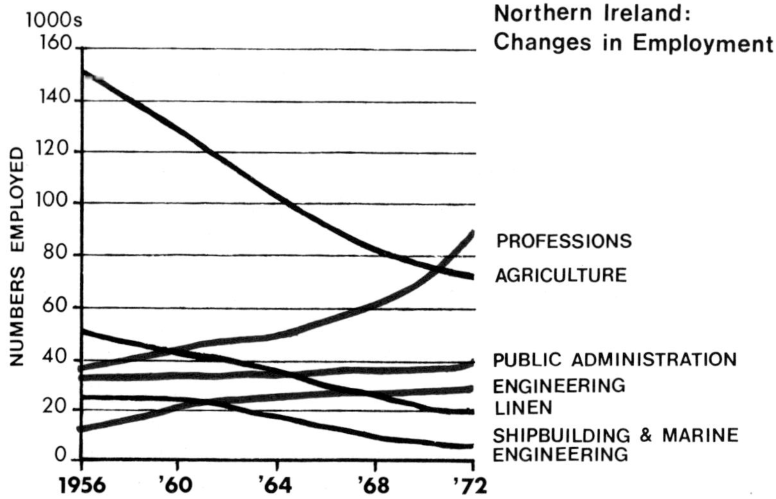

The most striking feature of the map is the divide between the more heavily industrialised eastern counties and the rural west. This is particularly noticeable in the case of the engineering industry but it also applies to the other staple industry—textiles. As we have already seen, the recent decline of the textile industry has severely affected these western areas and attempts have been made to overcome the problem by attracting new industries to the area. This policy has met with little success and many people have moved to the Belfast region or beyond to look for work.

The situation in the east is more complicated, largely because of the dominance of Northern Ireland's one major city, *Belfast*. Belfast grew up at the mouth of the River Lagan, on the shores of Belfast Lough. Here there was a deep water harbour and a large area of surrounding lowland suitable for development. At an early date, therefore, it emerged as an important port, well situated for trade with Britain. This function has remained important and today Belfast handles the bulk of the region's trade. During the nineteenth century industrialisation began to take place and the population grew from thirty thousand in 1821 to three hundred and ninety thousand in 1911. Most of these people came from the rural areas, attracted by the opportunities of work in the shipyards and linen mills. This period of rapid growth has created many of the problems which face the city today, when the population has reached more than half a million, eg:

1. A large amount of housing which is old and ready for demolition.
2. Basic industries which are declining.
3. An out of date road system which has caused congestion.
4. A complex pattern of religious and political divisions which have encouraged civil strife.

In spite of these problems, Belfast continues to attract people from the surrounding areas and its influence in the province has tended to grow. As a result, attempts have been made to divert industry and population away from the city and this has led to the spread of new industrial development into the surrounding towns and villages, eg Ballymena and Craigavon, which have become important centres for the production of artificial fibres, tyres and engineering products. This policy has benefited some areas, particularly the lowlands surrounding Lough Neagh, but it has done little to improve the situation in the more remote areas of the south and west, where there is little industry and unemployment is very high.

Belfast, showing the position of the docks, shipyards and aircraft industry. Note the concentration of industry around the port.

Even more serious for the future of Northern Ireland, the number of new industries attracted to the region has started to decline, partly because of the recent economic weakness of Britain and partly because of political problems which have become more severe during the 1970's.

The Political Problem

The political problems of Northern Ireland stem from the fact that, from the sixteenth century onwards, protestant immigrants from Scotland and other parts of Britain settled in a country which was predominantly Roman Catholic. The greatest concentration occurred in the province of Ulster which today makes up Northern Ireland. Here the protestants formed an elite which owned most of the land and controlled industry and trade. When Ireland obtained independence from Britain in 1920 these protestants demanded to remain within the United Kingdom. As a result, Ireland was partitioned and the province of Ulster remained British, being governed partly from Westminster and partly by an assembly which met at Stormont.

NORTHERN IRELAND: Religious problem

Unfortunately within Ulster, there were large numbers of catholics who were given little say in the government of the province and who were treated as second class citizens. In 1967 they began a movement for civil rights and in the following years this led to violence between the two religious groups. Since then the British army has been called in and the violence has escalated.

The situation today is that, out of a population of one and a half millions, one third is Roman Catholic and generally in favour of a united Ireland; and two thirds are protestant and generally in favour of remaining within the United Kingdom. The two appear to be irreconcilable and, if this is so, there appears to be little hope of improving the economic situation within the province.

A car bomb explosion in Ireland.

Notes

The Correlation Coefficient

The correlation coefficient is a measure of the relationship between two factors. It can be measured in several ways. One of the simplest is the rank order method.

For example a farm comprises four fields. Each has a different angle of slope and each produces a different yield of wheat. Is there any relationship between the two factors?

1. Arrange the information in a form which makes comparison easy.

SLOPE IN°	FIELD	YIELD
5	1	30
1	2	28
2	3	38
18	4	19

2. Number each field according to its rank order in each factor (largest =1, smallest = 4).

SLOPE	FIELD	YIELD
2	1	2
4	2	3
3	3	1
1	4	4

3. Work out the difference (d) between these rankings.

FIELD	d
1	0
2	0
3	3
4	−3

4. Square these numbers to obtain d^2

FIELD	d^2
1	0
2	0
3	9
4	9

5. Add together the values in the d^2 column to obtain Σd^2.
6. Substitute these values into the following formula:

$$\text{Correlation coefficient} = 1 - \frac{6\,\Sigma d^2}{n(n^2 - 1)}$$

Where n = the number of cases, ie fields.

$$= 1 - \frac{6 \times 18}{4(16 - 1)}$$

$$= 1 - \frac{108}{60}$$

$$= 1 - 1{\cdot}8 = -0{\cdot}8$$

7. Correlation coefficients range from + 1 to −1.

8. +1 indicates a perfect positive relationship between two factors, ie the values for one factor increase as those for the other increase.

9. −1 indicates a perfect negative relationship between the factors, ie the values for one factor increase as those for the other decrease.

10. 0 indicates no correlation.

In our example, therefore, −0·8 indicates a strong negative correlation between slope and yield, ie yield decreases as slope increases.

Index of Directness

a) To measure the directness of a link between two places A and B:

1. Measure the straight line distance between A and B.

2. Obtain the actual distance travelled between A and B, using the method of communication being studied, eg road.

3. The Index of Directness = $\frac{\text{Actual distance}}{\text{Straight line distance}} \times 100$.

4. The higher the number, the less direct the link.

b) To obtain the index of directness for a network:

1. Construct a matrix to show all the journeys possible within the network. Enter the straight line distances. Repeat for actual distances.
2. Add together *all* straight line distances.
3. Add together all actual distances.

4. The Index of Directness = $\frac{\text{Actual distance}}{\text{Straight line distance}} \times 100$

5. The higher the number, the less direct and less efficient the network.

Index of Accessibility

1. Draw a map of the network being studied.
2. Simplify this map by:
   a) Marking the towns with dots.
   b) Drawing the links which make up the network as straight lines.
   c) Marking any junctions which occur outside towns with dots.
3. Draw a matrix large enough to contain all the centres included in the network.
4. On it enter the number of links used to travel, by the shortest possible route, from each centre to every other centre in the network. A link in this case is the stretch of routeway between any two dots.
5. Calculate the total for each centre.
6. The centre with the smallest total is the most accessible. For example:

A road network contains four centres—Silverdale, Goldenhill, Ledbury and Ironbridge.
The matrix would be:

	SILVER-DALE	GOLDEN-HILL	LED-BURY	IRON-BRIDGE	TOTAL
Silverdale	—	3	4	2	9
Goldenhill	3	—	5	6	14
Ledbury	4	5	—	2	11
Ironbridge	2	6	2	—	10

Since the centre with the smallest total is the most accessible, the order of accessibility would be:

1. Silverdale
2. Ironbridge
3. Ledbury
4. Goldenhill.

The Location Quotient

The location quotient measures the concentration of an activity in a given area.

Example

1. Obtain the employment statistics for the activity being studied.	eg Employment in the pottery industry in NORTH STAFFS 45,000
2. Calculate this as a percentage of the total labour force in the area.	25%
3. Obtain comparable statistics for the United Kingdom as a whole.	1%
4. Divide the local figure by the national figure to obtain the Location Quotient.	$LQ = \frac{25}{1}$ $= 25$

5. Any figure over 1 shows a concentration of activity in that area.

Acknowledgements

Thanks are due to the following for the use of photographs:-

Port of Felixstowe Dock and Railway Co p 11; Townsend Thoresen p 13; West Air Photography p 15; Port of Bristol Authority pp 16, 17; British Airports Authority p 22; Meteorological Office p 34; Aerofilms Ltd pp 39, 45, 46, 68, 70, 99, 121, 142, 155, 161, 162, 163, 166, 184, 186, 188, 194, 202, 205, 208, 217, 225, 228, 232, 240, 248, 251, 259, 262, 268 (below), 283, 299, 305; Farmers Weekly pp 54, 55; White Fish Authority p 73; Derek Crouch (Contractors) Ltd p 82; British Steel Corporation p 91; JKS St Joseph, Committee for Aerial Photography University of Cambridge pp 108, 109; Central Electricity Generating Board p 115; UKAEA p 116; British Petroleum Co Ltd pp 126, 128; BP Chemicals International Ltd p 131; British Railways Board p 140; British Leyland Ltd p 147; British Transport Dock Board p 170; Freeman Fox & Partners p 171; K M Andrew p 192; British Aluminium Co Ltd p 197; National Coal Board p 254; Durham County Council p 256; Anglian Water Authority p 268 (above); Blackpool Gazette & Herald Ltd p 279; Peak Park Joint Planning Board p 289; Department of Commerce, Belfast p 302; Popperfoto p 307.

Index